Lecture Notes of the Institute for Computer Sciences, Social Informatics and Telecommunications Engineering   71

Giovanni Giambene   Claudio Sacchi (Eds.)

Personal Satellite Services

Third International ICST Conference, PSATS 2011
Malaga, Spain, February 17-18, 2011
Revised Selected Papers

 Springer

Volume Editors

Giovanni Giambene
University of Siena
53100 Siena, Italy
E-mail: giambene@unisi.it

Claudio Sacchi
University of Trento
38123 Trento, Italy
E-mail: sacchi@disi.unitn.it

ISSN 1867-8211   e-ISSN 1867-822X
ISBN 978-3-642-23824-6   e-ISBN 978-3-642-23825-3
DOI 10.1007/978-3-642-23825-3

Springer Heidelberg Dordrecht London New York

Library of Congress Control Number: 2011935599

CR Subject Classification (1998): C.2, H.4

Typesetting: Camera-ready by author, data conversion by Scientific Publishing Services, Chennai, India

Printed on acid-free paper

Springer is part of Springer Science+Business Media (www.springer.com)

Preface

Satellite systems have an important role in providing information services and Internet access everywhere. Next-generation satellite services will cater to the demands of personal services by bringing satellite terminals directly to the hands of the user, thus making them really 'personal.' Particularly future-proof services via satellite will be broadband Internet access, e-health, transportation and environment monitoring, provision of innovative services in underdeveloped areas, and 3D TV broadcasting.

With the unique broadcast nature and ubiquitous coverage of satellite networks, the synergy between satellite and terrestrial networks provides immense opportunities for disseminating multimedia services to a wide range of audiences over large numbers of geographically-dispersed people. It is evident that the satellite will play a complementary, but essential, role in delivering multimedia and telecommunication services to infrastructure-less regions where the terrestrial high-bandwidth communication infrastructures are practically unreachable.

New R&D frontiers appear to be the development of terabits per seconds satellite networks, MIMO antenna techniques, relaying and cooperation and the adoption of cognitive radio.

The PSATS 2011 conference (February 17-18, 2011, Malaga, Spain) represented a very interesting opportunity to meet and discuss these future challenges, thus promoting the exchange of ideas.

PSATS 2011 brought together academic participants (mainly from Europe and Korea) and industries (e.g., Avanti Group, Astrium, ESA, Telespazio, Thales, and TriaGnoSys) interested in future techniques on satellite communications, networking, systems, and applications.

This year keynote talks were provided by Paul Febvre (INMARSAT) dealing with "Personal Satellite Access Terminals: Observations with a 40-Year Perspective" and by Haitham Cruickshank (University of Surrey, UK) on "Satellite Communications Security—Current Status and Future Expectations."

In addition to this, three special sessions were organized on current hot topics: (a) "Multiservice IP Next-Generation Satellite Networks" dealing with the DVB-RCS NG standardization work by Thales Alenia Space – Spain; (b) "Aeronautical Communications for Air Traffic Management" by the University of Bradford; (c) "Delay Tolerant Networking (DTN)" by the ESA SatNEx III project (www.satnex3.org). In particular, the recent standardization work on the DVB Interactive Satellite System specification was addressed by the first special session. Moreover, Aeronautical Communications for Air Traffic Management (ATM) were addressed in a two-part special session dealing with the Iris Program funded by the European Space Agency and the SANDRA (Seamless Aeronautical Networking Through Integration of Data Links Radios and Antennas) EU FP7 project, respectively. Iris proposes a satellite-based solution to support the evolution from "voice-based" communications to digital data links in continental and oceanic airspace. The following projects were presented within the Iris program: (a) THAUMAS (Tailored and Harmonised Satcom for ATM Uses, Maximising Re-use of Aero Swiftbroadband) working on extending the Inmarsat SwiftBroadband system; (b) ANTARES (Aeronautical Resources Satellite) project, focusing on the development of a new satellite-based communication system based on low-cost user terminals through the realization of a new satellite communication standard; (c) SIRIO project dealing with a feasibility analysis of satellite communication service provision in terms of complexity, risks and financial benefits for aeronautical services. Moreover, the SANDRA EU FP7 project was presented in part II of this special session, dealing with the definition, the integration, and the validation of a reference communication architecture for aeronautical communications via satellite. As for the DTN special session, issues were addressed referring to "challenged networks" where the usual TCP/IP protocol suite is unable to provide a satisfactory performance.

This was the third year that the PSATS conference was organized and it has now acquired a stable size and for the first time outside Italy, in the nice setting of Malaga. This year we also obtained the technical co-sponsorship of the Integral Satcom Initiative (ISI) platform (thus contributing to raising the industrial interest in this event) and an invited talk by Eutelsat focusing on the innovative service of Internet access for high-speed trains.

Before concluding this preface, we would like to thank the TPC Co-chairs, Igor Bisio (University of Genoa, Italy) and Fun Hu (University of Bradford) for having defined a very attractive technical program with 35 accepted papers. Moreover, we would like to thank the conference manager Brian Bigalke (Create-Net) and the organizer Aza Swedin (EAI) for their continuous support. A special thank you is for the local organizer, Mari Carmen Aguayo Torres of the University of Malaga, for her work and for selecting the conference location. Finally, special thanks are due for the Web Chair Susanna Spinsante of the University of Ancona for providing effective and timely support of the relevant tasks related to Web setup and information update.

We hope that next year in Bradford, PSATS 2012 will continue the success of the past editions.

Giovanni Giambene
Claudio Sacchi

Organization

PSATS 2011 was organized by the European Alliance for Innovation (EAI) and hosted by the University of Malaga, Spain. The PSATS 2011 technical program was defined with the co-sponsorship of the Integral SatCom Initiative (ISI).

Steering Committee

Imrich Chlamtac	President, Create-Net, Italy – Chair
Kandeepan Sithamparanathan	Create-Net, Italy
Stefano Agnelli	ESOA/Eutelsat, France
Mario Marchese	University of Genoa, Italy
Marina Ruggieri	University of Rome "Tor Vergata", Italy

Program Committee

General Co-chairs

Claudio Sacchi	University of Trento, Italy
Giovanni Giambene	University of Siena, Italy

Technical Program Co-chair

Igor Bisio	University of Genoa, Italy
Yim Fun Hu	University of Bradford, UK

Organizing Chair

Mari Carmen Aguayo-Torres	University of Malaga, Spain

Website Chair

Susanna Spinsante	Marche Polytechnic University, Italy

Publication Chair

Lorenzo Mucchi	University of Florence, Italy

Conference Coordinator

Aza Swedin	European Alliance for Innovation, Italy

Publicity Chair

Marco Lucente	University of Rome "Tor Vergata", Italy

Special Session on Multiservice IP Next-Generation Satellite Networks

Organizer

Ana Yun Garcia   Thales Alenia Space España, Spain

Special Sessions on Aeronautical Communications for Air Traffic Management – Parts I and II

Organizer

Yim Fun Hu   University of Bradford, UK

Special Session on Delay-Tolerant Networking

Organizer

Erich Lutz   German Aerospace Center, DLR, Germany

Technical Program Committee

Fatih Alagoz	Bogazici University, Turkey
Matteo Berioli	German Aerospace Center, DLR, Germany
Carlo Caini	University of Bologna, Italy
Marco Cello	University of Genoa, Italy
Haitham Cruickshank	University of Surrey, UK
Philip A. Dafesh	Aerospace Corporation, California, USA
Franco Davoli	University of Genoa, Italy
Tomaso de Cola	German Aerospace Center, DLR, Germany
Anton Donner	German Aerospace Center, DLR, Germany
Fabio Dovis	Turin Polytechnic, Italy
Alban Duverdier	Centre National d'Etudes Spatiales, CNES, France
Carles Fernandes Prades	CTTC, Barcelona, Spain
Laurent Franck	ENST Bretagne, France
Istvan Frigyes	Budapest University of Technology, Hungary
Wilfried Gappmair	TUG, Graz, Austria
Thierry Gayraud	LAAS, Toulouse, France
Gonzalo Seco Granados	Universitat Autonoma de Barcelona, Spain
Takis Mathiopoulos	Institute for Space Applications and Remote Sensing, Greece
Maurizio Mongelli	University of Genoa, Italy
Fernando Perez-Fontan	University of Vigo, Spain
Francesco Potortì	Institute of Information Science and Technologies, Italy

Gianluca Reali   University of Perugia, Italy
Sandro Scalise   German Aerospace Center, DLR, Germany
Andrea Sciarrone   University of Genoa, Italy
Tuna Tugcu   Bogazici University, Turkey
Marìa Àngeles Vàzquez-Castro   Universitat Autonoma de Barcelona, Spain
Markus Werner   TriaGnoSys GmbH, Germany

Sponsoring Institutions

CREATE-NET, Trento, Italy
University of Siena, Italy
University of Trento, Italy
University of Malaga, Spain
Integral SatCom Initiative (ISI)

Table of Contents

DVB-S2

Hybrid Networks

Delay Tolerant Networking

Channel Estimation and Interference Management

Aeronautical Communications for Air Traffic Management - Part II

Satellite Antenna Design

Localization Systems

DVB-RCS New Generation towards NGN Convergence

Borja de la Cuesta, Isaac Moreno Asenjo,
Ana Yun Garcia, and Juan Manuel Rodriguez Bejarano

Thales Alenia Space España, Tres Cantos (Madrid), Spain

Abstract. The convergence between fixed and mobile networks is the new paradigm in the world-wide telecommunications network. Satellites are striving to become an important actor in the Next Generation Networks (NGN) by offering seamless integration with the terrestrial networks. This provides an opportunity to the Satellite industry to advance and adapt this framework to the new DVB-RCS (Digital Video Broadcast - Return Channel Satellite) standard.

Keywords: DVB-RCS NG, convergence, NGN, standardization.

1 Introduction

The network of the future is founded on the development of a converged all-IP communications and services infrastructure that will gradually replace the current Internet, mobile, fixed and broadcasting networks. These next generation network infrastructures [1] must support the convergence and interoperability of heterogeneous mobile and broadband network technologies, in order to support seamless services that are accessible everywhere and to support mobile usage. There is a need for a shared network solution, based on the convergence of fixed, mobile and broadcasting environments, capable of exploiting legacy, evolved and new infrastructure components. NGN promise to be multiservice, multiprotocol, multi-access and IP based infrastructure. The evolution of existing broadband satellite systems towards an NGN infrastructure will be crucial for the successful integration of the satellite systems in the new converged framework.

The DVB-RCS New Generation (DVB-RCS NG) comprises a series of draft documents produced by the DVB TM-RCS group in response to requirements delivered by the analogous DVB Commercial Module [2]. First time, DVB specifies a complete interactive system through a system document which is composed of two main documents, one referring to the Lower Layers (LL), and a second one focussing on the Higher Layers of the Satellite (HLS) interactive system. The current version of these documents will be available by the first quarter of 2011. The objective of the DVB-RCS New Generation is to elaborate a new version of the standard mainly suitable for the consumer market. With respect to the previous version of DVB-RCS standard [3], the new standard will not only define aspects from physical and MAC/link layers but will cover IP and Upper layers

G. Giambene and C. Sacchi (Eds.): PSATS 2011, LNICST 71, pp. 1–9, 2011.

aspects. It is in this frame where a contribution based on the identification of functionalities, protocols and mechanisms necessary to be compliant with NGN framework will ensure the merge of new generation DVB-RCS into the NGN.

This paper provides an overview of the Next Generation Network architecture and how this concept is introduced in the new DVB-RCS NG standard. The paper starts with a brief introduction to the DVB-RCS NG standard and continues with the implications of the NGN convergence in the standard.

2   DVB-RCS NG Standard

The next generation of DVB-RCS standard is compound by layered specifications (Lower and Higher Layers). Both specifications are closely related, the MAC mechanisms in the LL specification must follow the requirements from the higher layer functions, while the signalling and resources must flow to the higher layers to be used appropriately in the network.

- The Higher Layers Satellite (HLS) specification covers the control, management and traffic areas.
- The Low Layers Satellite (LLS) specification covers the physical layer (definition of transmission parameters and frame constructions) and data link layer (definition of the logical link control and the medium access control protocols).

It is also foreseen that these main documents will be accompanied by an Implementation Guidelines and a System document.

2.1   Reference Architecture

The Reference DVB-RCS NG architecture is an overall Satellite Interactive Network with a large number of Return Channel Satellite Terminals (RCST) that comprises the following functional blocks:

- The Network Control Centre (NCC) which provides Control and Monitoring Functions (CMF). It generates control and timing signals for the operation of the Satellite Interactive Network to be transmitted by one or several Feeder Stations.
- The Network Management Centre (NMC) which provides management capabilities. It is responsible for the management functions (FCAPS) of Fault, Configuration, Accounting, Performance, Security management.
- RCSTs which provides star or mesh connectivity to end users

Two reference architectures will be considered: transparent and regenerative. The actors in both interactive systems are the same:

- Satellite Operator (SO), who manages the whole satellite, and sells capacity at the transponder level to one or several SNOs.

– Satellite Network Operators (SNO), who are assigned one or more satellite beams. The Master SNO owns the Hub/NCC of the interactive network and configures the frequency plan for the regenerative satellite or the Hub. Other SNOs only control their own capacity. SNOs distribute their own physical and logical resources to SVNOs.

– Satellite Virtual Network Operators (SVNO), who manage the Operator Virtual Networks (OVN). The OVN divides the capacity into several logical and independent networks. Each OVN is assigned a set of RCSTs in the management plane, is allocated physical resources in the control plane and is associated with Satellite Virtual Networks (SVN) in the traffic plane. They sell connectivity services to their subscribers. For the regenerative architecture, they may also manage one or several RSGWs.

– Subscribers (RCST), which are the set of user stations that receive service from the SVNOs. An RCST is assigned to one OVN, through whom the RCST is informed of the set of SVNs that it shall use.

– End-users, who are the final actors enjoying the satellite services and who are connected to the RCSTs LAN interfaces.

The two system architectures can be characterized as follows:

– Transparent system (see Figure 1)
  • Transparent satellite(s). One or more transparent satellites provide the link between terminals and the Hub, or among terminals for the transparent mesh system. DTP payloads can also give multi-beam connectivity.
  • Hub/NCC. Performs the control (NCC), management (NMC) and traffic plane (Traffic GW) functions.
  • Star or mesh terminals (RCSTs). Two types of terminals are considered: the star transparent terminal, that complies with the specifications of the DVB-RCS-NG standard, providing star connectivity, or mesh connectivity using a double satellite hop; the mesh transparent terminal is more complex, since it includes at least two demodulators (DVB-S2, DVB-RCS2) to provide single-hop mesh and star connectivity.

– Regenerative system (see Figure 2)
  • Regenerative satellite. Performs demodulation, demultiplexing, decoding (and probably decapsulation) functions at the receiver side, on-board switching (at different layers) for multi-beam systems, and the corresponding transmission functions after signal regeneration.
  • Management Station. It provides the management (NMC) and control (NCC) plane functions to the satellite network users.
  • RSGWs. The RSGW stations provide regenerative RCST users with access to terrestrial networks via the LAN interface. There may be one RSGW giving service to a small number of terminals, or to hundreds of terminals. Essentially, the gateway comprises one RCST, plus an SLA Enforcer and an Access Router, but may also include voice, traffic acceleration servers, or a backhauling module.
  • Regenerative terminals. These RCSTs are identical in terms of hardware to the star transparent terminals. Their software includes C2P functionalities to support dynamic mesh connectivity.

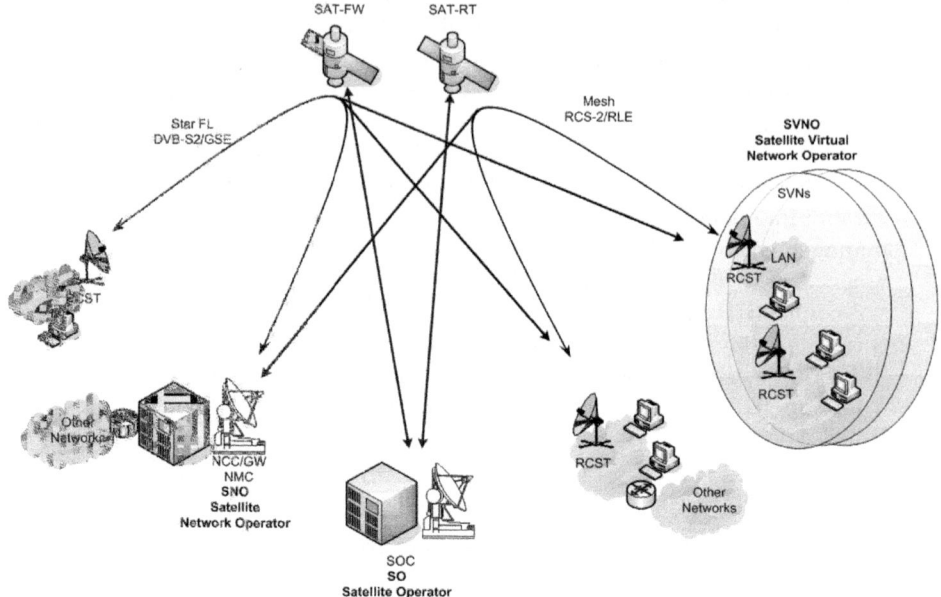

Fig. 1. Transparent network architecture

The reference architecture for the Higher Layers of DVB-RCS NG standard is divided into three different planes (User, Control and Management Plane). Each of the higher layers' functions can be mapped to one of the planes. Figure 3 represents this mapping of functions between planes and their different elements in a mesh regenerative system.

3   Convergence towards NGN

The evolution of DVB-RCS standard towards NGN will allow service providers to reach all their potential customers by creating a single telecommunications network environment. In this context, broadband multimedia satellite systems, working as the access network technology, should play a key role in NGN definition. Satellite systems, due to their characteristics, can be the most suitable technology to assure the independence between theservice provisioning and location.

The convergence requires interoperability between DVB-RCS NG satellite system and terrestrial systems at user, control and management planes, in order to allow network operators to deliver to end users the same experience as terrestrial access networks.

Network interoperability requires: the use of IP-based methods; interfaces for standard network-layer protocols; and enhancements when required to support the unique features of satellite. Operator interoperability demands the

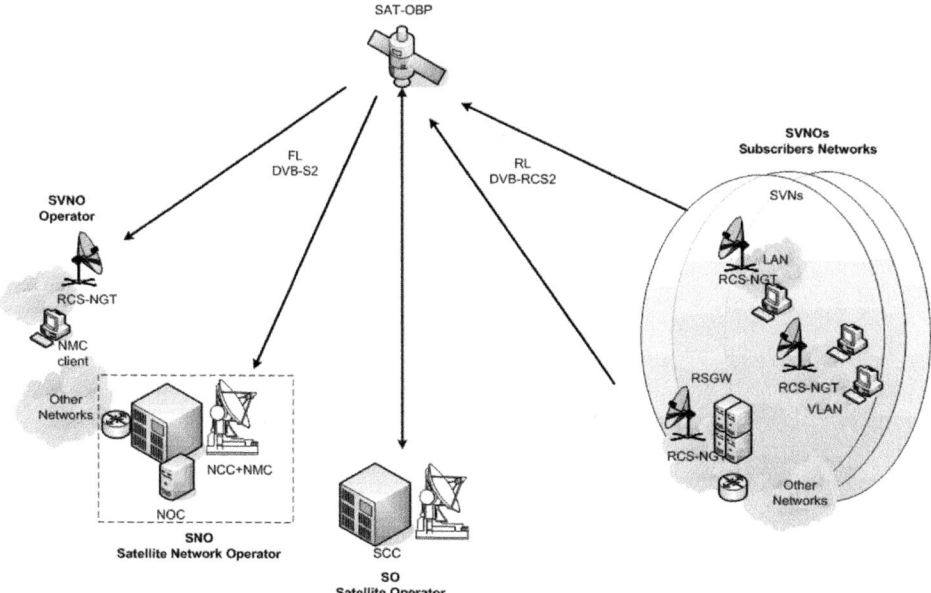

Fig. 2. Regenerative network architecture

specification of an OSS interface, for compatibility with ISP/Telcos, allowing the integration of several access networks in a common management platform.

The interoperability among different manufacturers, alongside the seamless integration with terrestrial networks, help to achieve a user experience close to those terrestrial access networks ones, whilst reducing costs reduction to make satellite services affordable for consumers. The features will be appropriate for a range of satellite systems (not only DVB-RCS) and will be made available to the wider community.

According to the CM-RCS, the DVB-RCS NG system shall be compatible with access technologies to allow it to be integrated into and interoperate with current broadband networks.

The proposed technologies in the Higher Layer Satellite specifications for the DVB-RCS NG will define a new framework for satellite networks capable of inter-operate and converge with the new terrestrial networks and Service Providers. It is crucial to ensure satellite systems become an integral part of the all-IP wave.

The integration of NGN and DVB-RCS NG implies the definition of interfaces, protocols and procedures in the transport, control and management plane.

3.1 Transport and Control Plane

In the integration of the DVB-RCS NG system with NGN, the following interfaces need to be defined:

– TNI (Transport Networks Interface).
– RCI (Resource Control Interface).

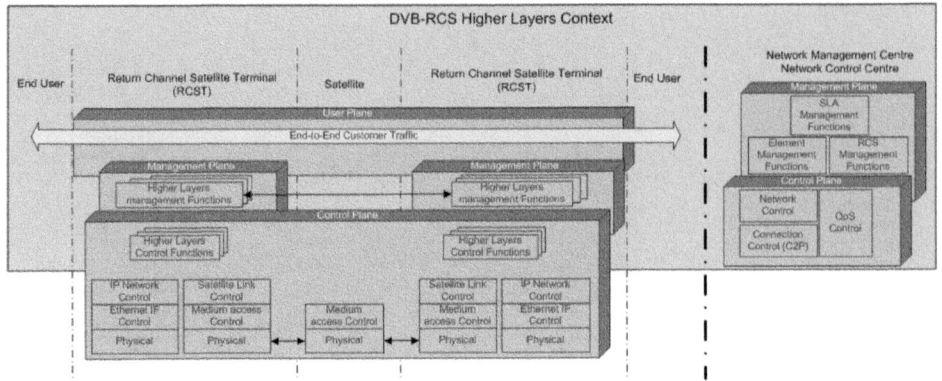

Fig. 3. Elements Functional Architecture

The Resource Control Interface (RCI) is used to allow resource requests from external actors (Other networks/ Telcos) into the RCS-NG systems. The main purpose of this interface is related to the provisioning of end-to-end Quality of Service; optionally it could include functionalities related to the SLA management. The specification of RCI should include:

 − Method for registration and authentication of requesters.
 − Protocol for resource allocation.

The Transport Network Interface (TNI) is defined to help the implementation of data services over a DVB-RCS transport network. It should recommend requirements from other protocols in order to communicate with the DVB-RCS NG system in a consistent way. For example, if the DVB-RCS system is connecting two HFC networks, the interface should specify the basic parameters for a seamless communication.

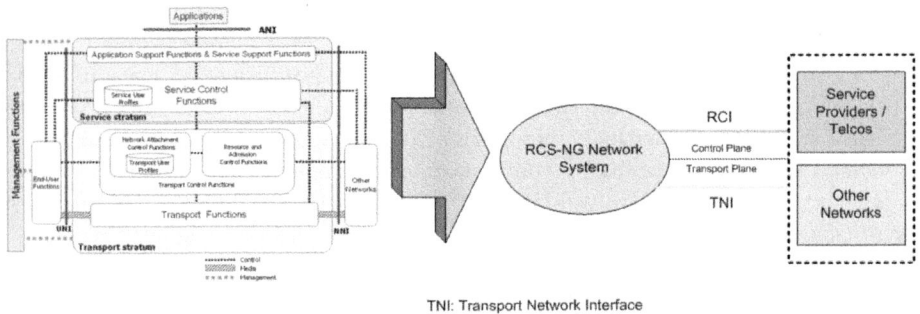

TNI: Transport Network Interface
RCI: Resource Control Interface

Fig. 4. Integration of NGN paradigm [4] into DVB-RCS NG system

3.2   Management Plane

The management functionality of the RCS NG network shall be envisaged in two environments (see Figure 5):

- Internal management: it shall be based on a central manager capable of interacting with network elements through common protocols and management information bases. The central manager is the NMC, which shall manage the following Network Elements: RCSTs, GWs, NCC, NMC, network devices and OBP for regenerative systems.
- OSS-NMC management (external): it compounds the functionalities related to the relationships with other networks and service providers.

The basic functionality of the NMC is the management of the elements of the network (RCST , GW , NCC). For this purpose, SNMPv3 [5] protocol and MIB [6] data bases (in the communication between NMC and network elements - Internal interface shall be supported). The NMC shall be interpreted as the SNMP manager, and the RCST , NCC or GW as SNMP agent. The activity of the NMC should be extended to include Service and Network related management functions and the interaction with external networks and Service Providers.

The interoperability among different vendors of DVB-RCS NG sub-systems is achieved through the harmonization of their management interfaces between higher layer OSS and individual Network Elements (the RCSTs). Thus, a more consistent management architecture is accomplished closer to the TMN framework of six-logical layers model (Business management, Service management, Network management, Element management, Network elements).

The Operation Support System combines the management functions that enable a Provider to monitor, control, analyze and manage systems, resources and services. The OSS should provide the development of a flexible service integration framework, which eases the introduction of new technologies and reduces the cost base. The OSS can be envisaged according to the functional areas FCAPS (Fault, Configuration, Accounting, Performance and Security). The functions of each area are summarized:

- Fault Management: The goals of fault management are to provide failure detection, diagnosis, and perform or indicate necessary fault correction. Fault identification relies on the ability to monitor and detect problems, such as error-detection events. Fault resolution relies on the ability to diagnose and correct problems, such as executing a sequence of diagnostic test scripts, and correcting equipment or configuration faults.
- Configuration Management:: it is concerned with adding, initializing, maintaining and updating network elements. The operation consists on modifying operating parameters associated to physical resources or logical objects (for example QoS).
- Accounting Management: it includes collection of usage data and permits billing the customer based on the subscriber's use of network resources.

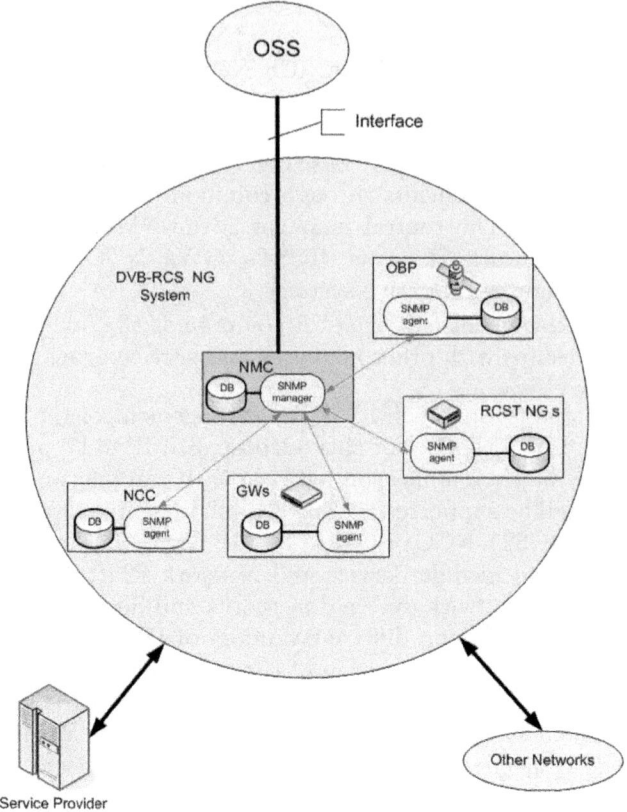

Fig. 5. Management Reference Framework

- Performance Management: Performance management functions include collecting statistics of parameters. These monitoring functions are used to determine the health of the network and whether the offered Quality of Service (QoS) to the subscriber is met.
- Security Management: It is concerned with both security of management information to protect the operations systems as well as managing the security information.

4   Conclusions

A correct definition of the Higher Layer Satellite specification for interactive satellite broadband satellite networks is crucial to ensure satellite systems become an integral part of the all-IP wave. DVB-RCS New Generation has already performed a "Call for Technologies" stage. The main objective of this new standard is to elaborate a new version of the standard mainly suitable for the consumer market, and it will include protocols and mechanisms to be compliant with NGN framework.

The key point and interest of NGN is to really find the solution for an easy integration and interworking of satellite networks with the rest of terrestrial networks. The DVB-RCS NG will define the main functional modules related to NGN principles in the management, control and traffic planes.

References

1. MITU-T Recommendation Y.2011: General principles and general reference model for next generation networks (2011)
2. DVB Technical Module has approved the release of a Call for Technologies for Next Generation DVB-RCS, http://www.dvb.org/technology/dvbrcs
3. ETSI EN 301 790 v1.5.1, Digital Video Broadcasting (DVB); Interaction channel for satellite distribution system (January 2009)
4. ETSI Telecoms and Internet converged Services and Procolos for Advanced Networks. Web, http://www.etsi.org/tispan/
5. Case, J., Mundy, R., Partain, D., Steward, B.: Introduction and Applicability Statements for Internet Standard Management Framework, SNMP Research, Network Associates Laboratories, Ericsson, RFC 3410 (December 2002)
6. McCloghrie, K., Rose, M.: Management Information Base for Network Management of TCP/IP-based internets: MIB-II, STD 17, RFC 1213 (March 1991)

IP Satellite Services and Applications Evolution towards the Future Internet

Juan Manuel Rodriguez Bejarano, Ana Yun Garcia,
Borja de la Cuesta de Diego, and Isaac Moreno Asenjo

Thales Alenia Space España, Tres Cantos (Madrid), Spain

Abstract. Satellite networks are evolving and changing its elements to converge with the NGN concept. The use of IP in satellite networks facilitates convergence, but also opens the door to integrate from the beginning the satellite IP services which would be part of the Future Internet. It is therefore necessary to analyze the applications and services that are typically susceptible of being used in a satellite network, so that its evolution goes online with the Future Internet. Actors and triggers of the evolution are analyzed as well as the critical points in the provision of services are identified.

Keywords: IP, Satellite, Services, Applications, Future Internet, NGN.

1 Introduction

For some time satellite systems have been evolved such systems capable of delivering interactive IP services. These satellite networks have changed its elements to converge with the NGN[1] (Next Generation Network) when necessary.

This shift to IP data networks facilitate the convergence with other networks, but so far has been carried out by adapting the system elements that typically have been used for broadcasting and not for interactive networks. Now it is open a door where from the beginning it is possible to integrate IP services via satellite, taking an active part in what would be the Future Internet.

It is therefore necessary to identify the applications and services that are likely to be used in a satellite network to promote its use and those who are most alien to it, to facilitate its convergence. This would evolve interactive satellite networks in line with the design of the Future Internet.

This paper analyzes some actors and triggers that would foster the development of the Future Internet and would identify the critical points in the taking off of services and applications over satellite networks. There are not only considered the Network aspects as the required connection bandwidth required or the necessary traffic engineering are considered, alongside higher-level factors that impose requirements to the lower levels.

1.1 NGN vs Future Internet

Integration in NGN does not necessarily imply a convergence with the Future Internet[2]. Therefore, firstly it is necessary to define the main differences

G. Giambene and C. Sacchi (Eds.): PSATS 2011, LNICST 71, pp. 10–18, 2011.
© Institute for Computer Sciences, Social Informatics and Telecommunications Engineering 2011

between the two concepts in order to analyze the new requirements that satellite networks are aimed to meet. While Future Internet initiatives are more long-term oriented, NGN ones are solutions that may already be realistically incorporated in current networks.

The Future Internet initiatives[3], focused on the long term, should consider new requirements and principles to improve the experience of the new applications. At the same time, short-term initiatives as NGN are designed to improve the architecture of the Internet in such way that the requirements and principles of the Future Internet would still be met.

The efforts of the NGN initiative have focused so far on the network aspects needed to provide triple play services (i.e. telephone, high-speed Internet access and television services). However, recent studies on NGN do not include requirements related to the content-centric networks. It is expected that future studies on NGN may address some of these requirements, and their outputs will become a key point in the Future Internet development. Even so, at this point of the progress it is possible to outline the expected services and applications evolution needs towards the Future Internet.

1.2   High Level Requirements for Satellite Networks

High-level requirements include scalability features or the ease of access to the network as well as security, connectivity and mobility services. Alongside these high high level services new applications are expected to be provided by the Future Internet networks as well. These future expected applications would be the ones that would make a more social and collaborative network where users are both generating and consuming content; new content such as 3D, augmented reality, artificial intelligence or virtual reality. To support these new applications it is necessary to define from the start the new requirements that new networks would need.

Satellite networks are likely a key element in the new networks compliant with the future internet principles [4]. Their ubiquity makes satellite communications networks capable of providing the essential mobility or backhauling when terrestrial networks can not afford them. The Internet of the future must be therefore built from the "broadband for all" concept and regarding to satellite networks it is summarized in two concepts:

- Firstly, the ability satellite networks have to provide broadband services to everyone and everywhere. Satellite networks are an alternative for rural internet and the final solution to provide broadband Internet access to most unapproachable areas or in disaster relief operations.
- Secondly a network designed for competition is one of the most important elements of the Future Internet that shall be designed. That is, networks supporting flexible business models where multiple parties can participate in an open framework that supports participation and encourages innovation.

The interactive communications satellite networks are actually a solution for the "everyone and everywhere" Future Internet promulgated concept, but it is necessary to evolve some technical aspects and the business model towards the concept of open competition between networks. For this end, it is not only necessary to provide the same applications and services that the terrestrial networks offer but also the generation of new features that only satellite networks would be able to grant, which all together would give the satellite business its own niche market. Potential services are: smart remote healthcare systems, smart systems for transport and mobility, environmental information systems and security systems.

This paper analyzes the technical enablers for emerging Future Internet capabilities such as:

- Increased service performance[5], by increasing data rates which depend mostly on physical layer aspects (RF power, G/T, waveform efficiency).
- Market segmentation: LEO satellites for handheld services and GEO for high broadband access.
- Polymorphism features to supply several services transparently to the user.
- Interoperability and integration with terrestrial networks and NGN services, that is to enforce terminal profiles and usage conditions
- Network virtualization to supply a common service access point with terrestrial networks.
- Information and content-centric Networking to provide contents from everywhere to everywhere and everyone.
- Innovative service coverage configurations with flexibility in the allocated power per beam, and in the beam bandwidth. Different and manageable connectivity scenarios with flexible channels routing. Support of direct connectivity between terminals and multicast mesh capabilities. Cellular network concept, with new frequency bands covering multiple spots in a coordinated way or covering selected areas that other networks cannot reach.
- Terminal handovers capabilities under different satellite or different networks.
- Full integration with small and handheld devices and VSAT antennas.

2   Future Internet Enablers for Satellite Networks

This section analyzes different R&D topics that are currently being studied within the Future Internet initiatives framework, in particular those related to the topics of the satellite networks world. Topics are analyzed from the perspective of the needed aspects to reach a new generation satellite network that meets the application or service requirements, trying to identify the enablers that would potentially achieve it.

2.1   Increased Service Performance

One of the key aspects in the Future Internet take-off is the next generation mobile radio technologies. An increased service performance is related to the increase of the data rate. This depends mostly on physical layer aspects (RF power, G/T and waveform efficiency).

On one hand, new technologies should be designed to be cost-competitive, spectrum and energy efficient. But on the other hand technologies must be adapted for their use in mobile systems. This results in two points: the satellite is expected to be able to play an important role within the future internet; and the satellite must meet certain requirements.

Satellite communications are a natural way to provide mobility services to devices. However, problems mostly related to the antennas integration imply, prevent satellites from a full integration with mobile devices. However the Future Internet initiative is also a source of new ideas and opportunities to detect new action items. Thus, to solve those restrictions, some technologies must be improved:

− Adaptive modulation and coding schemes
− Multiple antenna and user detection systems
− Cross layer design
− Low latency transmission regimes

It is also expected within a Future Internet Framework that complementary technologies cooperate at nodes and terminal levels. That means an improvement in all nodes technologies (i.e. satellite payloads, terminals and hubs), network topologies (i.e. new system models with flexible architectures), and efficiency (i.e. is dynamic power and capacity estimations).

2.2 Market Segmentation

The Future Internet is the Internet for people. This means that each user must be part of it and feel their needs are met. So, the Internet of the future must be able to deliver in every location what the user wants and it must meet the user needs in terms of application, services, Service Level Agreement (SLA) and cost. A free open market based on segmented network services will foster undoubtedly the Internet ecosystem. There are two levels of market segmentation, internal and external.

For external market segmentation, satellite technologies must offer the same services than the ones offered by terrestrial networks. SATNET must be converted into both access networks and transport networks. Satellite technologies must converge and offer the same services as the rest of the networks (adopting the NGN structure) and must become transparent by user's choice. There is no doubt that the technology election will be based primarily on offering something unique. Thus, it is necessary a transformation into platforms (for the new generation applications and services) that only SATNET are able to offer. For example, satellites have a potential benefit in mobile communications, in long distance communication and in providing services in certain areas where other technologies would have difficulties (because of cost or complexity). These features are then the ones to be exploited for the competition with other networks.

On a second level, internal market segmentation would allow a potential growth of the satellite market. This is to generate a service differentiation allowing users to choose the most appropriate service to their needs. For example satellite orbit could differentiate some services (LEO satellites for handheld services and GEO for high broadband access) or different technologies could differentiate applications (interactive satellite network for internet access and transparent communications for content streaming). Network topologies can also differentiate the satellite network use. For example in GEO satellites, mesh topologies are the only ones able to provide real-time services such as VoIP or videoconference.

2.3   Poliymorphism

Market segmentation does not necessarily mean offering a unique service. It means providing the most appropriate service according to a need. However, the network of the Future Internet should be able to offer all kind of services, by transforming its physical or logical architecture to meet the service and application requirements.

Network virtualization helps to provide common services in a transparent way with respect to the physical network used. This gives the satellite networks of an external polymorphism feature that allows them to offer both their native services and interactive communications.

On the other side it is also necessary to encourage internal polymorphism, i.e. the topological transformation of the communications features as required. For example, the same satellite system can provide mesh or star topologies depending on the user requirements. Satellite networks can change their topology to implement hybrid satellites, hubs, terminals and gateways.

A hybrid satellite is one having transponders that can curse traffic in both transparent and regenerative modes. A hybrid hub would control and manage the two types of payloads and assign the capacity from the users to one or the other depending on the transmission needs. For example, if mesh connectivity is needed, the hub would assign capacity on the regenerative channels, while other traffic would use transparent capacity resources. A hybrid GW would be able to assist in establishing mesh connections directed to terminals acting as an entrance to other networks, but also to derive the transparent traffic received to other networks.

2.4   Interoperability and Integration

The integration with terrestrial networks and NGN services is totally necessary for success. The provision of any service must be possible in any network, therefore also in satellite networks. Applications must run transparently over the access network. In other words, an application must not be adapted to the network features, but instead the network is the one that must to meet the application and user profiles. A common SLA framework for terrestrial and satellite networks has to be developed. This would allow for a best-service selection and would foster networks and services interoperability and integration, offering the user the most suitable service.

A Service Level Agreement Framework consists in a common set of rules, agreements and templates concerning on Service Level Agreement (SLA) that all network elements of all networks share and understand. Within this framework, two interconnected networks are aware of the equivalence and reciprocity of all SLAs applied from the original network to other networks. Therefore, it is easier for them to serve as both access or transport networks, as well as it facilitates the handover and backhauling procedures between networks better meeting and matching the required SLAs. To achieve these requirements, there must be an exchange of control and management signaling concerning to SLAs. Signalling is requiered to include information for requesting an SLA, to spread the possible SLAs within a network or to other networks and to indicate the status of compliance of these.

2.5   Network Virtualization

In order that the competition between networks becomes real and that the selection of one or another network does not become an impediment, satellite networks should be abstracted as a service layer with a common structure with other networks. This concept is referred to a network virtualization.

Satellites typically have a unique network architecture because of its transmission features. However, it is due time to convert their components to a new network generation structure, by converting the components of the old broadcast communications to interactive elements that meet the NGN specifications.

The service access points should be common to other networks and should be an SLA negotiation mechanism which enables satellite networks to: compete with other networks, perform handovers with them both to cover the segments terrestrial networks cannot reach or where better SLA is expected.

Stratification and virtualization of the network structure, would convert its elements (terminal, satellite, hub and Gateways) on simple nodes that can interconnect with other networks. The future of these elements would be to operate in a native IP form.

2.6   Information and Content Centric Networking

The inherent ubiquity of satellite systems allows us to clearly reduce the gap in the access to information. On one hand, it allows to send information and contents that otherwise would be very difficult to access. This becomes a reality in sensor networks such as Supervisory Control And Data Acquisition (SCADA) located in remote sites.

On the other hand, the satellite can provide access to information and content in places where other systems fail or whose deployment cost-effective. Thus the application layers in the satellite systems should also be transformed to an NGN structure fully integrated with other networks.

2.7   Connectivity Opportunities

Future Internet must allow, to connect applications where no connections were possible before and should allow innovative ways to connect users. Satellite is, in fact an innovative method to connect users in areas where otherwise would not be possible, or would at least be very difficult. Within this feature inherent to the technology, the satellite networks must be redesigned to allow connections and topologies demanded transparently from the users and applications. If so, they would provide the same services as terrestrial networks, but with the added value of features such as mobility or ubiquity.

New opportunities in configurations, topologies and connectivities include innovative service coverage configurations with flexibility in the allocated power per beam, flexibility in the beam bandwidth, different and manageable connectivity scenarios with flexible channel routing, support for direct connectivity between terminals and multicast mesh capabilities.

Satellite network innovative concepts may also allow new market opportunities. This is the case of the new designed Ka band cellular network concept. With new frequency bands covering multiple spots in a coordinated way or covering selected areas, a frequency reuse is possible and a total coverage in areas where other networks cannot reach.

2.8   Handover

Handover is the essence of the networks integration. Whether it implies a change in the access network or a certain handling of a transport network, it should be performed seamlessly, i.e. in such way the user does not feel it. Therefore a global mechanism that governs the SLA exchange process is necessary. Handover is a key question in satellite networks as multiple handovers due to different technologies (beam handover, device handover, other networks handover) could exist.

Handover affects mainly to the ground segments. The hub gateway or control part of the network has to connect some signaling functions with other networks to allow transparently handovers. It is necessary to implement the control functions which make the SLA of the service remain stable or at least, changes in the SLA should only be allowed under request. Finally, it is also necessary that the user terminals can provide a transparent service. In this sense, transparency enables satellite networks to become a real network capable of providing the same services with the same SLA as terrestrial networks do, but with improved features when necessary.

2.9   Devices Integration

Future Internet is the internet of things. Internet is expected to be shared by million of sensors and devices. Full integration with small sensors, handheld devices and VSAT antennas is necessary. This conflict with the actual concept of satellite networks, where there is a fixed user terminal with a large antenna. While the current terminals should be kept, it is also necessary to develop new technologies that foster the use of satellite in handheld devices and tiny sensors. For this purpose there are two fronts of action.

On one hand, it is necessary to reduce the size of antennas used by the devices. Smaller devices cannot accommodate large antennas and cannot implement the required processing in interactive terminals. It is therefore also necessary to develop new chipsets with smaller modulators and processors.

This applies to both sensors and handheld devices, but in the case of the latter it is important to give them the mobility features. In this case the devices antennas need to be very directive whilst properly handling and controlling the possible replicas. It also implies that satellites potentially used for mobile communications must be situated closer to the earth and transmit enough power so that the devices can easily capture the signal (eg, LEO satellites).

On the other hand, it is possible to continue using general-purpose satellite technology, but in this case feeders that distribute the signal to small devices would be necessary .

All in all, the use of satellite in small devices, requires a revolution in technology, such revolution affects chipset improvements, physical devices, modulators and devices, as well as new techniques that allow greater power concentration in certain areas, such as those based on cellular beams.

3   Conclusion

The potential of IP satellite communications systems is very high, and it is necessary to carry out some technological developments to get all the benefits from Future Internet. This paper summarizes the role of IP satellite communications systems' solutions in terms of applications and services to be offered within the Future Internet concept, as well as the technical enablers that would be necessary to evolve satellite communications towards the new model.

Some adaptations to the Future Internet are easy to achieve, but others require an special effort. In particular, it is necessary to improve the effort in R&D and certain satellite technology elements that would enable the gradual converging towards NGN-compatible models, therefore meeting the requirements of the Future Internet.

References

1. ETSI TISPAN (Telecommunications and Internet converged Services and Protocols for Advanced Networking), http://www.etsi.org/tispan/
2. Future Internet Assembly, http://www.future-internet.eu/
3. Stuckmann, P., Zimmermann, R.: Eur. Comm., IEEE Wireless Communications 16(5), 14–22 (2009), doi:10.1109/MWC.2009.5300298

4. European Future Internet Initiative (EFII), White paper on the Future Internet PPP Definition (February 2 2010), http://www.futureinternet.eu
5. Chuberre, N., Piccinni, M., Boutillon, J., Sanchez, A.A., Bejarano, J.M.R., Liolis, K.: Advanced Satellite Multimedia Systems Conference (ASMA) and the 11th Signal Processing for Space Communications Workshop (SPSC), pp. 162–168 (2010), doi:10.1109/ASMS-SPSC.2010.5586857

Modeling Dynamic Satellite Bandwidth Allocation for Situation-Awareness Applications

Laura Galluccio, Alessandro Leonardi, Giacomo Morabito,
Sergio Palazzo, and Corrado Rametta

Dipartimento di Ingegneria Elettrica, Elettronica ed Informatica,
University of Catania, Italy
{lgalluccio,aleonardi,gmorabi,spalazzo,crametta}@dieei.unict.it

Abstract. Dynamic allocation is a crucial issue when coping with satellite-provided services where many demanding users share a limited bandwidth. This is the case of situation-awareness applications, for example, where several users generate information (possibly multimedia) that must be delivered to a control center through a shared satellite link. In this context, bandwidth allocation should be adaptive so as to support differentiation in the treatment offered by the communication system to distinct data flows, depending on the specific criticality of the individual flows. In this paper we provide an analytical framework for modeling the performance of different approaches which can be adopted to face such a problem. As compared to existing literature in the field, the original contribution of this work is providing a general mathematical framework to support dynamic bandwidth allocation as a function of the criticality of the information being managed.

Keywords: Satellite, Dynamic Bandwidth Allocation, Criticality.

1 Introduction

Satellite services give coverage to remote and scarcely connected areas. This is a very important contribution to foster situation-aware and military operations, as well as to connect faraway villages in emerging countries. However, typically there are only a few access points to the satellite network, denoted in the following as *gateways*. If many users want to access the Internet, they should thus share the same set of gateways. So bandwidth sharing mechanisms are needed. The problem drastically worsens when users request for support of highly demanding real-time applications, such as video or audio, since the satellite bandwidth is costly and should be allocated according to the different traffic priority, i.e. the data criticality. Accordingly, highly dynamic and adaptive bandwidth allocation schemes should be designed. Traditional architectures for services differentiation like IntServ and DiffServ [8] are not sufficient since neither dynamic bandwidth reallocation is supported nor a feedback exists, so that source coding parameters can be tuned run-time. In this paper, differently from previous literature in the field, we mainly focus on mathematical modeling of a satellite service access

G. Giambene and C. Sacchi (Eds.): PSATS 2011, LNICST 71, pp. 19–32, 2011.

system where multiple sources are grouped in clusters. Sources are characterized through their emission process and the criticality of the information they carry.

Numerous models of traffic sources have emerged in the last years, both multimedia and not. Also, some of these models assume satellite communication channels [4,5,11] as likely in situation-awareness operations. However, none of these schemes considers dynamic bandwidth allocation when related to multiple sources transmitting on the same channel, and the priority that traffic could have due to the criticality of the area where it was generated. Similarly to our work, in [9] is proposed a scheduler aimed at sharing the limited uplink resources of a satellite system. In particular, the discussed solution refers to the case where many bursty users emit traffic with different QoS requirements, and different data rates are provided to each terminal to differentiate the QoS priority levels. The approach provides fairness and maximizes the exploitation of the capacity of the system, also in case of time-varying channels.

In this paper we provide a mathematical model to compare the performance of different schemes for situation-awareness operations. More specifically, we present a system where multiple clusters of non gateway devices, not connected to the Internet, are deployed and want to send information through a cluster gateway equipment. All the cluster gateways share a satellite link towards the headquarter of the operations. We assume that the traffic issued by each of these gateways could be characterized by a priority associated to the criticality of the area where it was generated. To this purpose, an intelligent mechanism for the management of the transmission at the gateway output should be devised. This obviously depends on the specific scheme being employed.

Specifically, we develop a framework of the overall system and we use it to test different schemes for situation-awareness operations. We provide this comparison by exploring parameters to support traffic differentiation at the queue in terms of allocated bandwidth, adaptive queue management and adaptive source coding as a function of the criticality level of both the cluster area and the overall system. However, supporting adaptive protocols as a function of the criticality level of the different areas in the system could be complex and costly. In fact, as a consequence of the adaptability of the system, the complexity of the procedures could also imply an increase in the delivery time. This could become a detrimental aspect in scenarios where responsiveness is of primary importance. So, the main contribution of this work is to provide network designers with a tool to design an efficient system by also taking into account cost constraints, reliability and effectiveness issues.

2   Overview

We consider that sources are distributed in a geographic area. Sources located nearby and served by the same gateway are grouped into a cluster C_i. Each sub-area where a cluster is located and, thus, each cluster, are characterized by a certain criticality level as a consequence of a given event occurring in the related area, e.g. a earthquake, a mine explosion, etc. Also, some sources in each area

could become particularly critical. The identification of this criticality can lead
to different approaches:

- *Plain approach* (PA): in this case the behavior of the traffic sources, as well
  as the bandwidth share assigned to each gateway, do not depend on the
  criticality levels. At the gateways, all packets are treated in the same way,
  regardless of the criticality of the information they carry.
- *Packet differentiation at the GW approach* (PDA): differently from the PA
  case, packets treatment at the gateway depends on the criticality of the
  information they carry. We assume that a D-RED buffer management scheme
  is utilized [1], i.e., the gateway applies a RED buffer management scheme in
  which the characteristic parameters depend on the criticality of each source.
  In this way, packets characterized by low criticality level are discarded with
  higher probability than packets with high criticality level.
- *Aggregate criticality-aware approach* (ACA): in this case the bandwidth share
  that will be used by each gateway depends on the criticality of the area
  where the cluster is located and not on the individual criticality levels of the
  sources. Buffer management at the gateway will be the same as in the PDA
  case.

3   Model

In this section we will model the system status. Accordingly, we should consider
both the cluster to which the source node under consideration belongs and the
remaining clusters. Moreover, the criticality of the area where the entire system
is located will impact on the criticality of the specific sub-area where the source
node under consideration belongs. This will also influence its cluster criticality
level. However, observe that the time scale of criticality variations is significantly larger than the one associated to packet transmission or buffer queueing.
Accordingly, in the next sections, we will consider the number of clusters with
high criticality level as assigned and we will perform an analysis of the system
status focused only on a single cluster. Then, final results will be obtained by
weighting the results provided in this case with the probability to have a given
number of critical clusters $NC_{C_1} = l$ over the NC^{max} clusters in the network.

In the following sections we will model the various components of the system.
More specifically, we consider the following states:

- $S^{(Q)}(t)$: state of the server process at the gateway node;
- $S^{(TS.E)}(t)$: state of the emission process at a given source (in the following
  identified as tagged source, TS);
- $S^{(TS.CL)}(t)$: state of the criticality level process at a given source;
- $S^{(C_i-TS.E)}(t)$: state of the emission process at sources in the cluster C_i, other
  than the given source;
- $S^{(C_i-TS.CL)}(t)$: state of the process representing the number of critical
  sources in cluster C_i other than the given source;
- $S^{(C_i.ACL)}(t)$: state of the process representing the criticality level of the area
  the cluster belongs to;

3.1   Process Representing the Criticality Level of the Area

As a consequence of the occurrence of an event, e.g. the explosion of a mine or the invasion of a region, a certain area where various sources are located can become critical. The criticality of an area where a cluster of nodes is located can be represented as a two-states Markov chain. More specifically we identify as $S^{(C_i.ACL)}(t)$ the state of this process. Its state space $\Im^{(C_i.ACL)}$ consists of only two possible states: not critical (0) and critical (1). Also, the transition state matrix can be identified as $Z^{(C_i.ACL)}$ and the stationary state probabilities can be obtained by solving the set of equations as in [10]

$$\begin{cases} \mathbf{\Pi}^{(C_i.ACL)}\mathbf{Z}^{(C_i.ACL)} = \mathbf{0} \\ \sum_{j \in \Im^{(C_i.ACL)}} \Pi_j^{(C_i.ACL)} = 1 \end{cases} \tag{1}$$

where $\mathbf{0}$ is an array of 2 elements all equal to zero. The characterization of the criticality level of the area where the cluster C_i is located is of paramount importance. In fact, in the following, we will see how the emission rates at the sources are conditioned and impacted by this criticality.

3.2   Process Characterizing the TS

We now focus on a generic source in a cluster C_i denoted as *tagged source* (TS). There is a large literature demonstrating that a traffic source can be modeled accurately in the time by means of *Markov Modulated Poisson Processes* (MMPP). According to such models, the process characterizing the behavior of a traffic source depends on the current state of an underlying Markov chain.

The state space of this process can be written by considering that $S^{(TS)}(t) = (S^{(TS.E)}(t), S^{(TS.CL)}(t))$ where $S^{(TS.E)}(t)$ and $S^{(TS.CL)}(t)$ have been introduced above and represent the state of the emission process at the tagged source and the state of the criticality level process at the TS, respectively. Accordingly, the state space $\Im^{(TS)}$ can be expressed as $\Im^{(TS)} = \Im^{(TS.E)} \times \Im^{(TS.CL)}$ where $\Im^{(TS.E)}$ and $\Im^{(TS.CL)}$ represent the state spaces of the two underlying processes. For worth of simplicity in the following we will assume that, again, there are only two states in the process characterizing the criticality of the TS, i.e. $\Im^{(TS.CL)} = \{0, 1\}$. The behavior of the TS depends on the criticality level of the information generated by the TS with respect to the number of sources in the same cluster C_i and the criticality level of the area where nodes of the same cluster C_i are located. Accordingly, if we denote as N_{C_i} the number of traffic sources in the cluster C_i, then the emission rate $\lambda^{(TS)}(t)$ of the TS source will be

$$\lambda^{(TS)}(t) = f\left(N_{C_i}, S^{(TS.CL)}(t), S^{(C_i.ACL)}(t), S^{(TS.E)}(t)\right) \tag{2}$$

where $f(\cdot)$ is a function which can be rewritten as the product of two terms:

- $f_{\text{State}}\left(S^{(TS.E)}(t)\right)$ which is a term that depends only on the current state of the tagged source emission process. This is the emission rate that would be used by TS in case there is no adaptation to the current system condition, i.e. no consideration of the criticality level.

- $f_{\text{TS\_Share}}\left(N_{C_i}, S^{(TS.CL)}(t), S^{(C_i.ACL)}(t)\right)$ which is a term that accounts for the share of bandwidth allowed to the TS source. This function takes into account both the criticality level of the TS when compared to the criticality level of the cluster area and the number of nodes in the cluster C_i. Note that, in case the emission process of TS does not change its behavior depending on the criticality level, then $f_{\text{TS\_Share}}\left(N_{C_i}, S^{(TS.CL)}(t), S^{(C_i.ACL)}(t)\right)$ is only a function of the number of nodes in the cluster C_i, i.e. $f_{\text{TS\_Share}}(N_{C_i})$. On the contrary if the behavior of the TS source depends on the criticality levels, then $f_{\text{TS\_Share}}\left(N_{C_i}, S^{(TS.CL)}(t), S^{(C_i.ACL)}(t)\right)$ is expected to increase as the criticality level of the TS increases with respect to the criticality level of the area where the cluster is located. Similarly, the value of $f_{\text{TS\_Share}}\left(N_{C_i}, S^{(TS.CL)}(t), S^{(C_i.ACL)}(t)\right)$ is expected to increase as the criticality level of the cluster area increases.

Observe that the emission rate values, in general depend on the criticality level of the TS although in some cases it could happen that the emission rate array is independent of it. Also consider that a change in the criticality level of a TS could not result in a variation in the emission rate depending on the state, but only lead to a change in the type of packets being sent (i.e. the packet priority), although the rate could remain the same.

To characterize the emission process at the TS source, we need a transition rate matrix $\mathbf{Z}^{(TS.E)}$ and an array of the stationary state probabilities $\mathbf{\Pi}^{(TS.E)}$ which should satisfy a relationship analogous to the one in eq. (1).

Concerning the TS criticality level, it could be represented using a two-states Markov chain. In order to characterize this Markov chain, we use a state transition rate matrix $\mathbf{Z}^{(TS.CL)}$ which generic element $z_{h,j}^{(TS.CL)}$ represents the rate of transition of the criticality state at each TS from state h to j. Let us note that the rate of transitions from a state to another will depend on the criticality of the area the cluster belongs to. More specifically,

$$
z_{h,j}^{(TS.CL)} = \begin{cases} \alpha_h^{(TS.CL)} & \text{if } j = h + 1 \\ \beta_h^{(TS.CL)} & \text{if } j = h - 1 \\ -\sum_{\substack{j \in \Im^{(TS.CL)} \\ j \neq h}} z_{h,j}^{(TS.CL)} & \text{if } h = j \\ 0 & \text{otherwise} \end{cases} \tag{3}
$$

The terms in the transition rate matrix, $\alpha_h^{(TS.CL)}$ and $\beta_h^{(TS.CL)}$ are related to the process describing the criticality of the area where the cluster is located.

$$
\alpha_h^{(TS.CL)} = \begin{cases} \alpha_0^{(TS.CL)} \\ \alpha_1^{(TS.CL)} \end{cases} \qquad \beta_h^{(TS.CL)} = \begin{cases} \beta_0^{(TS.CL)} & \text{if } S^{(C_i.ACL)}(t) = 0 \\ \beta_1^{(TS.CL)} & \text{if } S^{(C_i.ACL)}(t) = 1 \end{cases} \tag{4}
$$

To completely characterize this process, we also need the array of the steady state probabilities. The latter, together with the transition rate matrix will satisfy a relationship similar to the one shown in eq. (1). Once the underlying Markov chains have been introduced, the state of the TS source can be characterized through the transition rate matrix. If we identify as $s_1^{(TS.E)}$ and $s_2^{(TS.E)}$ two states of the TS emission process and $s_1^{(TS.CL)}$ and $s_2^{(TS.CL)}$ two states of the TS criticality level process, the terms of the transition rate matrix associated to the process describing the global behavior of the TS, can be written as:

$$
z_{h,j}^{(TS)} = \begin{cases} z_{(s_1^{(TS.E)},s_2^{(TS.E)})}^{(TS.E)} & h \neq j \ , j = s_1^{(TS.CL)} = s_2^{(TS.CL)} \\ z_{(s_1^{(TS.CL)},s_2^{(TS.CL)})}^{(TS.CL)} & h \neq j \ , j = s_1^{(TS.E)} = s_2^{(TS.E)} \\ -\sum_{j \in \Im^{(TS)}} z_{h,j}^{(TS)} \text{ with } h = j \\ \qquad j \neq h \\ 0 & \text{in other cases} \end{cases} \tag{5}
$$

The steady state probabilities of the TS can be calculated as a solution of the system in eq. (1) with TS instead of $C_i.ACL$. Consider that with such an approach for modeling the TS source, we are able to limit the explosion in the number of states; in fact, with this MMPP modeling of the TS process, we reduce the number of states to $\Im^{(TS)} = \Im^{(TS.E)} \times \Im^{(TS.CL)}$. This represents a primary advantage if compared to traditional literature in the field where ON-OFF sources require only 2 states but do not allow characterization of the sources with high detail [10]. On the contrary, traditional modeling of video and audio sources implies a drastic increase in the number of states as discussed in [7,10].

3.3 Process Representing the Number of Critical Sources in Cluster C_i Other Than the TS

This process, denoted as $S^{(C_i-TS.CL)}(t)$, can be modeled through a birth-death chain with N_{C_i} states where N_{C_i} has been defined as the number of traffic sources in cluster C_i, including the tagged source. Accordingly, the state space of this process is $\Im^{(C_i-TS.CL)} = \{0, 1, \ldots N_{C_i} - 1\}$. This Markov chain could be described through a transition state matrix $\mathbf{Z}^{(C_i-TS.CL)}$ and an array of the stationary state probabilities $\mathbf{\Pi}^{(C_i-TS.CL)}$. The terms in matrix $\mathbf{Z}^{(C_i-TS.CL)}$ can be written as in eq. (3), with $C_i - TS.CL$ instead of $TS.CL$ The transition rate matrix terms, $\alpha_h^{(C_i-TS.CL)}$ and $\beta_h^{(C_i-TS.CL)}$, depend on the process of the criticality level of the area where cluster C_i is located.

$$
\alpha_h^{(C_i-TS.CL)} = \begin{cases} \alpha_0 \cdot (N_{C_i} - 1 - h) \\ \alpha_1 \cdot (N_{C_i} - 1 - h) \end{cases} \quad \beta_h^{(C_i-TS.CL)} = \begin{cases} \beta_0 h & \text{if } S^{(C_i.ACL)}(t) = 0 \\ \beta_1 h & \text{if } S^{(C_i.ACL)}(t) = 1 \end{cases} \tag{6}
$$

Finally, let us remember that the transition rate matrix and the array of the steady state probabilities should again satisfy a system similar to eq. (1).

3.4 Process of Emission at Sources in Cluster C_i Other Than the TS

Once the emission process at a source TS in the cluster C_i has been identified, we should model the emission process at the other $N_{C_i} - 1$ sources. In order to make the analysis independent of the number of sources multiplexed, we introduce an approximation which consists of focusing only on the TS source and modeling the aggregate of the other sources loading the buffer by means of a single arrival process. In order to minimize the effects of the approximation, we model the aggregate traffic with a MMPP process matching its pdf and autocorrelation function. In fact, it is well known that buffer performance depend on first and second order statistics (i.e. pdf and autocorrelation function) of the source feeding the buffer. So the target is to build an MMPP with specific pdf and autocorrelation function. This is the well known inverse eigenvalues problem [3] which has been solved in the continuous time and discrete time [7] cases. We will refer in the following to the approach for the continuous time case. To this purpose, the SMAQ tool [6] can be employed to obtain a Markov chain by the probability density function and the autocorrelation functions of the process. To apply this strategy, we observe that, when assuming $N_{C_i} - 1$ identical and incorrelated sources, the pdf can be obtained as the convolution of the pdf of the emission processes of the composing sources and the autocorrelation is $N_{C_i} - 1$ times the autocorrelation of the individual emission process.

$$R_{[C_i - TS.E, C_i - TS.E]}(\tau) = (N_{C_i} - 1) \cdot R_{[TS.E, TS.E]}(\tau)$$
$$Pdf_{[C_i - TS.E, C_i - TS.E]}(\delta) = \otimes_{N_{C_i} - 1} Pdf_{[TS.E, TS.E]}(\delta) \tag{7}$$

where $\otimes_{N_{C_i} - 1}$ denotes $N_{C_i} - 1$ times convolution of the pdf of the emission process at the individual TS.

The state of this process is denoted as $S^{(C_i - TS.E)}(t)$ and the state space is identified as $\Im^{(C_i - TS.E)}$ where $\Im^{(C_i - TS.E)} = \{0, \dots K^{(C_i - TS.E)}\}$ having identified as $K^{(C_i - TS.E)}$ the maximum number of states of this process. The state transition rate matrix, $\mathbf{Z}^{(C_i - TS.E)}$, and the array of the emission rates, $\mathbf{\Lambda}^{(C_i - TS.E)}(t)$, can be identified. Observe that the emission rates array depends on the criticality level of the area where the cluster is located. More specifically,

$$\lambda_j^{(C_i - TS.E)} = \begin{cases} \lambda_0^{(C_i - TS.E)} & \text{if } S^{(C_i.ACL)}(t) = 0 \\ \lambda_1^{(C_i - TS.E)} & \text{if } S^{(C_i.ACL)}(t) = 1 \end{cases} \tag{8}$$

Finally the array of the steady state probabilities can be calculated by considering that the state transition rate matrix and the steady state probability array should satisfy a system similar to the one in eq. (1).

3.5 System Process

Let us now model the cluster system. In particular we are interested in modeling a finite buffer of maximum length q_{max} fed by the traffic generated by multiplexing a tagged source TS and the aggregate of the other sources $C_i - TS$ belonging to the same cluster C_i. To this purpose we will model the system as an MMPP/MMPP/1/q_{max} process.

The state of the system MMPP, which has been estimated given that the number of clusters in the system exhibiting a high criticality level NC_{C_1} is l, is denoted as $S^{(\Sigma_l)}(t)$, and can be written as

$$S^{(\Sigma_l)}(t) = (S^{(Q)}(t), S^{(\Sigma_l - Q)}(t)) \quad where \quad \Sigma_l = \{\Sigma | NC_{C_1} = l\} \tag{9}$$

The state space $\Im^{(\Sigma_l)}$, consequently, could be written as the Cartesian product of the state spaces of the server process Q and the complimentary system, except for the server, i.e. $\Sigma_l - Q$. Accordingly $\Im^{(\Sigma_l)} = \Im^{(Q)} \times \Im^{(\Sigma_l - Q)}$.

Observe that the process denoted as $\Sigma_l - Q$ is obtained by exploiting the processes identified before. More specifically, by saying $S^{(\Sigma_l - Q)}(t)$ the state of the $\Sigma_l - Q$ process, this state could be described through a transition rate matrix $\mathbf{Z}^{(\Sigma_l - Q)}$ and an array of the stationary state probabilities, $\mathbf{\Pi}^{(\Sigma_l - Q)}$.

$$S^{(\Sigma_l - Q)}(t) = (S^{(TS)}(t), S^{(C_i - TS.E)}(t), S^{(C_i - TS.CL)}(t), \tag{10}$$

According to the theory of MMPPs [10], the transition rate matrix of the Σ_l MMPP could be represented as the Kronecker sum of the matrices of the underlying processes Q and Σ_l. More specifically $\mathbf{Z}^{(\Sigma_l)} = \mathbf{Z}^{(Q)} \oplus \mathbf{Z}^{(\Sigma_l - Q)}$.

Now let us focus on the buffer which is one of the components of the overall system. We assume a "Late arrival system with immediate access" [2]. Observe that the service rate is not constant but varies as a function of the criticality level of the area where is located the cluster to which the considered TS belongs. Also, the service rate depends on the criticality level of the other clusters in the overall system, i.e. $\mu = \omega(C_i.ACL, S.CL)$ where $S.CL$ is the process representing the criticality of the other clusters in the overall system other than the considered one to which the TS belongs. More specifically, as will be thoroughly discussed in the following, the buffer management procedure uses a threshold such that, when the queue length is lower than this threshold, no packets are dropped. Then, only packets with low criticality level will be discarded; finally, upon reaching the maximum queue length, all packets, independently of their criticality level, will be dropped. Such a server process could be represented by a transition rate matrix $\mathbf{Z}^{(Q)}$ and a steady state probability array $\mathbf{\Pi}^{(Q)}$ which should again satisfy a system analogous to the one in eq. (1).

If we identify as $s_1^{(\Sigma_l - Q)}$ and $s_2^{(\Sigma_l - Q)}$ two states of the $\Sigma_l - Q$ process and $s_1^{(Q)}$ and $s_2^{(Q)}$ two states of the server process, the state transition rate matrix $\mathbf{Z}^{(\Sigma_l)}$ of the overall system can be written as: $z^{(\Sigma_l)}_{(s_1^{(\Sigma_l - Q)}, s_1^{(Q)}),(s_2^{(\Sigma_l - Q)}, s_2^{(Q)})} =$

$$= \begin{cases} z^{(Q)}_{s_1^{(Q)}, s_2^{(Q)}} & s_2^{(Q)} = s_1^{(Q)} + 1, s_1^{(\Sigma_l - Q)} = s_2^{(\Sigma_l - Q)} \\ z^{(Q)}_{s_1^{(Q)}, s_2^{(Q)}} & s_2^{(Q)} = s_1^{(Q)} - 1, s_1^{(\Sigma_l - Q)} = s_2^{(\Sigma_l - Q)} \\ [z^{(\Sigma_l - Q)}]_{(s_1^{(\Sigma_l - Q)}, s_2^{(\Sigma_l - Q)})} & s_1^{(Q)} = s_2^{(Q)}, s_1^{(\Sigma_l - Q)} \neq s_2^{(\Sigma_l - Q)} \\ -\sum[z^{(\Sigma_l)}]_{(s_1^{(Q)}, s_1^{(\Sigma_l - Q)}),(s_2^{(Q)}, s_2^{(\Sigma_l - Q)})} & s_1^{(Q)} = s_2^{(Q)}, s_1^{(\Sigma_l - Q)} = s_2^{(\Sigma_l - Q)} \\ 0 & \text{otherwise} \end{cases} \tag{11}$$

Observe that matrix $\mathbf{Z}^{(\Sigma_l)}$ is the following block matrix:

$$Z^{(\Sigma_l)} = \begin{bmatrix} \mathbf{Z}'^{(\Sigma_l - Q)} & \mathbf{Z}'^{(Q)} & \mathbf{0} & \cdots & \mathbf{0} \\ \mathbf{Z}''^{(Q)} & \mathbf{Z}'^{(\Sigma_l - Q)} & \mathbf{Z}'^{(Q)} & \cdots & \mathbf{0} \\ \cdots & \cdots & \cdots & \cdots & \cdots \\ \mathbf{0} & \cdots & \mathbf{0} & \mathbf{Z}''^{(Q)} & \mathbf{Z}'^{(\Sigma_l - Q)} \end{bmatrix} \tag{12}$$

where

- matrix $\mathbf{Z}'^{(\Sigma_l - Q)}$ is the transition rate matrix of the process $\Sigma_l - Q$ where elements on the main diagonal are substituted with the opposite of the sum of the elements of matrix $Z^{(\Sigma_l)}$ per each row.
- matrix $\mathbf{Z}'^{(Q)}$ is a diagonal matrix where elements on the main diagonal are the transition rates of the process Q from state i to $i + 1$.
- matrix $\mathbf{Z}''^{(Q)}$ is a diagonal matrix where elements on the main diagonal are the transition rates of the process Q from state i to $i - 1$.

By solving the usual set of equations in eq. (1), the value of the stationary state probabilities can be obtained. However, observe that solution of such a system could be very complex due to the excessive number of states. Accordingly, we exploit the Neuts theory [10] to solve the problem by reducing the complexity of a factor proportional to the maximum queue size, i.e. q_{max}. More specifically,

$$
\begin{cases}
\mathbf{\Pi}_0^{(\Sigma_l)} = \mathbf{\Pi}^{(\Sigma_l - Q)} \cdot [\sum_{q \in \Im^{(Q)}} V_q]^{-1} \\
V_1 = Z_{0,0}^{(\Sigma_l)} \cdot [Z_{1,0}^{(\Sigma_l)}]^{-1} \\
V_q = \left(-V_{q-2} \cdot Z_{q-2,q-1}^{(\Sigma_l)} - V_{q-1} \cdot Z_{q-1,q-1}^{(\Sigma_l)} \right) \cdot [Z_{q,q-1}^{(\Sigma_l)}]^{-1}
\end{cases}
\tag{13}
$$

where the array of the stationary state probabilities for the states of the system depends on the stationary state probabilities of the process identified as $\Sigma_l - Q$.

Also, an emission rate array $\Lambda^{(\Sigma_l)}$ can be used to describe the rate of emissions depending on the criticality state of the cluster. Now, we could finally calculate the array of the steady state probabilities for the entire system:

$$
\begin{aligned}
\Pi_y^{\Sigma} &= \sum_{l=1...NC^{max}} \Pi_y^{(\Sigma_l)} \cdot \Pr\{NC_{C_1} = l\} \quad where \\
Pr\{NC_{C_1} = l\} &= \binom{NC^{max}}{l} Pr\{\text{Cluster}_C\}^l \cdot (1 - Pr\{\text{Cluster}_C\})^{NC^{max} - l}
\end{aligned}
\tag{14}
$$

and $Pr\{\text{Cluster}_C\}$ is the probability of the criticality state of the cluster and NC^{max} is the maximum number of clusters in the area.

4   Performance Analysis

In this section we investigate the loss probability and the delay distribution.

Loss Analysis. The buffer at the gateway can both queue or drop packets. To this purpose, the following scheme is employed for buffer management:

- The buffer queues packets independently of their criticality level as soon as the queue length remains below an appropriate threshold q_{Th};
- When the buffer queue length is between q_{Th} and q_{max}, low criticality level packets are dropped with probability 1;
- When the queue length is equal to the maximum q_{max}, all packets are dropped with probability 1, independently of their criticality level.

According to the above mentioned scheme for buffer management, the loss probability at a gateway buffer can be calculated as the sum of the probability to drop packets when the queue length is in $[q_{Th}, q_{max}]$ and when the queue length is the maximum possible q_{max}. More specifically,

$$P_{Loss} = \sum_{s^{(\Sigma)} \in \Im^{(\Sigma)} \text{ s.t.} s^{(TS.CL)}=0, q_{Th} < s^{(Q)} < q_{max}} [\Pi^{(\Sigma)}]_{s^{(\Sigma)}} + \\ + \sum_{s^{(\Sigma)} \in \Im^{(\Sigma)}, s^{(Q)}=q_{max}} [\Pi^{(\Sigma)}]_{s^{(\Sigma)}} \tag{15}$$

Delay Analysis. In order to characterize the delay, we apply the definition of average delay, i.e. $E\{\Delta\} = \int_0^{+\infty} \delta \cdot f_\Delta(\delta) d\delta$, where $f_\Delta(\delta) = \frac{\partial}{\partial \delta} F_\Delta(\delta)$. Consequently we will derive the cumulative distribution of the delay at a gateway node. More specifically, the probability that the delay Δ is lower than δ could be written as:

$$F_\Delta(\delta) = Pr\{\Delta \le \delta\} = \Pi^{(C_i.ACL, Q)} \cdot e^{-\mathbf{Z}^{(C_i.ACL, Q)} \cdot \delta} \cdot \mathbf{v} \tag{16}$$

with $\Pi^{(C_i.ACL, Q)} = Pr\{S^{(C_i.ACL)}(t) = s_{(C_i.ACL)}, S^{(Q)}(t) = s_q\}, \mathbf{v} = [1, 0, \ldots 0]^T$ and $\mathbf{Z}^{(C_i.ACL, Q)}$ is the transition rate matrix whose generic element represents the transition rate from state $(s_{C_i.ACL_1}, s_{q_1})$ to $(s_{C_i.ACL_2}, s_{q_2})$.

5   Model of the System for the Different Approaches

In this section we will explicit the values of some system parameters depending on the addressed approach.

PA Approach. When considering this approach, the behavior of the sources does not depend on the criticality level. Accordingly, at the gateway packets will only be dropped based on a drop tail scheme when the maximum of the buffer length q^{max} is reached. More specifically, processes $C_i.ACL$, and $TS.CL$ will not be needed to characterize the system. So, the array of the emission rates at the TS source can be written as $\lambda^{(TS)}(t) = f_{State}(S^{(TS.E)}(t)) \cdot f_{TS\_Share}(N_{C_i})$.

Moreover, $C_i - TS.CL$ reduces to a process having a single state since all sources in the cluster will be characterized by having the same criticality level. The emission process at other sources in the cluster other than TS reduces to a single emission rate $\lambda^{(C_i - TS.E)}$. Finally, the state of the process $\Sigma_l - Q$ reduces to $S^{(\Sigma_l - Q)}(t) = (S^{(TS)}(t), S^{(C_i - TS.E)}(t))$.

Once the process $\Sigma_l - Q$ has been characterized, each term of the stationary state probability array can be written as $\mathbf{\Pi}^\Sigma = \mathbf{\Pi}^{\Sigma_l}$. The loss probability can be thus calculated as $P_{Loss} = \sum_{s^{(\Sigma)} \in \Im^{(\Sigma)}, s^{(Q)}=q_{max}} [\Pi^{(\Sigma)}]_{s^{(\Sigma)}}$ The average delay can be written as $E\{\Delta\} = \mathbf{\Pi}^{(Q)} \cdot \mathbf{T}^{-1} \cdot \mathbf{H}''' \cdot \mathbf{T} \cdot \mathbf{v}$, where $\mathbf{H}'''$ is defined as follows:

$$\mathbf{H}''' = \begin{bmatrix} -\frac{1}{h'_1} & 0 & \cdots \\ \cdots & \cdots & \cdots \\ 0 & \cdots & -\frac{1}{h'_B} \end{bmatrix} \tag{17}$$

and h'_i with $i \in [1 \ldots B]$ is the generic eigenvalue of matrix $\mathbf{Z}^{(Q)}$.

PDA Approach. In this case, the behavior of the sources depends on the individual criticality level. Accordingly, at the gateway, packets will be dropped with a policy which depends on their critical level. When considering this approach, process $C_i.ACL$ is not needed to characterize the system. So, the array of the emission rates at the TS source can be written as $\lambda^{(TS)}(t) = f_{State}(S^{(TS.E)}(t)) \cdot f_{TS\_Share}(N_{C_i}, S^{(TS.CL)}(t))$. Finally, the state of the process $\Sigma_l - Q$ reduces to $S^{(\Sigma_l-Q)}(t) = (S^{(TS)}(t), S^{(C_i-TS.E)}(t), S^{(C_i-TS.CL)}(t))$. Once the process $\Sigma_l - Q$ has been characterized, each term of the stationary state probability array can be written as $\mathbf{\Pi}^{\Sigma} = \mathbf{\Pi}^{\Sigma_l}$. The loss probability can be thus calculated as $P_{Loss} = \sum_{s(\Sigma)\in\Im(\Sigma),s(Q)=q_{max}} [\Pi^{(\Sigma)}]_{s(\Sigma)}$. The average delay can be written as $E\{\Delta\} = \mathbf{\Pi}^{(Q)} \cdot \mathbf{T}^{-1} \cdot \mathbf{H}''' \cdot \mathbf{T} \cdot \mathbf{v}$, where $\mathbf{H}'''$ is defined as eq. (17) with h'' instead of h'.

ACA Approach. When considering this approach, the behavior of the sources depends on the criticality level of the area where the cluster is located. Accordingly, buffer management at the gateway will be the same ad in the PDA cases. In this case, process $TS.CL$ is not needed to characterize the system since it automatically comes from the aggregate criticality of the cluster. So, the array of the emission rates at the TS source can be written as $\lambda^{(TS)}(t) = f_{State}(S^{(TS.E)}(t)) \cdot f_{TS\_Share}(N_{C_i}, S^{(C_i.ACL)}(t))$. Moreover, $C_i - TS.CL$ reduces to a process having a single state since all sources in the cluster will be characterized by having the same criticality level. The emission process at other sources in the cluster other than TS reduces to a single emission rate $\lambda^{(C_i-TS.E)}$. Finally, the state of the process $\Sigma_l - Q$ reduces to $S^{(\Sigma_l-Q)}(t) = (S^{(TS)}(t), S^{(C_i-TS.E)}(t), S^{(C_i.ACL)}(t))$. Once the process $\Sigma_l - Q$ has been characterized, each term of the stationary state probability array can be written as from eqs. (1). The loss probability can be thus calculated as $P_{Loss} = \sum_{s(\Sigma)\in\Im(\Sigma),s(Q)=q_{max}} [\Pi^{(\Sigma)}]_{s(\Sigma)}$. The average delay becomes: $E\{\Delta\} = \mathbf{\Pi}^{(C_i.ACL,Q)} \cdot \mathbf{T}^{-1} \cdot \mathbf{H}'' \cdot \mathbf{T} \cdot \mathbf{v}$, where $\mathbf{H}''$ is defined as eq. (17), where h'_i with $i \in [1 \ldots B]$ is the eigenvalue of matrix $\mathbf{Z}^{(C_i.ACL,Q)}$.

6   Performance Results

We consider a scenario where $NC^{max} = 10$ clusters share an overall satellite capacity equal to 2 Mbps. In accordance to DVB-RCS communication standard, data sent by the gateways is transmitted in frames of 188 bytes containing a header of 4 bytes and 184 bytes of payload. We assume that all the sources in the clusters maintain active an audio call and use the GSM half rate encoding scheme, based on the *Vector Self-Excited Linear Predictor* (VSELP) codec at bit rate of 5.6 kbit/s. Also we assume that in a given area, a critical event occurs on the average each 10 hours and that the average duration of such critical event is one hour. Regarding the criticality level process, we assume that

- when the area criticality is low, on the average, each source becomes critical after 30 minutes and remains such for 10 minutes.
- when the area criticality is high, on the average, each source becomes critical after 10 minutes and remains such for 30 minutes.

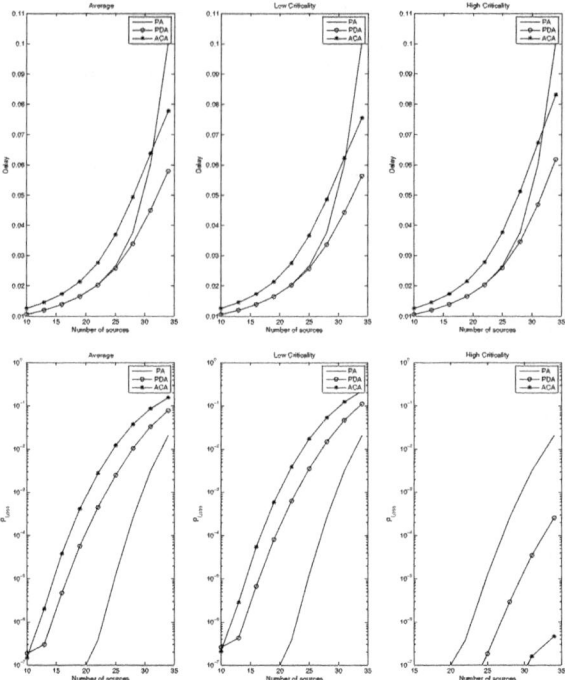

Fig. 1. Delay and loss probability as a function of the number of sources in the a) Average case b) Low criticality case c) High criticality case

Data packets coming from the sources at each gateway are buffered in a queue that can accommodate a maximum of $q_{max} = 30$ packets. In Figures 1, we show the average delay (up) and the packet loss probability (down) versus the number of sources in each cluster. The curves in the above figures have been obtained assuming that the threshold of the buffer size above which low criticality packets are dropped is set equal to $q_{Th} = 15$. In each figure we have three different plots:

- *Left plots* show the average delay and the packet loss probability for any packet (independently of its criticality).
- *Central plots* show the average delay and the packet loss probability for low criticality packets.
- *Right plots* show the average delay and the packet loss probability for high criticality packets.

Finally, observe that in each plot we provide three curves: one for each of the approaches analyzed. In Figure 1 we observe that the average delay increases as the number of sources increases and that PDA approach achieves the lowest delay, always. This is because it drops low criticality packets when the buffer size is higher than 15 and therefore, on the average, packets find smaller buffers; which results in shorter delays. Instead, by comparing ACA and PA we observe that the delay achieved by ACA is higher for most of the values of the number of

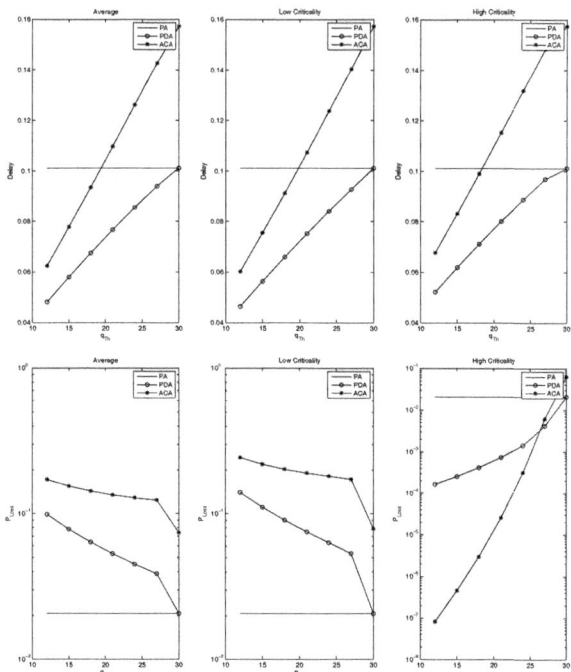

Fig. 2. Delay and loss probability as a function of the buffer threshold q_{Th} in the a) Average case b) Low criticality case c) High criticality case

sources. This is because ACA assigns more capacity to high criticality areas even when such additional capacity is not needed. When the number of sources in each cluster is high (and the traffic load is high as well) ACA achieves lower delay than PA. This is because the additional capacity given to high criticality areas is fully utilized and the delay is lower because ACA drops low criticality packets when the buffer size is higher than 15. In the left plot of Figure 1 we observe that the PA approach achieves the lowest packet loss probability. This is because the other approaches drop packets when the buffer size is higher than 15. However, in the right plot of the same figure we observe that the loss probability of high criticality packets is much lower in the case of the PDA and ACA approaches.

Similar observation can be drawn by observing Figures 2 where we show the average delay and the loss probability versus the threshold q_{Th} assuming that the number of sources in each cluster is equal to 34.

7 Conclusions

Satellite communications provide support to situation-awareness applications. However, areas where such services are deployed are usually equipped with only few gateway ports which should be shared among operators. Such operators usually move in groups, so clustering could be a natural solution to allow gateway sharing. However, due to the high cost of the satellite link especially when users request support of bandwidth demanding applications, bandwidth allocation should be adaptive and dynamically take into account the dynamics of the clusters and the importance, i.e. criticality, of the traffic being issued by the sources. In this paper, differently from the previous literature in the field, we provided a mathematical model of an aggregate of cluster sources and studied the dynamics of the bandwidth allocated based both on the criticality of the individual sources and of the area where the sources are located.

Acknowledgment. This work was partially supported by the European Commission in the framework of the FP7 Project IMSK (contract n. 218038).

References

1. Aweya, J., Ouellette, M., Montuno, D.Y.: A Control Theoretic Approach to Active Queue Management. Elsevier Computer Networks 36(2-3) (July 2001)
2. Bae, J.J., Suda, T., Simha, R.: Analysis of individual packet loss in a finite buffer queue with heterogeneous Markov modulated arrival processes: A study of traffic burstiness and a priority packet discarding. In: IEEE Infocom 92, Florence, Italy (May 1992)
3. Chu, M., Golub, G.: Inverse eigenvalue problems: Theory, Algorithms and Applications. Oxford Scholarship (September 2007)
4. Galluccio, L., Licandro, F., Morabito, G., Schembra, S.: An Analytical Framework for the Design of Intelligent Algorithms for Adaptive-Rate MPEG Video Encoding in Next Generation Time-Varying Wireless Networks. IEEE Journal on Selected Areas of Communications 23(2) (February 2005)
5. Galluccio, L., Morabito, G., Schembra, G.: Transmission of Adaptive MPEG Video over Time-Varying Wireless Channels: Modeling and Performance Evaluation. IEEE Transactions on Wireless Communications 4(6) (November 2005)
6. Li, S.-q., Park, S., Arifler, D.: SMAQ: A Measurement-Based Tool for Traffic Modeling and Queuing Analysis Part I: Design Methodologies and Software Architecture. IEEE Communications Magazine 2(7) (August 1998)
7. Lombardo, A., Morabito, G., Schembra, G.: A Discrete-Time Paradigm to Evaluate Skew Performance in a Multimedia ATM Multiplexer. IEEE/ACM Trans. on Networking 7(1) (February 1999)
8. Mahadevan, I., Sivalingam, K.M.: Quality of Service Architectures for Wireless Networks: IntServ and DiffServ Models. In: Fourth International Symposium on Parallel Architectures, Algorithms, and Networks, I-SPAN (June 1999)
9. Narula-Tam, A., Macdonald, T., Modiano, E., Servi, L.: A dynamic resource allocation strategy for satellite communications. In: Proc. of IEEE Milcom 2004 (October-November 2004)
10. Neuts, M.F.: Matrix-Geometric Solutions in Stochastic Models: An Algorithmic Approach. J. Hopkins University Press, London
11. Xu, J., Shen, X., Mark, J.W., Cai, J.: Adaptive Transmission of Multi-Layered Video over Wireless Fading Channels. Transactions on Wireless Communications 6(6) (June 2007)

Optimum Beam Bandwidth Allocation Based on Traffic Demands for Multi-spot Beam Satellite System

Unhee Park, Hee Wook Kim, Dae Sub Oh, and Bon Jun Ku

ETRI, 138 Gajeong-dong, Yuseong-gu, Daejeon, Korea
{unipark,prince304,trap,bjkoo}@etri.re.kr

Abstract. Multibeam satellite networks can extend the service coverage as deploying its spot beam. It is important to allocate the appropriate resources to downlink multibeams to prevent the unnecessary waste of resources in the satellite system. This paper presents an optimum beam allocation scheme for multi-spot beam satellite system, as beam bandwidth to be allocated is controlled dynamically. We apply the Lagrange theory to obtain the optimization formula for bandwidth allocation of each spot beam in order to meet the total bandwidth constraint. Eventually we can find out the optimum beam profile respect to bandwidth.

Keywords: multi-spot beam, satellite, beam allocation, optimization.

1 Introduction

It is crucial to manage the satellite downlink communication resources effectively in order to maximize the utilization of limited on-board resources over satellite networks.

A future satellite will generate its wide service coverage area by using multiple spot beams. The goal of the satellite system with narrow spot-beams is to support high data rates to user terminals and thus achieve maximum throughput for satellite systems. For this, it needs to study on dynamic resource allocation method among the each beam according to different traffic demands and other channel conditions.

Some attempts have been proposed to adjust the power for optimal resource allocation in [1][2][3]. These techniques have the inherent drawback of high-cost system since it has controllable multi-port travelling wave tube amplifiers (TWTAs) with nonlinearity. To alleviate this problem, we consider the adjustment of the beam bandwidth to maximize the spectral efficiency according to the operation condition instead of the each spot beam power.

In this paper, we propose the adaptive beam bandwidth allocation scheme based on traffic demands and channel condition. The rest of the paper is organized as follows. In section 2, we describe a system configuration of the multi-spot beam satellite network. Section 3 presents how to calculate the optimum bandwidth allocation using a Lagrangian function theory. In the section 4, simulation result shows the validity of the proposed scheme. Finally, the conclusion is drawn in section 5.

G. Giambene and C. Sacchi (Eds.): PSATS 2011, LNICST 71, pp. 33–39, 2011.
© Institute for Computer Sciences, Social Informatics and Telecommunications Engineering 2011

2   System Configuration of Multi-spot Beam Satellite

Multi-spot beam antenna techniques can achieve the beam pattern having a high directional gain as well as narrow beam width. Also it is possible to switch among the beams into several areas according to high speed phase conversion. Therefore, it can make a flexible construction of service via effective operation of limited communication resources. In addition, the total system capacity increases by providing appropriate resources according to traffic distribution and channel conditions, and thus we need a means to allocated reasonable beam resources such as power, bandwidth and spot beams.

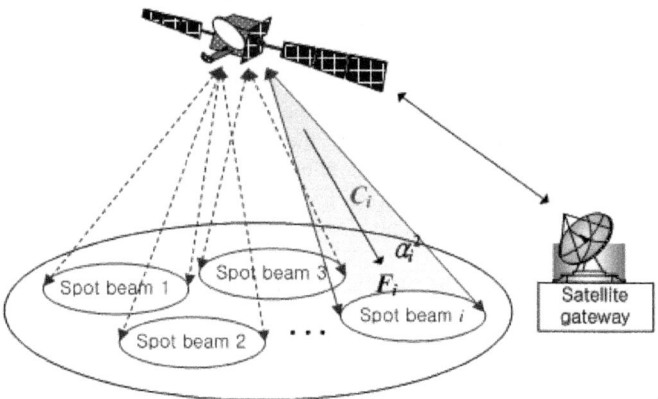

Fig. 1. A multi-spot beam satellite that provides capacity C_i for the ith cell of traffic demand F_i

Figure 1 shows a system configuration of satellite with multiple spot beams. In the network, a multi-spot beam satellite in geostationary orbit and an ensemble of satellite cell sites are deployed. ith beam requires some traffic demand F_i to be served, and multi-beam satellite allocate the capacities to cells. On-board transmission resource (bandwidth in this scheme) is divided and allocated to meet required traffic demand for each spot beam. Using the time sharing scheme for Gaussian broadcast channels, we can obtain the Shannon bounded capacity C_i for ith spot beam as follow [4].

$$C_i = W_i \log_2 \left(1 + \frac{\alpha_i^2 P}{W_i N_0} \right), \tag{1}$$

where α_i^2 represent the signal attenuation and N_0 is noise power density for ith beam. We assume uniform signal attenuation across each narrow spot beam. And also P is the allocated power and uniform for all beams. As the W_i is the ith beam bandwidth to be allocated, it is a considered factor to achieve the reasonable resource allocation and total capacity improvement in this paper. In the next section, we derive the beam profile for bandwidth based on traffic demands and channel conditions to minimize the waste of the resource and to maximize the spectral efficiency, thereby achieving optimum beam allocation method.

3   Optimum Bandwidth Allocation (OBA)

3.1   Derivation of Optimum Beam Bandwidth Profile

As one of the metrics to evaluate the system performance of resource allocation over satellite downlinks, the authors in [2] addressed some tradeoff between different objects for system optimization. They derived the downlink multibeam capacity optimization problem and proposed a schematic method. Motivated by this paper, we formulate an optimum beam bandwidth profile of the parallel multibeams with respect to traffic distributions to achieve the reasonable fairness among users.

In this paper, we only focus on the problem of bandwidth allocation when the total traffic demands exceed the total system capacities. To be the best optimum case, it ought to match the traffic demand F_i and capacity C_i and means the gap between the two should be minimized as possible across the all spot beams, for $i=1,2,\ldots,n$. In view of this, we adopt a square deviation cost function between capacities and traffic demands and formulate our beam bandwidth allocation problem as given here.

$$\text{Minimize } \sum_{i=1}^{n}(F_i - C_i)^2. \tag{2}$$

We have some constraints to solve this optimization problem. First, we do not exceed the more bandwidth required by traffic demand from each beam (or $C_i \leq F_i$) to prevent the unnecessary waste of resources. Second condition is subject to total bandwidth supply such as $\sum_{i=1}^{n} W_i \leq W_{total}$.

Applying the Lagrangian function as $L(W_i, \Lambda) = \sum (F_i - C_i)^2 + \Lambda(\sum W_i - W_{total})$, we have the optimum beam bandwidth profile W_i, which should satisfy as follow equation (3).

$$F_i - W_i \log\left(1 + \frac{\alpha_i^2 P}{W_i N_0}\right) = \frac{\frac{\Lambda W_i \ln 2}{2}\left(1 + \frac{\alpha_i^2 P}{W_i N_0}\right)}{W_i \ln 2\left(1 + \frac{\alpha_i^2 P}{W_i N_0}\right)\log\left(1 + \frac{\alpha_i^2 P}{W_i N_0}\right) - \frac{\alpha_i^2 P}{N_0}}, \tag{3}$$

where Λ is a Lagrange multiplier which is determined by total bandwidth constraint. Nonnegative Λ means that it satisfies the constraint for $C_i \leq F_i$.

We need a verification process to confirm whether the beam bandwidth W_i which can be obtained from (3) is the optimum case or not, since it does not closed-form solutions can be found out W_i in terms of F_i. For this, we provide the solution for W_i through the approximation process in the case of low and high signal-to-noise ratios (SNRs) region.

3.2 Approximation Formula to Verify Optimization Bounds

At the low SNR region of $\alpha_i^2 P/W_i N_0 \ll 1$ using the property $\log_2(1+x) \approx x/\ln 2$ for x is very small, we can derive the first order approximation formula from (3) as follow.

$$
W_i = \begin{cases} \left(\dfrac{2\alpha_i^2 P}{\lambda N_0 \ln 2}\right)\left(F_i - \dfrac{\alpha_i^2 P}{N_0 \ln 2}\right), & if \;\; F_i > \dfrac{\alpha_i^2 P}{N_0 \ln 2} \\[4mm] 0, & if \;\; F_i \le \dfrac{\alpha_i^2 P}{N_0 \ln 2} \end{cases}
\tag{4}
$$

Next, for the high SNR region of $\alpha_i^2 P/W_i N_0 \gg 1$, we use a truncated part of the Taylor series of the $\log_2(1+x) \approx (x/\ln 2) - (x^2/2\ln 2)$, and then we also get the third order approximation, given as

$$
\frac{\lambda(\ln 2)^2}{(\alpha_i^2 P)^3}W_i^3 + \left[\frac{2}{(\alpha_i^2 P)N_0^2} - \frac{F_i(\ln 2)}{(\alpha_i^2 P)^2 N_0}\right]W_i^2 + \left[\frac{F_i(\ln 2)}{(\alpha_i^2 P)N_0^2} - \frac{2}{N_0^3}\right]W_i + \frac{(\alpha_i^2 P)}{2N_0^4} = 0.
\tag{5}
$$

In general, there are many methods to solve cubic polynomials and the form of the roots of cubic equation is determined by discriminant. After investigating the discriminant of (5), we know that it has a real root and two imaginary roots. Here we adopt the only real root to decide the optimization boundary. In the section for simulation part, we will compare these two approximations of (4), (5) and numerical solution (3).

3.3 Updating the Lagrangian Multiplier

From (3) and the constraint for Lagrangian multiplier which is determined by total bandwidth constraint, we have a formula for the Lagrangian multiplier Λ as follow.

$$
\Lambda = \frac{2}{\ln 2}\left[F_i - W_i \log\left(1 + \frac{\alpha_i^2 P}{W_i N_0}\right)\right] * \frac{\ln 2\left(1 + \dfrac{\alpha_i^2 P}{W_i N_0}\right)\log\left(1 + \dfrac{\alpha_i^2 P}{W_i N_0}\right) - \dfrac{\alpha_i^2 P}{W_i N_0}}{1 + \dfrac{\alpha_i^2 P}{W_i N_0}},
\tag{6}
$$

As mentioned earlier, since the beam bandwidth profile is not yield close-form solutions, it needs an intuitive approximation method to find a closed form solution for W_i by using the relationship between total traffic demand and beam bandwidth profile. In

[3], the heuristic method to search the Lagrangian multiplier for the optimal power allocation is presented, and we use this method to find the optimal bandwidth W_i and Λ. First, if a beam requires sum of the traffic demands, F_{sum}, and then, total bandwidth W_{total} will be allocated the beam, we can calculate an initial value Λ_0. Using binary search as rule of thumb, we set $\Lambda_{min}=\Lambda_0/10$ and $\Lambda_{max}=\Lambda_0*10$. We undergo a process to find the optimal Λ in the range of $\Lambda_{min}, \Lambda_{max}$ according to several different simulation scenarios.

$$\Lambda_0 = \frac{2}{\ln 2}\left[F_{sum} - W_{total}\log_2\left(1+\frac{\alpha_i^2 P}{W_{total}N_0}\right)\right]* \frac{\ln 2\left(1+\frac{\alpha_i^2 P}{W_{total}N_0}\right)\log_2\left(1+\frac{\alpha_i^2 P}{W_{total}N_0}\right)-\frac{\alpha_i^2 P}{W_{total}N_0}}{1+\frac{\alpha_i^2 P}{W_{total}N_0}}$$

(7)

Inserting the Λ_0 to (3), the initial optimum beam bandwidth W_i^{opt} for each spot beam can be calculated. However, it is not definitive optimal values, and thus, we need the process to update these allocated bandwidths. A set of updating process is as follow.

i)   Save the sum of the allocated bandwidth $\sum W_i^{opt}$.
ii) If $\sum W_i^{opt} > W_{total}$, then set $\Lambda_{min}=\Lambda$ and let $\Lambda=(\Lambda_{min}+\Lambda_{max})/2$. Find W_i^{opt} again using (3).
iii) If $\sum W_i^{opt} < W_{total}$, then set $\Lambda_{max}=\Lambda$ and let $\Lambda=(\Lambda_{min}+\Lambda_{max})/2$. Find W_i^{opt} again using (3).
iv) The updating process is repeated iteratively until $\sum W_i^{opt}=W_{total}$.

The optimum solution W_i^{opt} achieves the reasonable proportional fairness according to traffic demand for all spot beams. Eventually, we expect to improve the total capacity.

4   Simulation Result

This section presents the simulation results. For the simulation, we assume signal attenuation is uniform across each narrow spot beam and $\alpha_i^2=1$. In addition, the number of beams to be allocated is 28, and total bandwidth constraint is 100Hz. The signal power to noise power spectral density(P/N_0) for each beam is 200w. The minimum traffic demand is 5bps at the first beam, and it goes up by 5bps at each beam up to 140bps(at the 28th beam) like $\{F_i \mid F_1=5, F_2=10,\ldots, F_{17}=135, F_{18}=140\}$.

Table 1. Total allocated bandwidth after optimum bandwidth allocation algorithm

j	1	2	3	4	5	6
$\sum W_i^{opt}$	327.4	77.8	130.5	98.6	112.4	105
j	7	8	9	10	11	12
$\sum W_i^{opt}$	101.7	100.1	99.2	99.5	99.7	99.9
j	13	14	15	16		
$\sum W_i^{opt}$	99.9	100.1	100.1	100		

Table 1 shows the results of total allocated bandwidth after optimum algorithm. It means we obtain the final optimum result at the 16th step($\sum W_i^{opt} = W_{total}$). Next table 2 and figure 2 shows the updating process to search the optimal Λ finally. It looks unstable in the early stage, and then it has stable value gradually.

Table 2. Mortification process to find the optimal Lagrangian multiplier Λ

j	1	2	3	4	5	6
Λ	25.604	140.82	83.213	112.02	97.616	104.82
j	7	8	9	10	11	12
Λ	108.42	110.22	111.12	110.67	110.44	110.33
j	13	14	15	16		
Λ	110.27	110.25	110.26	110.27		

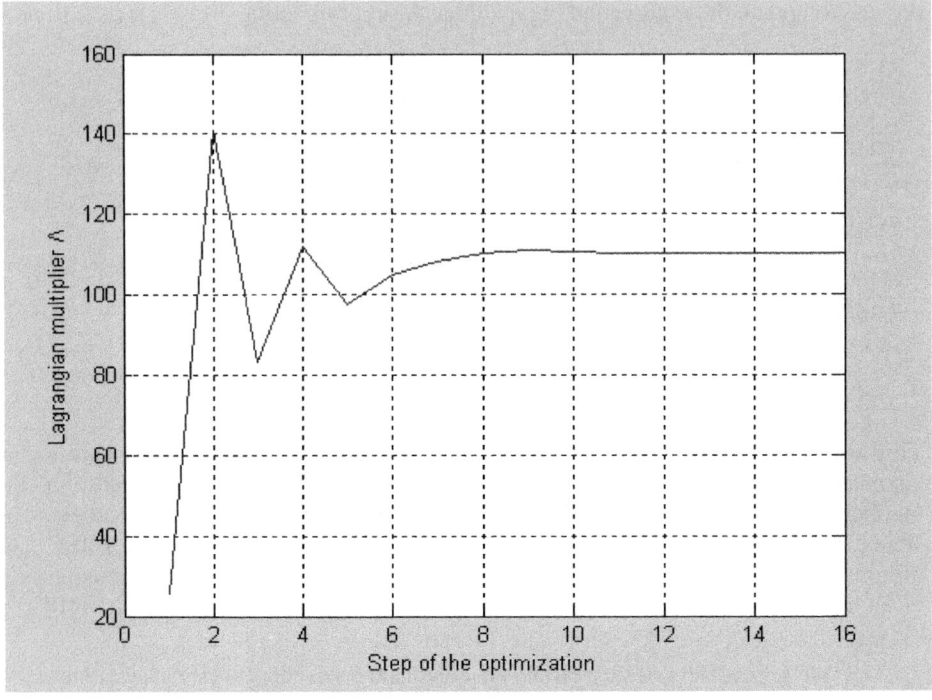

Fig. 2. Mortification process to find the optimal Lagrangian multiplier Λ

We will compare the approximated close-form answers for low and high SNR in (4) and (5) with the numerical solution of (3) to confirm the optimum distribution of beam bandwidth W_i as follow figure 3. We can confirm the bandwidth allocation of (3) is under the optimization boundary regions.

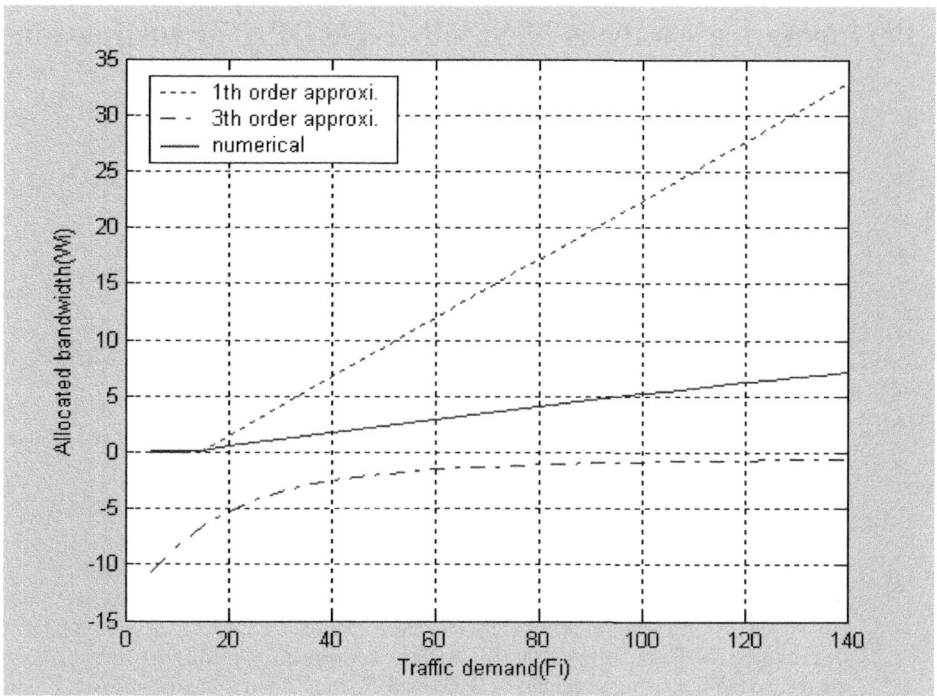

Fig. 3. Optimum beam bandwidth allocation W_i for demand F_i in (3) and its approximated close-form solutions in (4), (5)

5   Conclusion

In this paper, we formulate an optimum beam bandwidth profile of the parallel multibeams with respect to traffic distributions to achieve the reasonable fairness among users. In addition, we show the process to search the optimal beam bandwidth updating the Lagrangian multiplier by heuristics method. The simulation result show the proposed resource allocation scheme is under optimized boundary region to minimize the gap of the traffic demands and total capacity.

References

1. Neely, M.J., Modiano, E., Rohrs, C.E.: Power Allocation and Routing in Multibeam Satellites with Time-Varying Channels. IEEE/ACM Trans. On Networking 11(1), 138–152 (2003)
2. Choi, J.P., Chan, V.W.S.: Optimum power and beam allocation based on traffic demand and channel conditions over satellite downlinks. 10th IEEE Transactions on Wireless Communications 4(6), 2983–2993 (2005)
3. Hong, Y., Srinivasan, A., Cheng, B., Hartman, L., Andreadis, P.: Optimal Power Allocation for Multiple Beam Satellite Systems. In: IEEE Radio and Wireless Symposium (2008)
4. Cover, T.M., Thomas, J.A.: Elements of Information Theory. Wiley, New York (1991)

Performance Analysis of Satellite-HSDPA Transmissions in Emergency Networks

Alessandro Raschellà[1], Giuseppe Araniti[2], Anna Umbert[1], Antonio Iera[2], and Antonella Molinaro[2]

[1] Signal Theory and Communication Dept., Universitat Politecnica de Catalunya (UPC)
Jordi Girona, 1-3, 08034 Barcelona, Spain
{alessandror,annau}@tsc.upc.edu
[2] ARTS Laboratory - Dept. DIMET - University "Mediterranea" of Reggio Calabria
Reggio Calabria - 89100, Italy
{araniti,antonio.iera,antonella.molinaro}@unirc.it

Abstract. In this work we are interested to investigate how the Geostationary (GEO) satellite can be used to employ efficiently High Speed Downlink Packet Access (HSDPA) technology supporting Multimedia Broadcast/Multicast Service (MBMS) in a scenario in which the terrestrial network is not available. An exhaustive simulation campaign has been carried out with the aim to determinate the Satellite-HSDPA (S-HSDPA) performances in different radio channel conditions. In particular, a Good/Bad Channel model has been taken into account to achieve a specific radio information and to be able to evaluate the maximum date rate that can be obtained by S-HSDPA.

Keywords: GEO satellite, HSDPA, MBMS, Good/Bad channel.

1 Introduction

Nowadays, Third Generation (3G) cellular wireless networks, such as Universal Mobile Telecommunication System (UMTS), are able to provide support to mobile videoconferences, multimedia streaming (i.e. TV on mobile phones), broadband transmissions and downloading services [1]. Notwithstanding, a terrestrial-only segment may not be adequate to environments exhibiting high exacting communication requirements. It is the case of so called disadvantaged areas: either rural areas or areas involved in unpredictable catastrophic events. This is why the integration of space segments into the UMTS architecture deserves a special attention from the scientific and industrial communities involved in the design of *Emergency Networks*.

UMTS network, already, foresees also a satellite component, called Satellite-UMTS (S-UMTS), aiming to provide services to mobile users in such scenarios. According to [2], [3] the following important purposes can be highlighted that justify the need for integration of S-UMTS in the terrestrial system: *(i) Geographical complement,* as an extension of covered area where the only Terrestrial-UMTS (T-UMTS) is not adequate or not available; *(ii) Personal universal communications,* as an omnipresent scenario in order to realize an *anyone, anywhere* and *anytime* cellular applications distribution.

G. Giambene and C. Sacchi (Eds.): PSATS 2011, LNICST 71, pp. 40–49, 2011.

In an *Emergency Scenario*, the limited radio resources of satellite component have to be efficiently utilized, in order to allow the information exchange among rescue teams. Indeed, many groups of first responders may need to be established and to share logistic information of various nature, e.g. dispatching, maps, video of the incident area, etc.. The employment of High Speed Downlink Packet Access (HSDPA) technology [4] supporting Multimedia Broadcast/Multicast Services (MBMS) in satellite network can enhance the overall system performance, since on one side HSDPA increases the date rate with respect to standard UMTS and on the other hand, multicast emergency transmissions can be delivered to groups of receivers at the same time; thus avoiding data duplications both in the core network and over the air interface.

Several research works have been conducted, targeted to study the HSDPA for MBMS applications taking into account both terrestrial network and High Altitude Platforms (HAPs). For instance, in [5] authors demonstrated that the maximum number of users served by the High Speed Downlink Shared Channel (HS-DSCH) hardly depends on the User Equipment (UE) speed. Instead, in [6], [7] authors made some performance study of HSDPA for PtM MBMS applications; while in [8], [9] authors introduced the use of HAPs for supporting the terrestrial networks in disadvantaged areas.

This paper aims at investigating the possibility to provide connectivity via Geostationary (GEO) satellite to rescue teams and first responders involved in the incident area. In particular, a preliminary study on Satellite-HSDPA (S-HSDPA) performances have been conducted with the purpose to evaluate the maximum amount of resources (in term of *Data Rate*) that a S-HSDPA can provide. In case of Terrestrial-HSDPA (T-HSDPA) the Adaptive Modulation and Coding (AMC) technique allows changing the kind of modulation depending on the radio channel conditions. In particular, the Channel Quality Information (CQI) includes this information; in fact, the transport block size, the number of used physical channels and the kind of modulation are evaluated from the CQI value [4].

Such information depends on the channel conditions sensed by the UE, exchanged with the Terrestrial Radio Network Controller (T-RNC). In fact, the T-RNC receives the CQI reported from the UE every 2 ms; for instance, if such a CQI indicates that the Quality of Service (QoS) is worsening, the T-RNC can select a more robust ACM level. In case of satellite transmission scenarios different propagation features must be considered to be able to investigate the performance of the S-HSDPA, used in *Emergency Networks*.

Therefore, considering that we took into account both fixed and mobile users, in the case of fixed users is assumed that terminals have a Line-of-Sight (LoS) path loss to the GEO satellite, while for mobile scenario a classic *Good/Bad channel* model has been utilized to take into account the different radio channel conditions. In the *Good/Bad channel*, state fluctuations are due to shadowing phenomena that mainly depend on the mobile environment. Basically, the *Good channel* is characterized by an unshadowed LoS path loss, while the *Bad channel* is related to shadowed periods of Non LoS (NLoS) path loss [10].

The paper is organized as follows. Section 2 provides a general overview of HS-DSCH channel. Then, in Section 3 we illustrated the experiment scenario, including the description of the considered S-UMTS scenario and the radio channel conditions highlighting the differences between *Good* and *Bad channel*. The results of an exhaustive simulation campaign are the focus of Section 4. While, conclusive remarks are given in Section 5.

2 HS-DSCH Description

HS-DSCH is the shared transport channel carrying the user data with HSDPA that is the downlink transmission component of the High Speed Packet Access (HSPA) technology. In a wireless network environment, the selected Energy per Bit-to-Spectral Noise Density (E_b/N_0) is equivalent to a certain Block Error Rate (BLER) for a particular application bit rate. Nevertheless, the E_b/N_0 is not a suitable metric for HSDPA because the bit rate on the HS-DSCH can change every Transmission Time Interval (TTI), by using different modulation schemes, code rates and physical channel codes. Therefore, in the HS-DSCH scenario, the E_b/N_0 is taken over from the Signal to Interference Noise Ratio (SINR) representing a more suitable measurement metric as it is independent of the considered modulation. Equation (1) illustrates this parameter.

$$SINR = SF_{16} \frac{P_{HS-DSCH}}{pP_{own} + P_{other} + P_{noise}} \tag{1}$$

In the equation $P_{HS-DSCH}$ represents the transmission power assigned to the HS-DSCH, P_{own} is the own cell interference, P_{other} is the neighbouring cells interference, P_{noise} is the Additive White Gaussian Noise (AWGN), p is the orthogonality factor (that can be zero in the case of perfect orthogonality), while SF_{16} is the SF value fixed to 16 for the HS-DSCH [4].

The main goal of HSDPA is to raising user data rates enabling a utilization of different applications characterized by diverse QoS requirements. The shorter TTI compared to the WCDMA one and the implementation of the scheduling functionalities in the Node-B, allow a fast adaptation to the channel state variations [4]. Such state variations affect the SINR that in turn, how it will be clarify in Section 4, changes the CQI values.

The state of the satellite channel information is obtained thanks to the uplink data (CQI measurements) sent from the UE in according to the *Good/Bad Channel* model. The long propagation delays could make CQI measurements useless, notwithstanding, different delay compensation techniques are proposed in literature   to solve this kind of problem, for instance in [11] a possible solution is proposed for Ku and Ka band satellite links.

In the next sections the impact of the new radio channel conditions to the SINR will be explain.

3 Experiment Scenario

In this section we explain the main features of the GEO satellite radio interface suggested by the Satellite Earth Stations and Systems Technical Committee of the European Telecommunications Standards Institute (ETSI) and taken into account in our research work. The satellite channel varies from the terrestrial one in terms of the following propagation features: *(i)* longer propagation delay (120 ms satellite to earth delay for GEO satellite); *(ii)* Doppler effects; *(iii)* path loss; *(iv)* multipath fading; *(v)* interferences; *(vi)* effect of the distance from the centre of the area covered by the satellite, being equidistant from all users. Moreover, the satellite radio interface is compatible with the 3GPP WCDMA radio one; hence, it is also compatible with the band at 2 GHz for the 3G devices. Such an issue clearly gives the advantage to smooth over the impact of the *satellite enabled terminals* cost, guarantying interoperability with the 3G environment [12].

In our research study, we considered a multi beam GEO satellite; the scenario is shown in Fig. 1.

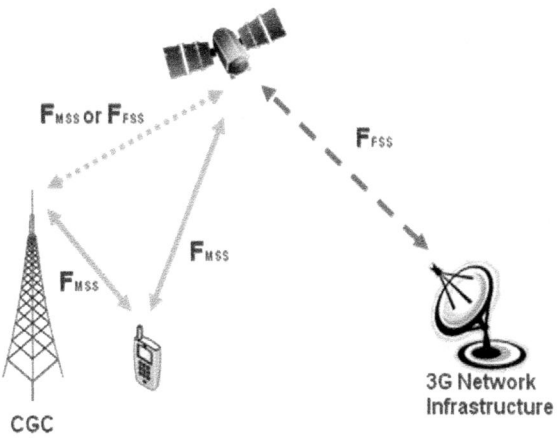

Fig. 1. S-UMTS architecture including a GEO satellite

All the features of the Radio Access Network (RAN) representing the network part are located in the gateway; then we considered that all the UEs distributed in the domain of a particular gateway stay in the coverage area managed by it. The gateway configures the physical layer of the radio interface using the protocol of the 3GPP radio interface, to smooth over the satellite channel features. Hence, the UE radio access parameters are configured by the RAN and sent to UE over the radio interface. From the figure we can notice that there exist two different ways to send data to the UEs; in fact, either a direct link between satellite and UEs and a Complementary Ground Component (CGC), which relays signals to/from the satellite, can be used. In our study we deal with transmission data by the link between the satellite and the UEs.

Different kinds of UEs can work in case of satellite coverage such as handset or vehicular. In the uplink direction the codes orthogonality is kept in 99% of the cases. In our study we considered 62.5 dBW Equivalent Isotropically Radiated Power (EIRP) per spot, a satellite antenna C/I of 12 dB, a GEO satellite located at an altitude of 35800 km covering an area with a radius equal roughly to 600 km.

As we already said, a *Good/Bad Channel* model has been taken into account in our research work; the *Good channel* is related to an unshadowed LOS path loss, while the *Bad channel* is characterized by shadowed intervals of NLOS path loss. In particular, in case of NLOS we considered a further *loss element* of the signal transmitted by the satellite, named *excess path loss*. This contribution of the path loss depends on the following parameters: *(i)* local environment; *(ii)* vehicle heading; *(iii)* link frequency; *(iv)* satellite elevation angle; *(v)* street side. In [13] is highlighted how the first two contributions dominate and how the *excess path loss* is equal to a value between 10 dB (in suburban areas) and 25 dB (in urban situations). In our research work we considered the worse case (i. e. 25 dB) in case of *Bad Channel*.

Moreover, time intervals of *Good* and *Bad channel* are characterized by an exponential distribution. The average time period in the *Good state* (T_g, i.e. when LOS path loss is experienced) and the one in the *Bad state* (T_b, i. e. when instead a NLOS path loss including the *excess path loss* is experienced) depend on the power value sensed by the terminals and on the UE speed varying from LOS to NLOS condition and vice versa. Equations (2) and (3) illustrate T_g and T_b [12].

$$T_g = \frac{e^{\eta} - 1}{f_D \sqrt{2\pi\eta}} \tag{2}$$

$$T_b = \frac{1}{f_D \sqrt{2\pi\eta}} \tag{3}$$

Where f_D is the Doppler shift obtained as:

$$f_D = \frac{V f_p}{c} \tag{4}$$

Where V represents the average UE speed, c is the light speed and f_p is the carrier frequency. While, η defined by (5) is the ratio between the threshold power level P_t, which enables to distinguish amongst the *Good channel* and the *Bad channel*, and the Root Mean Square (RMS) value of the interference power value, P_{rms}.

$$\eta = \frac{P_t}{P_{rms}} \tag{5}$$

The fading model that we have taken into account in our simulations is also strictly related to the channel condition. In particular, in case of *Good channel* we considered the state with the Rician probability Density Function (PDF) that represents unshadowed areas with relatively high received signal power; while in case of *Bad channel* we took into account the state with Rayleigh-Lognormal PDF, representing shadowed areas with low received signal power [14].

4   Obtained Results

The purpose of the realized simulation campaign is to investigate how the satellite can be utilized to employ efficiently HSDPA technology supporting MBMS in

Emergency Network. To do that we studied how the *Good* and *Bad channel* conditions affect the SINR that has to be assured and then the maximum *Data Rate* that the Satellite-HS-DSCH (S-HS-DSCH) can support. Furthermore, we took into account the UE speed equal to 3 km/h.

Firstly, as the SINR is strictly linked to the assigned S-HS-DSCH transmission power and on the state of the channel, in Fig. 2 we illustrate the SINRs that have to be guaranteed when changing the cited parameters. As expected the assured SINR improves when the S-HS-DSCH assigned power value increases and in case of *Good channel*.

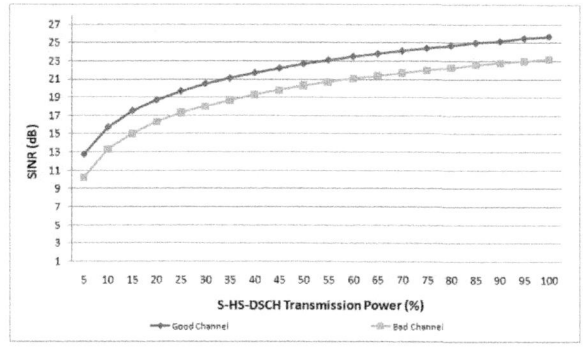

Fig. 2. SINR vs S-HS-DSCH Transmission Power

Moreover, in HSDPA, the maximum *Data Rate* depends on the CQI that, in turn, is closely connected to the SINR values. Indeed, the CQI parameter varies in according to channel conditions experienced by the UE. For each CQI value are associated the transport block size, the number of used physical channels, the modulation technique and, finally, the *Data Rate* (see Table 1).

It is worth noting that values reported in Table 1 are applicable for both terrestrial and satellite systems. In this work, the Satellite-HSDPA coding and modulation chains has been implement by using Simulink to determinate the maximum *Data Rate* that satellite system can provide. An exhaustive simulation campaign has been carried out.

Fig. 3 and 4 illustrate the SINRs to be guaranteed for different values of CQIs and BLERs, when the state of channel is *Good* and *Bad* respectively. Obviously, in case of *Good channel* the better radio channel condition allows obtaining a higher CQI keeping the same SINR and as a consequence a higher *Data Rate*. From the matching of obtained results and the values reported in Table 1, it is possible to determinate the maximum *Data Rate*, that satellite systems can provide in *Good* and *Bad* channel conditions for a desired Block Error Rate (BLER) and a given HS-DSCH transmission power.

Table 1. HS-DSCH Parameters

CQI	Modulation	#Physical Channel	Data Rate (kbps)
1	QPSK	1	68.5
2	QPSK	1	86.5
3	QPSK	1	116.5
4	QPSK	1	158.5
5	QPSK	1	188.5
6	QPSK	1	230.5
7	QPSK	2	325.5
8	QPSK	2	396
9	QPSK	2	465.5
10	QPSK	3	631
11	QPSK	3	741.5
12	QPSK	3	871
13	QPSK	4	1139.5
14	QPSK	4	1291.5
15	QPSK	5	1659.5
16	16-QAM	5	1782.5
17	16-QAM	5	2094.5
18	16-QAM	5	2332
19	16-QAM	5	2643.5
20	16-QAM	5	2943.5
21	16-QAM	5	3277
22	16-QAM	5	3584
23	16-QAM	7	4859
24	16-QAM	8	5709
25	16-QAM	10	7205.5
26	16-QAM	12	8774
27	16-QAM	15	10877
28	16-QAM	15	11685
29	16-QAM	15	12111
30	16-QAM	15	12779

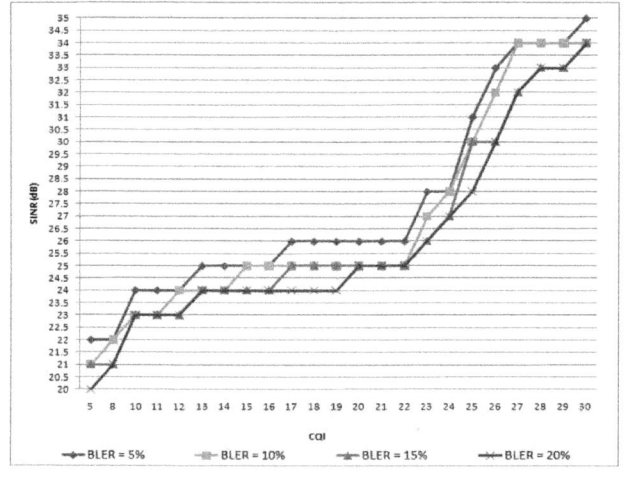

Fig. 3. SINR vs CQI in case of Good Channel

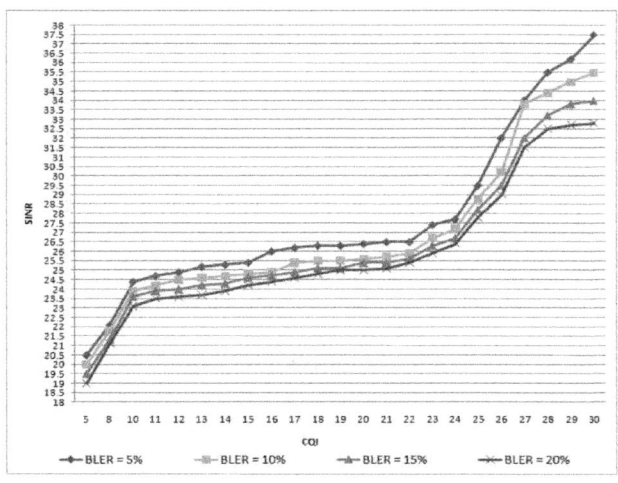

Fig. 4. SINR vs CQI in case of Bad Channel

Fig. 5 is a graphical representation of this *Data Rate*, obtained when varying S-HS-DSCH transmission power and channel state, given a target BLER of 10%. In case of T-HSDPA the transmission power assigned to the T-HS-DSCH can vary between 35 and 60% of the overall transmission power, nonetheless we assumed that the values of S-HS-DSCH transmission power can reach out the overall transmission power because we are supposing a GEO satellite used only for S-HSDPA applications.

From the figure clearly emerges how the transit from *Good channel* to *Bad channel* lead to a considerable reduction of the *Data Rate*. This means that in case of change of the radio condition channel some rescue team could lose suddenly the connection. Therefore, obtained results besides giving a performance analysis of S-HSDPA supporting MBMS applications, can be further utilized to introduce the main requirements of possible RRM policies that aim to maximize the system capacity and the number of services that a satellite network can provide in a disaster area.

Hence, to avoid losing the connection of a considerable group of rescue teams, the following approaches could be taken into account: *(i)* during the low intervals of *Bad channel* the BLER target can be made worse, then changed to 20%, decreasing the QoS but at the same time safeguarding the connection of several UEs; *(ii)* the greater *Data Rate* of *Bad Channel* (i.e. 631 kbps) could be used to provide the PtM services. So doing it will be possible to have one (or more than one) multicast group of rescue teams able to receive streaming services without loss of connectivity employing the PtM transmission of HSPA. The remaining resources will be used to provide delay tolerant services, such as file downloading, messaging and so on. Hence, rescue teams can obtain continuously MBMS data with a lower but guaranteed *Data Rate* (in the *Bad Channel* conditions) and they can take advantage of the additional *Data Rate*, provided from the better conditions of the *Good state*, for downloading delay tolerant files such as maps or messages from other teams; *(iii)* the power assigned to S-HS-DSCH can be dynamically modified in according to the channel conditions. Indeed, from Fig. 5 one can notice that the greater *Data Rate* can be obtained using only the

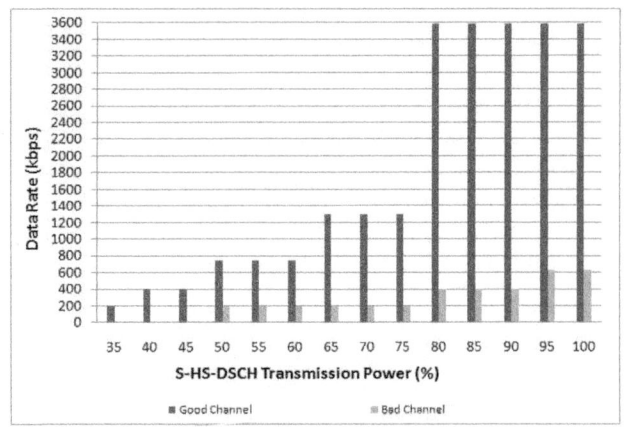

Fig. 5. Data Rate vs S-HS-DSCH Transmission Power, with BLER 10%

80% of the overall S-HS-DSCH transmission power in case of *Good channel* and the 95% in case of *Bad channel*. Hence, the saved power could be used to provide services on dedicated channels.

5 Conclusions

In the last few years the UMTS is supported by a satellite component, defined as S-UMTS. The goal of this research work was to investigate the HSDPA performance in a Satellite environment (S-HSDPA) in which the terrestrial network is not available. The maximum Data Rate has been evaluated when a Good/Bad Channel model was taken into account for the satellite link. Furthermore, as a future work obtained results can be used to introduce RRM policies with the purpose to improve the performance of S-HSPA system.

References

1. Holma, H., Toskala, A.: WCDMA for UMTS – Radio Access for Third Generation Mobile Communications. John Wiley and Sons, Chichester (2004)
2. Narenthiran, K., Karaliopoulos, M., Evans, B.G., De-Win, W., Dieudonne, M., Henrio, P., Mazzella, M., Angelou, E., Andrikopoulos, I., Philippopoulos, P.I., Axiotis, D.I.: S-UMTS access network for broadcast and multicast service delivery: the SATIN approach. Int'l. J. Satellite Commun. and Net (January–February 2004)
3. Huber, J.F., Weiler, D., Brand, H.: UMTS, the Mobile Multimedia Vision for IMT-2000: A Focus on Standardization. IEEE Communications Magazine (September 2000)
4. Holma, H., Toskala, A.: HSDPA/HSUPA for UMTS – High Speed Radio Access for Mobile Communications. John Wiley and Sons, Chichester (2004)
5. Raschellà, A., Umbert, A., Araniti, G., Iera, A., Molinaro, A.: On the Impact of the User Terminal Velocity on HSPA Performance in MBMS Multicast Mode. ICT, Marrakech (2009)

6. Chaudhry, A., Khan, J.Y.: A Group Based Point-To-Multipoint MBMS Algorithm over the HSDPA Network. ISWPC, Melbourne (2009)
7. Vartiainen, V., Kurjenniemi, J.: Point-To-Multipoint Multimedia Broadcast Multicast Services (MBMS) Performance over HSDPA. PIMRC, Athens (2007)
8. Raschellà, A., Araniti, G., Iera, A., Molinaro, A.: High Altitude Platforms: Radio Resource Management Policy for MBMS Applications. PSATS, Rome (2009)
9. Raschellà, A., Araniti, G., Iera, A., Molinaro, A.: Radio Resource Management Policy for Multicast Transmissions in High Altitude Platforms. ICSSC, Edinburgh (2009)
10. Lutz, E., Cygan, D., Dippold, M., Dolainsky, F., Papke, W.: The Land Mobile Satellite Communication Channel-Recording, Statistics, and Channel Model. IEEE Transactions on Vehicular Technology 40(2) (May 1991)
11. Párraga, C., Kissling, C.: Delay Compensation Strategies for an Efficient Radio Resource Management in DVB-S2 Systems. ISWCS, Siena (2005)
12. Martin, B., Chuberre, N., Lee, H.J.: IMT-2000 Satellite Radio Interface for UMTS Communications over Mobile Satellite Systems. ASAM, Bologna (2008)
13. Hess, G.G.: Land-Mobile Satellite Excess Path Loss Measurement. IEEE Transactions on Vehicular Technology VT29(2) (May 1980)
14. Kaufman, O., Lyandres, V.: Model of the Land Mobile Satellite Channel Based on the Stochastic Differential Equations Approach. EEEI, Tel-Aviv (2002)

THAUMAS: Study on Swift Broadband Improvement for Compliance with SESAR Requirements

Laurent Bouscary, Jean-Christophe Dunat, and Ababacar Gaye

Astrium Satellites, 31 rue des Cosmonautes, 31402 Toulouse, France
{laurent.bouscary,jean-christophe.dunat,
khalifa-ababacar.gaye}@astrium.eads.net

Abstract. Future Air Traffic Management (ATM) communications will rely on advanced satellite and ground based communication means. THAUMAS studies the possibility of using, for the satellite based component, an extension of the Inmarsat SwiftBroadband (SB) satcom service, the current version being already widely deployed and successfully operated by many airlines. This document summarizes the overall THAUMAS approach for the system design. Starting from the analysis of the current limitations of SB, this approach is based on a gradual evolution starting from the key upgrades while starting as early as possible feasibility studies on topics judged as less urgent regarding today's ATM needs. The final result will be the so called SwiftBroadband-Safety (SB-S).

Keywords: ATM, SESAR, SwiftBroadband, safety, sitcom.

1 Introduction

To modernize the existing solutions and to anticipate on future needs, the European Union is moving towards the implementation of the Single European Sky ATM Research (SESAR), a new Air Traffic Management (ATM) framework (combining technological, economic and regulatory aspects). Satellite communications for Air/Ground communications (voice and data exchanges between cockpits and flight control centres) have an important role to play in this future ATM infrastructure, both in Europe and in the rest of the world.

In coordination with the European Commission, Eurocontrol, Air Navigation Service Provider and the SESAR consortium, the ESA[1] ARTES 10 Programme ("Iris") intends to define and develop the use of Satcom for ATM communications in the future ATM system defined by SESAR. As part of the activities endorsed by ESA, THAUMAS is working on extending the Inmarsat SwiftBroadband system. Still within Iris, several operator studies are on-going to define business cases for the use of satcom for ATM.

Within THAUMAS, the system proposed to support future ATM communications is based on an extension of the Inmarsat Swift Broadband satcom service, the current version being already widely deployed and successfully operated by many airlines.

---

[1] SA : European Space Agency.

G. Giambene and C. Sacchi (Eds.): PSATS 2011, LNICST 71, pp. 50–59, 2011.
© Institute for Computer Sciences, Social Informatics and Telecommunications Engineering 2011

The current Swift Broadband system is already compliant with a large portion of the Iris System Requirements Document (SRD) [2] requirements, but not fully compliant, thus explaining the rationale for its extension. It is to be noted that the overall features required by the SRD correspond to an objective (in terms of performances, safety, coverage, etc) that could be required in a real operational environment by 2020+ (to let some time for the technologies to maturate; the aircrafts and human operators to adapt, and the market to develop).

The objective of THAUMAS is to propose a satellite communication system to SESAR compatible with mandatory International Civil Aviation Organisation (ICAO) provisions that best answers: the users' requirements, the end-to-end concept of operations, at the least possible cost for the airspace users and finally which can be operated by a certified Communication Service Providers (CSP). Ultimately this satellite communication infrastructure is intended to be included in the SESAR ATM Master Plan.

2   Review of SRD Requirements

A first important objective of the study consisted in reviewing the SRD requirements and in determining the areas of current compliance, and the areas of requiring upgrades to finally reach a full compliance with most of the SRD requirements.

The review of the requirements has shown that the current SB protocol, as it is defined today, does not fully comply with all the requirements, but still offers a very good baseline for a future system that could offer:

- Compatibility with low gain aircraft antenna: additional bearers (i.e. coding rate, modulation scheme and symbol rate) will be made available to ensure compatibility with low gain antenna, even in low elevation situations,
- Improved transmission latency for small messages, using improved random access mechanisms and satellite resources management between regional and narrow beams[2],
- Fast system recovery and efficient redundancy management to minimize unavailability of service,
- Options for a decentralised ground segment architecture,
- Interface with ATN[3]/OSI[4] and ATN/IPS[5] protocols, through the implementation of dedicated gateways.

It was identified that partial compliance may remain on transmission latencies, and service area.

---

[2]   Each Inmarsat 4 satellite provides one global beam, 19 regional beams and around 228 narrow beams. Although communications can be managed within the regional beams, traffic is usually handled in narrow beams to take advantage of better satellite performances. The management of the terminal (when in standby) is done using the regional beams.

[3]   ATN: Aeronautical Telecommunications Network.

[4]   OSI : Open Systems Interconnection.

[5]   IPS : Internet Protocol Suite.

Transmission latencies will be deeply investigated in the next steps, looking in parallel at the end-to-end performances (also considering proposed improvements), as well as potential updates in the definition of the latency requirements (more suitable with GEO[6] satellites constraints).

For the service area, issues come from high latitude where the visibility of the satellite may not be possible during manoeuvres. Various options could be considered. Like additional antenna could be installed on-board the aircraft (this is not the preferred option from an airliner perspective), additional satellites with non-GEO orbits (for instance HEO[7]) could be used, but the business case could be more tricky, or availability of the services could be relaxed during manoeuvres. This last option seems the most realistic one, although implies a decision from aviation to relax requirements in some specific operational cases.

Some critical points were raised during the requirements' review. They concern the geographical area over which the satcom service will be delivered, the definition of the concept of operation for future communication means, the targeted ground segment infrastructure and the communication standard performances.

2.1   Main Geographical Area

Looking at traffic estimations and predictions, ECAC[8] is not a homogenous domain since air traffic is mainly concentrated in Central Europe. It has been proposed to concentrate the analysis on the FABEC[9] area (the Functional Airspace Block composed of Belgium, France, Germany, Luxembourg, the Netherlands and Switzerland) as the requirements for this high density airspace will include the need of the other airspaces.

In parallel, investigations will be required on the need for service provision in high latitude countries would dramatically constrain the system design for a relatively small part of the traffic.

It was stressed that the use of satcom in TMA[10] airspaces is a challenging requirement, especially at high latitudes. In those types of airspace, aircraft are manoeuvring to maintain separation constraints. Should the requirement on the satcom system remain equivalent to the ones on the L-DACS[11] system (very high availability and short message latency), it will be necessary to install multiple antenna on-board the aircraft to keep the satcom link alive. This complex (and costly) installation will be required for TMA only, since ENR[12] and ORP[13] airspaces have less stringent requirements.

In addition, TMA area will be equipped with both VDL-2[14] and L-DACS capabilities, covering rather restricted areas around the airport, hence living room for

---

[6]   GEO: Geostationary Earth Orbit.
[7]   HEO: Highly Elliptical Orbit.
[8]   ECAC: European Civil Aviation Conference.
[9]   FABEC : Functional Airspace Block Europe Central.
[10]   TMA : Terminal Manoeuvring Area.
[11]   L-DACS : L-band Datalink Air-ground Communications System.
[12]   ENR: En –Route.
[13]   ORP: Oceanic, Remote, Polar.
[14]   VDL-2 : VHF Data Link mode 2.

efficient frequency re-use. Thus, the provision of a third data link over satcom in TMA could be considered less mandatory than in ENR (and obviously ORP) regions, especially in low/medium density airspaces.

All those aspects will have to be considered for the consolidation of the satcom usage in TMA airspaces.

2.2 Dual Data Link Concept and Boundaries of the System

The current concept of operation proposed by the SESAR Definition Phase is based on the idea that the future satcom system will work jointly with the future L-DACS systems, both of them working "in complement of VDL2/ATN to support the new most demanding data-link services" (refer to [3]).

It is understood that SESAR JU still have to clarify and to confirm both the dual-link concept (in the sense L-DACS + satcom) and the integration of the new dual-link with other communications means (especially VDL2).

The dual link concept could be interpreted either as a simultaneous transmission approach, the messages being duplicated over the two links or as a complementary approach with transmission over a nominal link, the other one being kept for hot redundancy. The comparison of the two concepts concluded that the latter approach seems more appropriated. The basic idea is to consider the satcom system and the terrestrial system as complementary in the sense that messages can be transmitted either by one or the other system, each system providing a backup mode in case of the failure of the other one.

2.3 Ground Segment Architecture

Considering the ground segment of the satcom system, one point to be confirmed is the need for a distributed architecture. According to the FAB evolution, in the coming years, air traffic flows will not be constrained anymore by national boundaries and ground communication infrastructures will thus be rationalized. Satcom infrastructures for ATM will have to follow this new trend. Hence, the probability of a need for one ground Earth station (GES) per country seems very low, and ground satcom infrastructure is more likely to be shared among all FABs.

The satcom system would have to be able to support the handover of an aircraft from one FAB to another. This would not necessarily mean a GES handover, but rather a handover between FABs performed at an upper layer.

2.4 Communication Standard Performances

The definition of the communication standard requirement is a tricky point. So far, performances requirements are defined for a single user, based on the information available in Communications Operating Concept and Requirements (COCR) [1]. However, there is a need to define the test conditions, i.e. load conditions of the network, under which those performances should be reached.

In other words, the implementation of the SRD requirements into system specifications is suitable to verify the performances required per aircraft.

For the full traffic aspects, the critical point remains to define the way to validate that the overall behaviour of the system is in line with the end-users' expectations.

Concerning voice services, for the time being, the use of satcom is mainly foreseen in oceanic and remote areas. In addition, voice will essentially be used as a safety back-up means of communication in case of a) incoherence/erroneous data transmission: in this case, point to point communication between the pilot and the controller could be sufficient or b) loss of the data link: in case of satcom, the voice communication will probably not be available either. In this case, considering satcom voice as back-up to satcom data link may not be of interest.

The operational conditions for the usage of voice shall be specified together with the associated performance requirements as soon as possible.

3   System Design Evolutions

The SB system already offers flexible adaptation to propagation channel, as well as flexible resources management. This system has been designed taking benefit from the last generation of Inmarsat 4 satellites. The satellite payload has a transparent, bent-pipe digital signal processor (DSP) that provides very fine granularity in the allocation of bandwidth to the various beams, and also enables the generation of a variety of different types of beams, with the required pointing.

The overall so called SB-S (Swiftbroadband-Safety) system is being defined to comply with "Iris System" definition specified in the SRD.

3.1   Decentralized Ground Segment

Some of the requirements expressed in the current SRD do not seem to come from technical aspects, but more from institutional issues. The requirement for defining a system compatible with a decentralized architecture is one of them. This point is critical, with major impacts on the telecommunication resources management.

The current BGAN resource management scheme is centralized by design and is based on the following components to ensure an optimized allocation:

o   Payload Control System (PCS): optimizes the payload configuration to accommodate the channels demand (received by the GRM).

o   Frequency Planning System (FPS): analyses the resource usage and tries to anticipate (long-term trend) the future demands to prepare adequate frequency planning in advance of peaks.

o   Global Resource Manager (GRM): manages the radio resource (channels) for the entire satellite. It answers channel requests from LRM and request for additional missing channels to PCS.

o   Local Resource Manager (LRM): allocates slots within its channels and analyses the demands from beams up to requesting for additional channels to GRM.

Using this architecture, the current capacity and resources allocation to LRM (located in GES) can be static, semi-static or dynamic, the current mode of operation being the dynamic one, in order to maximize the satellite capacity usage.

Several resource management architectures have been studied: from partially to fully decentralized. The current version of SB is partially compliant with ground segment Iris requirements as it does not currently support a fully decentralized architecture. However, moving to a decentralized architecture is nevertheless feasible, even if some components require staying centralized (e.g. the PCS is a central component linked to the satellite payload). Note that in the light of these requirements, a decentralized approach (not only applicable to the SB case) would have several impacts on the existing elements by reducing the system flexibility and the efficiency of the resource usage, thus leading to:

- o A lack of resource availability in one GES whilst other GES resources are underused
- o Over-dimensioning of communication resources capacity to mitigate this risk
- o Pre-emption is less efficient (a GES can only pre-empt resources within the local entity)

At this stage, resource management aspects have been qualitatively evaluated in front of the SRD requirements. Once the communications scenarios will be mature enough, quantitative study will complement this qualitative work on in the next phases, focusing on capacity analysis and sizing, reconfiguration time for each type of nominal scenario and reconfiguration time for each type of fault-back/recovery scenario.

3.2 Redundancy Management

The impact of the use of two satellites (one for redundancy purpose) and the need for a fast switching (meaning in less than a given duration defined at system level) and safe switching (meaning that all the ATM traffic on the nominal satellite shall be preserved and fit on the backup one) between the nominal and the backup satellite on the PCS has also been investigated. Several options have been envisaged such as:

- o The backup satellite is pre-configured with the same channels as the ones of the nominal
- o The backup satellite (when in backup mode) does not host any non ATM traffic or hosts only a limited non ATM traffic
- o The backup satellite (when in backup mode) hosts non ATM traffic without limitation

All options are technically achievable, and the final selection of the final solution is still to be done, considering the impact on the business model.

3.3 Interoperability with "Non THAUMAS" Elements

Non-THAUMAS elements (or non-Iris elements) correspond to other satcom systems than the ones specified by the SRD (ie GEO/HEO L-band satellites using SB-S protocol for THAUMAS) as well as terrestrial solutions. It's worth mentioning that the definition of the rules for choosing a communication means or another (for instance between terrestrial and satcom technologies) has been considered outside of the scope of the system design activities.

Interoperability between the designed system and "non-THAUMAS" elements concerns other satcom systems, as well as the terrestrial ATM backbone.

Interoperability with other GEO systems could be achieved by re-using the same aircraft terminals and protocol stack. In order to achieve interoperability, the performances of the space segments must be equivalent and the ground segments must be compliant with the future system specifications.

Concerning HEO systems, interoperability could be achieved either by defining a single protocol stack compatible with both GEO and HEO satellites or by using two protocol stacks taking benefits from Software Define Radio technologies, which is the preferred solution. Another option is to consider the HEO system as an independent one, similar to a terrestrial network.

This third approach is also valid for interoperability with LEO[15] systems and it is proposed to keep it for the future system since it seems difficult to define a single communication standard compatible with both GEO and LEO types of orbits. Defining a unique interface between the aircraft Communication Management Unit (CMU) and the various Satellite Data Units (SDU), as it is currently the case for the provision of ACARS services (Iridium units essentially emulate an Inmarsat SATCOM SDU), would overcome the difficulty.

When it comes to interoperability with the terrestrial ATM backbone, the THAUMAS ground segment connectivity will have to be aligned with the network concepts brought by the PENS project (Pan European Network Services). This initiative from EUROCONTROL and ANSPs aims at providing a common IP based network service across the European region covering voice and data communication and providing efficient support to existing services and new requirements that are emerging from future Air Traffic Management (ATM) concepts.

The key requirements for support of ATN-IPS relate to the support for IPv6 service, which is a native service offering of 3GPP[16] networks, on which the Family-SL communication standard is based.

The future ATN-IPS IPv6 network architecture, addressing and routing policies require further clarification; however, the implementation and configuration of an IPv6 service capability is not seen as a significant risk item for the Swift Broadband.

4 Protocol Upgrades

In order to reach compliance with the ESA Iris system requirements, the most required system improvements are:

- Support of low gain antenna
- Reduction of the message transmission delay through signalling optimization
- Improvement of the random access scheme

These upgrades, complementary to those explained in the previous section, would be built on top of the SB baseline protocol, leading to the SB-S service adapted to safety aeronautical operations.

---

[15] LEO: Low Earth Orbit.
[16] 3GPP : 3rd Generation Partnership Project.

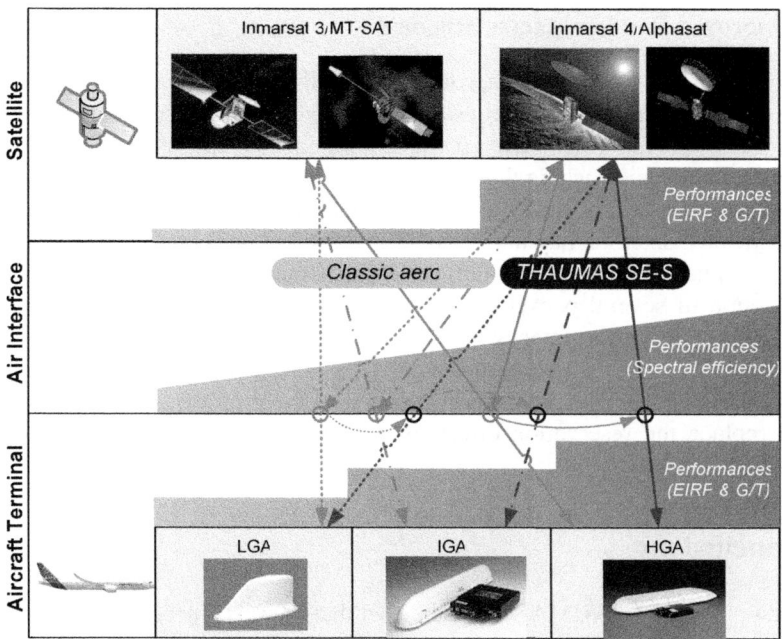

Fig. 1. Overview of Technical Improvements

4.1 Low Gain Antenna Support

To support the omni-directional mobile terminals, new physical bearers have been defined that allow robust operation for fixed-wing and rotary-wing aircraft at low elevation angles. The introduction of the new bearer types into both the Radio Access Network and the Mobile Terminals is the key development that needs to be undertaken to support this capability.

The support of low gain antenna is accompanied with a strategy of defining an optimized terminal architecture in terms of number of antennas, coverage, redundancy, price, dynamic configuration.

4.2 Reduced Message Transmission Latency

The default behaviour of the Swift Broadband system is to release the Radio Access Bearers (RAB, the logical connection across the radio interface) after a configurable period of non-use (typically 60 seconds). The reason for this is to enable the spot beam resources to be released and re-used in another beam if no traffic is being carried in a beam. With the low-latency requirements of the COCR [1] traffic this is considered unacceptable behaviour, so additional mechanisms need to be implemented to ensure RABs remain persistent and are supported in either spot beams or regional beams depending upon the loading within a spot beam.

Such a modification in the protocol signalling would reduce the protocol overhead of re-establishing a RAB after inactivity. This gain would be particularly important for bursty traffic.

4.3 Improved Random Access Scheme

Transmission latency will be improved for small messages, using improved random access mechanisms and satellite resources management between regional and narrow beams. It was identified that partial compliance may remain on very short transmission latencies, where the transmission delay toward the geostationary satellite is not negligible, and on service area, especially considering high latitude where the visibility of the satellite may not be possible during aircrafts manoeuvres.

The planned improved random access scheme would allow the simultaneous transmission of several packets into the same timeslot while coming from different terminals. For very short latency messages of a bursty traffic, there is no much time to spend in exchanging signalling messages to reserve and prepare for the transmission of this short latency message. As a consequence, the proposed random access scheme would replace the reservation effort for a slot by a random (but more successful) access into a specific-slot.

5 Conclusions

The first phase of THAUMAS has identified that there is a need to clarify the concept of operation and the future usage of the satcom link within the dual-link concept. The overall envelop of requirements considered so far certainly includes "nice to have" features that could dramatically increase the cost of the service for the end users. Concrete proposal have been done in this direction. In particular, it seems important to focus the effort on the key objectives that will support the improvement of air traffic situation in the dense area of the European airspace.

Despite these uncertainties in the usage landscape, it has been demonstrated that the SwiftBroadband system has been designed in a flexible way, using up to date techniques and as such it constitutes a very good basis to contribute to the consolidation of the requirements of the future satcom system for ATM. Indeed, it takes benefits of the strong background of Inmarsat in the aeronautical services provision.

For cost reason, the current implementation of the system does not take benefit of all potential options, but could be implemented. In addition, whenever required, further improvements and functionalities have been identified to achieve compliance with the end users requirements in the core area of Europe, leading to the SB-S system.

Remaining partial non compliance to the current requirements (that may be updated in the future) are linked to the use of geostationary satellite, and will therefore be the same for any system based on this kind of constellation. Compliance could be achieved using other types of constellations, but cost and business case issues have to be carefully considered before moving in this direction.

The next phase of this THAUMAS initiative will consolidate technical trade-off initiated in phase 0, and will confirm system design validation through simulation activities.

References

1. Communications Operating Concept and Requirements for the Future Radio System; COCR Version 2.0
2. IRIS PHASE 2.1 System Requirements Document; reference Iris-B-OS-RSD-0002-ESA; issue 1; revision 2 (November 5, 2009)
3. SESAR, D3, The ATM Target Concept, DLM-0612-001-02-00a (September 2007)

ANTARES System Design Options for Satellite-Based Air Traffic Management

Alessandro Di Stefano, Stefano Buratti, and Paolo Conforto

Thales Alenia Space Italia, Rome, Italy
alessandro.distefano@external.thalesaleniaspace.com,
{stefano.buratti,paolo.conforto}@thalesaleniaspace.com

Abstract. The future European Air Traffic Management (ATM) System is currently being defined by the Single European Sky ATM Research (SESAR) programme. Iris is the European Space Agency programme to develop an Air-Ground communication system for the SESAR programme. Within the Iris Programme, ANTARES focuses on the development of a new satellite-based communication system based on the use of low-cost user terminals through the realization of a new satellite communication standard. The present paper describes the system design process defined and adopted in the ANTARES project to cope with the uncertainties of the user requirements The proposed process is based on the definition of System Architecture Options allowing to evaluate the impact of requirements variability on the satellite system design. Each System Architecture Option will implement a set of Design Options which are system or technology solutions adopted to design the system or its elements. An insight on the Design Option relevant to the airborne user terminal features is provided in the paper.

Keywords: ANTARES, L-Band, ATM, SESAR, System design options.

1 Introduction

The need of a dual link to support air-ground communication in high-density continental airspace has been recognized of key importance by the Single European Sky ATM Research (SESAR) working groups. This dual link will rely on two separate means of communication to avoid common points of failure; one link relies on a new terrestrial line-of-sight technology in L-Band (LDACS), while a satellite communication system will also provide communication services over high-density continental areas to ensure the required availability.

Iris is the European Space Agency (ESA) programme to develop a new Air-Ground communication system for Air Traffic Management (ATM) as the satellite-based communication solution for the SESAR programme. The system design Phase B of the Iris Programme is called ANTARES (AeroNauTicAl REsources Satellite based) and has started in November 2009. ANTARES will contribute to the modernisation of ATM by providing more efficient connections for digital data to cope with the growing amount of information required and the increasing number of users.

G. Giambene and C. Sacchi (Eds.): PSATS 2011, LNICST 71, pp. 60–73, 2011.

The general ANTARES system architecture is depicted in Fig. 1 which highlights the physical system components – i.e. the space segment, the user terminals segment and the ground segment – along with the communication standard, including the whole set of system functionality (from the physical layer up to the network layer of the protocol stack) allowing the communication between system components with a given level of performance. The interfaces with the external European ATM (EATM) system are also depicted.

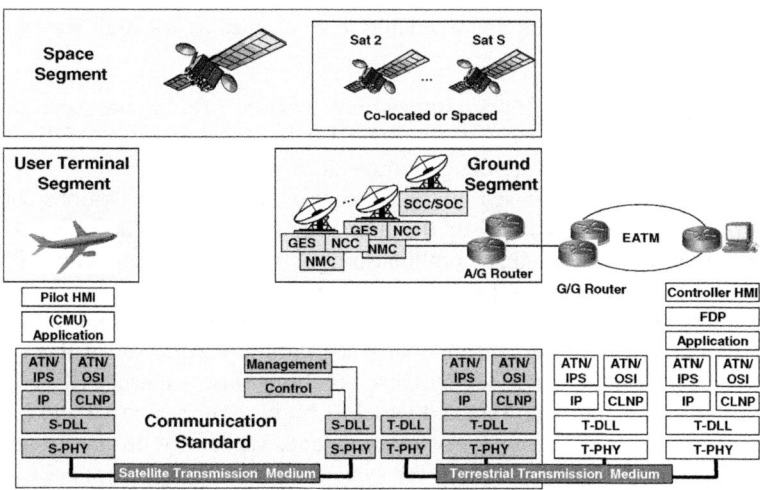

Fig. 1. ANTARES General System Architecture. In the figure above are depicted the network architecture consists of the mobile terminal, lo space segment e in ground segment comprendente le ground earth stations (GESs), the Satellite Control Centre (SCC), the Satellite Operation Centre (SOC), the Network Control Center (NCC) and the Network Management Centre (NMC). Under the respective protocol stack. It is possible identify the Physical layer (PHY), Data Link Layer (DLL), the Connectionless Network Protocol (CLNP), the Internet Protocol (IP) under the Aeronautical Telecommunication Network (ATN) both for Internet Protocol Suite (IPS) and Open System Interconnection (OSI). The stack go up to the Communications Management Unit (CMU), Foundry Discovery Protocol (FDP) for the pilot and controller respectively.

The ANTARES end-to-end system architecture has been defined by identifying the set of functionality to be supported, by allocating them to the physical elements and by suitably dimensioning the overall system so as to provide services to the estimated population of aircrafts with the required performance.

The ANTARES system aims at meeting a set of requirements which are generated by ESA considering (among others) the activities currently ongoing in the context of SESAR Joint Undertaking (JU) and the baseline for applications and communication performance defined in the FAA/Eurocontrol COCR (Communication Operating Concept and Requirements for the Future Radio System) [7]. Since these activities are running in parallel to the ANTARES project, some key requirements and their flow-down to the satellite communication system are not firmly defined yet. This entails that an uncertainty is still present in key requirements for the some specific areas such as:

1. Security: both information security and transmission security.
2. Number and type of applications/services.
3. Number of equipped aircraft as function of time.
4. Service provider configuration and associated ground segment architecture
5. User terminal performance which are achievable considering airframes installation constraints and available technologies as for high power amplifier and antenna etc.

In order to cope with these uncertainties, a systematic process has been defined by ANTARES system team to provide an understanding of the impact of the variability of key requirements on the satellite communication system design.

The proposed process is based on the concept of Requirement Options corresponding to different "interpretations" of the user requirements. In particular, a Requirement Option is defined as any combinations of attributes, i.e. technical variables which are used in order to qualify a given system feature.

Different system architectures may be defined by suitably combining the different alternatives of each Requirement Option. Even though a wide set of theoretical system architectures can be generated by these combinations, a subset of five alternative system architectures has been selected which is highly representative and particularly appropriate to evaluate the impact of requirements variability on the system design. This set of alternatives has been used for system dimensioning and specifications and each of the alternatives has been analyzed, dimensioned, specified and assessed on the basis of suitably selected figures of merit. The final result of this design process is that a system architectures "catalogue" is offered by the ANTARES project which includes five solutions with five complete sets of system and segment specifications each of them suitably assessed. The most suitable solution will be selected by operators and aeronautical stakeholders. As such, the ANTARES architecture depicted in Fig. 1 provides a general view of the system and a specific System Architecture Option is obtained by particularising this figure with the identified system choices.

As highlighted above, the Requirement Options are used in ANTARES in order to cope with uncertainties of requirements which are outside the ANTARES boundaries and therefore not under the control of the ANTARES team. Moving within the ANTARES system perimeter, the system design has been performed by identifying a wide set of possible Design Options (more than thirty) representing different technical choices which can be made to design the system or its elements and which are entirely under the definition and responsibility of the ANTARES team. All these Design Options have been analyzed and duly traded-off so as to produce the most appropriate technical solutions which are reflected in the system technical specifications and design.

Considering the user requirements relevant to the need to have very small user terminals on board aircrafts, which are devised so as to minimize the equipment weight, size, power dissipation and costs, it is worth pointing out that key Design Options refer to the user terminal features.

Several options have been investigated as for the number of user terminal antennas (single antenna or dual antenna with different inclination angles with respect to airframe body) taking into account their installation positions on the aircraft and their performance as well as the radiofrequency performance of the high power amplifier.

The present paper aims at describing the system design process for the definition of the ANTARES System Architecture Options and the selection of the system baseline. Moreover, an insight on the trade-off performed on the user terminal configurations is provided. In particular , the paper is organised as follows. The present section is devoted to introduce the main concepts to understanding the proposed system design process. Section 2 gives guidelines for the ANTARES system design process, focused on the Requirement Options and on the *System Architecture Options*. Requirement Options have been specifically distinguished from *Design Options,* shown in the Section 3 that will focus on the user terminal capability to maintain link during aircraft manoeuvring.

2   System Design Process

In order to cope with the uncertainties of the user requirements which are currently being defined in activity running in parallel to the ANTARES project, a systematic process has been defined by ANTARES system team to provide an understanding of the impact of the variability of the key requirements on the satellite communication system design. The proposed process is schematically depicted in Fig. 2 and consists of the following steps:

- Step 1:
    - Definition of the Requirement Options (ROs).
    - Definition of the System Architecture Options (SAOs) from all the combinations of the Requirement Options.
    - Definition of the Figures of Merit (FOMs) for the quantitative evaluation of the System Architecture Options.
- Step 2:
    - Identification of a reduced number of System Architecture Options.
- Step 3:
    - Analysis and design of the system configurations associated with the System Architecture Options identified in the Step 2.
- Step 4:
    - Quantitative assessment of the alternative System Architecture Options identified in the Step 2 and designed in the Step 3 on the basis of the Figures of Merit identified in the Step 1.
- Step 5:
    - Selection of the baseline System Architecture Option.
    - Identification of the set of baseline Requirement Options associated with the baseline System Architecture Option.

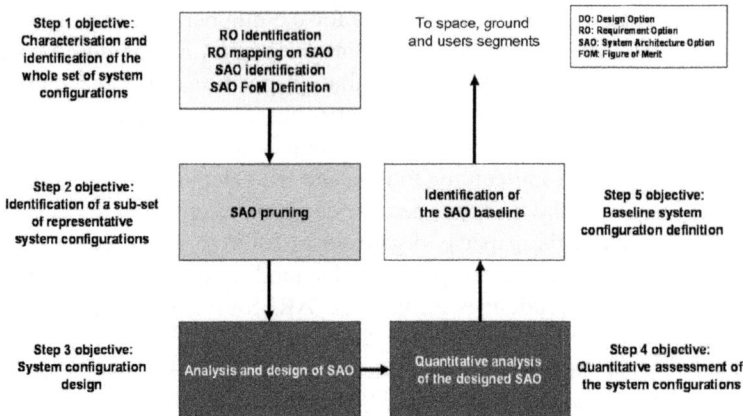

Fig. 2. Requirement Option Baseline Identification Overall Process

2.1 Requirement Options Definition

The Requirement Options are adopted in order to cope with the uncertainties of user requirements. As such, they correspond to different "interpretations" of the user requirements. In particular, a Requirement Option is defined as any combinations of attributes, i.e. technical variables which are used in order to qualify a given system feature. The ANTARES Requirement Options are presented in the Table 1.

Table 1. ANTARES Requirement Options Definitions

Requirement Options		Requirement Option Definition
RO1	Information and Communication Security (INFOSEC/COMSEC)	Information security and communications security capabilities that are applied within the ANTARES system boundaries in order to mitigate the threats on informatics system and on communication devices or other electronic systems. The attributes of RO1 are determined by the results of the threat analysis.
RO2	Transmission Security (TRANSEC)	Transmission security capabilities that are applied within the ANTARES system boundaries, in order to mitigate the threats deriving from RF interference sources that may affect the user plane for the provisioning of the AOC and ATC services, the control plane and the management plane.
RO3	System Capacity	Information volume (Mbps) entering the ANTARES System at network layer.
RO4	End User Terminal Capability	Radiofrequency transmission performance (EIRP) of the end user terminal on-board the Satcom equipped aircrafts.
RO5	Ground Segment Architecture	Topology of the ground segment architecture.

2.1.1 Requirement Option on INFOSEC/COMSEC (RO1)
This Requirement Option refer to the system capabilities aiming at mitigating the general threats and attacks to the ANTARES system and applying on the different interfaces such as the satellite air interface between GES and AES, the communication

interfaces between elements of the ground segment (GES, NCC, NMC, SOC/SCC etc.), the telecommand and telemetry interfaces.

Three attribute values have been defined for this Requirement Option based on different security implementation "levels":

- NULL refers to the case that INFOSEC/COMSEC mechanisms are implemented neither in the EATMN and nor by ANTARES (a part those to guarantee integrity);
- LOW refers to the case that satcom-specific authentication and integrity mechanisms are implemented by ANTARES;
- HIGH refers to the case that a full set of INFOSEC/COMSEC mechanisms are required to be implemented by ANTARES;

The following table describes the specific feature of the ANTARES system linked to the Requirement Option:

Table 2. Attributes and Values for the Requirement Option on INFOSEC/COMSEC

Attribute	Attribute Definition	Attribute Logical Value		
		NULL	LOW	HIGH
Confidentiality	Transmitted data are not disclosed to unauthorized entities or process	NO	NO	YES
Integrity	Messages are received with no modification. Data are not accidentally or maliciously modified, altered or destroyed	YES	YES	YES
Authentication	The identity of the entities involved in a communication is verified	NO	YES	YES
Access Control	The users privilege to access the resources of a system is granted only to authorized users, programs, processes, or other systems	NO	NO	NO
Non repudiation	The transmitter or the receiver is not allowed to deny its acts	NO	NO	NO

2.1.2 Requirement Option on TRANSEC (RO2)

This Requirement Option refers to the system capabilities supporting the ANTARES transmission security. Two attributes have been defined to cover two extreme cases corresponding two possible levels of immunity against known and unknown interference:

- LOW, entails that the ANTARES system is robust only to interferences from known equipment, as for coordination and design results in front of possible inter-system and intra-system interference sources (non malicious).
- HIGH entails that the ANTARES system is robust to interferences from both known and unknown equipment; this case covers the case of malicious jammers on the ATM satellite system.

As far as the immunity against known equipment interference, the Ku-band uplink the level of immunity to interference has been dimensioned considering the following two categories of interference sources:

- Out-of-band/spurious emissions from earth stations belonging to other satellite systems, according to the normative documents [1] and [2].

- Off-axis EIRP emission density within the band from earth stations belonging to adjacent satellite systems, transmitting within the band of interest (14-14.25 GHz), according to the normative documents [2] and [3].

For the Ku-band downlink, the level of immunity to known interference has been dimensioned considering the following two categories of interference sources:

- Out-of-band/spurious emissions from other satellite systems, according to [4]
- In-band emissions from adjacent satellite system, transmitting within the band of interest, according to [5].

For the L-band user up-link, the known interference on the ANTARES uplink frequency band due to the out-of-band emissions generated by other terminals has been evaluated by assuming as reference the normative documents [6].

The level of immunity against unknown equipment interference (both intentional and unintentional) has been evaluated in terms of possible thresholds for the acceptable I/S levels, where I indicates the interfering signal and S is the useful signal. The following cases have been considered as for the user uplink

- unknown, out-of-band interference where out-of-band indicates bands externals to the 10MHz bandwidth allocated to the aeronautical services.
- unknown in-band, out-of-channel interference where in-band indicates the 10MHz bandwidth allocated to the aeronautical services and out-of-channel means that the interference is not overlapped to the ANTARES system carrier.
- unknown, in-band, in-channel interference where in channel means that the interference is overlapped to the ANTARES system carrier

In any case, the telecommand and telemetry link will be protected from possible intentional and unintentional jammers by means of spread spectrum-based technique.

2.1.3 Requirement Option on System Capacity (RO3)

This Requirement Option is to provide a range of capacity values (i.e. offered load) within which the satellite communication system shall be designed. Table 1 defines three attributes of the Requirement Option and relates these attributes to the reference traffic scenarios they are associated with, considering the implications of both the growth of air traffic in the reference time frame and the applications being considered). The described reference traffic scenarios have been defined on the basis of the aircraft traffic profiles analyses performed in the context of the ANTARES framework. In this respect, several aircraft traffic profiles have been considered

Table 3. Attributes and Values for the Requirement Option on System Capacity

Attribute	Reference scenarios	Attribute Value (capacity)
ANTARES system traffic capacity	Low capacity scenario	FWD=0.6 Mbps RTN=0.34 Mbps
	Medium capacity scenario	FWD=2.3 Mbps RTN=0.4 Mbps
	High "plus" capacity scenario	FWD=4.65 Mbps RTN=0.75 Mbps

corresponding to different traffic growth rates for the predictions. Both the Air Traffic Services (ATS) and Aeronautical Operational Control (AOC) applications defined in [7] have been considered to dimension the overall data traffic over the ECAC.

2.1.4 Requirement Option on User Terminal Capability (RO4)

The attributes of this Requirement Option refer to two different cases of aeronautical user terminal design: one with "low performance" and one with "high performance". The different levels of user terminal performance have been quantified in terms of possible sizes of the airborne power amplifier, which may provide:

- 20W saturated power.
- 40W saturated power.

In both cases it is assumed that the user terminal should be designed so as to minimise the impact onto the aircrafts as for the installation. In this respect, the possibility to support nominal operational modes of the user terminal without need for active cooling is pursued as a key factor to reduce the above mentioned installation impacts.

These two cases are identified as representative of user terminal classes which may be installed on different typologies of aircrafts.

Analyses (e.g. link budgets, thermal analysis etc.) have been performed considering the actual operating point of the amplifier, taking into account the required output back-off (OBO) and the relevant physical losses due to installation constraints.

2.1.5 Requirement Option on Ground Segment (GS) Architecture (RO5)

This Requirement Option refers to the topology of the ground segment architecture. The ground segment elements considered to define this topology are:

- Ground Earth Station (GES).
- Satellite Service Provider (SSP).

Depending on both the number of GESs and SSPs, two different classes of ground segment topologies can be identified, which are captured by the attributes defined for this Requirement Option and listed in Table 4:

- Centralised, entails a topology which is fully centralised as for distribution of the physical elements and relevant functions.
- Decentralised, entails a topology presenting a certain level of decentralisation of the physical elements and/or relevant functions. The level of decentralisation may vary depending on the combination of attribute values.

Table 4. Requirement Option 5 (GS Architecture)

Attribute	Attribute Definition	Attribute Logical Value	
		Centralised	Decentralised
Number of SSP	Number of providers providing the SatCom service to the Communication Service Provider(s) (CSP)	1	3
Number of GES	Number of stations providing the Tx/Rx communication capability	1	15

2.2 System Architecture Options

By combining the Requirement Option values described in the previous section up to forty-eight System Architecture Options (SAO). These options are only theoretical. In practice, only System Architecture Options which are useful to highlight the impact of requirements variability need to be studied. In this respect, a subset of five alternative system architectures has been selected which is highly representative and particularly appropriate to evaluate the impact of requirements variability on the system design. They are defined as follows:

- System Architecture Option 1:
    - Supporting a low capacity traffic (RO3=LOW)
    - Presenting high resistance to the interference (RO2=HIGH)
    - Not implementing any specific information/communication security mechnisms (RO1=NULL)
    - Having a decentralised GS architecture (RO5=DECENTRALISED)
- System Architecture Option 2
    - Supporting a medium capacity traffic (RO3=MEDIUM)
    - Presenting high resistance to the interference (RO2=HIGH)
    - Not implementing any specific information/communication security mechanism (RO1=NULL)
    - Having a decentralised GS architecture (RO5=DECENTRALISED)
- System Architecture Option 3
    - As SAO 2, but with a centralised GS architecture (RO5=CENTRALISED)
- System Architecture Option 4
    - Supporting a medium capacity traffic (RO3=MEDIUM)
    - Presenting high resistance to the interference (RO2=HIGH)
    - Implementing specific information/communication security mechanisms (RO1=HIGH)
    - Having a decentralised GS architecture (RO5=DECENTRALISED)
- System Architecture Option 5
    - Supporting a very high capacity traffic (RO3=HIGH+)
    - Presenting high resistance to the interference (RO2=HIGH)
    - Not implementing any specific information/communication security mechanisms (RO1=NULL)
    - Having a decentralised GS architecture (RO5=DECENTRALISED)

As far as the Requirement Option on user terminal capabilities (RO4) is concerned, it may be observed that the two categories of UT performance will not drive the final choice on the system architecture. They may be regarded as proposed options to be selected on the basis of on aircraft installation constraints, operational conditions etc.

The above defined set of alternatives has been used for system dimensioning and specifications and each of the alternatives has been analyzed, dimensioned and specified. The final result of this design process is that a system architectures "catalogue" is offered by the ANTARES project which includes five solutions with five complete sets of system and segment specifications each of them suitably assessed. The most suitable solution will be selected by operators and aeronautical stakeholders. The system architecture in Fig. 1 include all the possible System Architecture Options which may result from the Requirements Option combination. A specific System Architecture Option is obtained by particularising this figure with the identified system choices.

2.3 Figure of Merit of System Architecture Options

The System Architecture Options are compared on the basis of a set of of quantitative variables aiming at capturing technical and programmatic elements.

These quantitative variables, also referred to as Figures of Merit (FOMs), are:

1. Cost of the user terminal (UT), including:
   - Cost for UT development.
   - Cost for installation and maintenance.
2. Cost of the overall system, including
   - Nonrecurring and recurring costs for space segment development (payload and platform).
   - Nonrecurring and recurring costs for ground segment development.
   - Cost for redundancy
   - Cost for launch
3. Cost of system operations, including:
   - Operational costs for the Satellite Operator relevant to in-orbit tests, payload monitoring and control, station keeping procedures execution, satellite relocation (if necessary), data archiving and retrieving, personnel and maintenance of the SCC and SOC etc.
   - Operational costs for the Satellite Service Provider (SSP) relevant to satellite bandwidth and resources procurement, personnel and NMC and NCC operations, maintenance of the SCC and SOC etc.
4. Spectrum (i.e. amount of required bandwidth).
5. System margins

3 Design Options

The Requirement Options have been introduced in ANTARES in order to cope with uncertainties of requirements which are outside the ANTARES boundaries and therefore not under the control of the ANTARES team.

Moving within the ANTARES system perimeter, the system design has been performed by identifying a wide set of possible Design Options (more than thirty) representing different technical choices which can be made to design the system or its elements and which are entirely under the definition and responsibility of the ANTARES team. All these Design Options have been analyzed and duly traded-off so as to produce the most appropriate technical solutions which are reflected in the system technical specifications and design.

A key requirements coming from the aeronautical stockholders is to have very small user terminals on board aircrafts. As a consequence, user terminal shall be devised so as to minimize the equipment weight, size, power dissipation and costs.

Toward this end, key Design Options has been defined which refer to the user terminal features. The following section focused on these several options.

3.1 Design Option on Single vs Multiple Antenna on Board the Aircraft

This DO aims at analysing, quantitatively evaluating and trading-offs the possible solutions to provide the required link availability, during aircraft manoeuvring and in particular banking and pitching. Several options have been investigated.

During the en-route phases of the aircraft, it is highly probable that the radio link between the aircraft user terminal and the satellite has line-of-sight characteristics, since no impairments are supposed to be interposed between the antennae. Nevertheless it is possible that, during aircraft manoeuvring (bank angle 35°) and for specific relative position between aircraft and satellite (especially at high latitudes / low satellite elevation angles ~5°), that the line-of-sight is lost. Fig. 3 shows examples of the UT antenna gain variation due to aircraft manoeuvring.

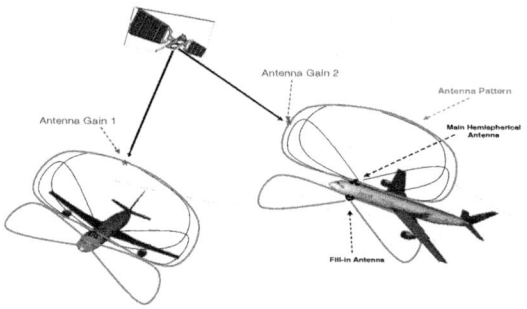

Fig. 3. User Terminal Antenna Gain Variation due to Aircraft Manoeuvring

The following implementation alternatives are considered for this DO:

- Alternative 1: Aircraft user terminal with single antenna (Reference).
- Alternative 2: Aircraft user terminal with two antennas located ~45 deg away from top of the fuselage on both sides.
- Alternative 3: Aircraft user terminal complemented by fill-in antenna.

The trade off among the different option has been based on (among others) link budget performance, antenna radiation pattern and geometrical link visibility analysis.

3.1.1 Trade-Off Criteria
The considered trade-off criteria for the evaluation of DO alternatives are:

- Geometrical availability of the link: the higher this availability, the higher the overall service availability provided the link margin is sufficient. This criterion is quantified by the percentage of time the satellite is visible over a statistical population of aircraft flight trajectories, i.e.

$$T_{\%} = \frac{\sum_i \sum_j t_{i,j}^{vis}}{\sum_i t_i^{vis}} \tag{1}$$

where:

$t_{i,j}^{vis}$ =time duration of j-th visibility period of the i-th aircraft flight trajectory.

t_i = time duration of the i-th aircraft flight trajectory.

$i \in$ [Set of aircraft flight trajectories].

$j \in$ [Set of visibility periods].

- Implementation feasibility: e.g. compatibility with aircraft installation constraints. This is a fundamental factor for acceptability of the Satellite Communication System overall. This criterion is quantified by the number of antennas installed on the aircraft, taking into account redundancy.
- Overall link margin at max banking angle. This parameter, together with the geometrical visibility defines the service availability during manoeuvring. This criterion is quantified by means of the margin of the signal-to-noise ratio variation with respect to the single antenna case:

$$\Delta\left(\frac{C}{N_{0,t}}\right) = \left(\frac{C}{N_{0,t}}\right) - \left(\frac{C}{N_{0,t}}\right)_{Single\ Antenna} \tag{2}$$

3.3 Trade-Off Justification and Conclusions

The quantitative assessment for the identified alternatives is reported in Table 5. The trade-off is based on two major analyses whose major results are briefly reported in the following:

- Link geometrical visibility analysis.
- Link budget analysis.

A third analysis has been performed on the antenna radiation, taken into account in the link budget analysis but are not explicitly reported in the following.

Table 5. DO Quantitative Assessment

Evaluation Criteria	Comment / Explanation		DO Quantitative Assessment		
			AES with Single Antenna	AES with two Antennas (at 45°)	AES complemented
Geometrical availability of the link	% of time the satellite is visible over a statistical population of aircraft trajectories= (1)		Airliner: 9999910747%	Airliner: 9999925334%	Airliner: 9999982938%
			Civil utility aircraft: 9998338439%	Civil utility aircraft: 9998956203%	Civil utility aircraft: 9999262673%
	Max Outage Time Occurence Duration		Airliner: 6.042 sec	Airliner: 6.486 sec	Airliner: 0.306 sec
			Civil utility aircraft: 67.885 sec	Civil utility aircraft: 6.692 sec	Civil utility aircraft: 67.885 sec
Installation feasibility	Number of antennas (taking into account redundancy)		2	4	4
Overall link performance	Link margin at max bank angle= (2)	Lat. 65°	ENR: 0 (reference) Bank: 0 (reference)	ENR: 0dB Bank: 4.3dB	ENR: 0dB Bank: 4.3dB
		Long. -20°			
		Lat. 65°	ENR: 0 (reference) Bank: 0 (reference)	ENR: 0dB Bank: 6.2dB	ENR: 0dB Bank: 6.2dB
		Long. 20°			
		Lat. 45°	ENR: 0 (reference) Bank: 0 (reference)	ENR: 0 Bank: 1.3 dB	ENR: 0dB Bank: 0.4dB
		Long. -20°			
		Lat. 45°	ENR: 0 (reference) Bank: 0 (reference)	ENR: 0 Bank: 1.3 dB	ENR: 0dB Bank: 0.4dB
		Long. 20°			

The data on geometrical availability of the link have been obtained by means of visibility analyses performed on antenna aircraft-to- satellite line of sight (LOS) link. These analyses, have been carried out taking into account the real aircraft manoeuvres, with the twofold objective of:

- Evaluate the link geometrical availability, contributing to the overall system availability which may be offered to the aviation end-users.
- Identify the solutions to increase the satellite link geometrical availability, such as suitable system dimensioning coping with real flight conditions, appropriate numbers of antennas on the aircraft and of satellites simultaneously operating.

A sample of 1000 realistic flight trajectories departing and arriving in airports inside the ECAC coverage has been generated in close co-operation with the University of Salzburg through the NAVSIM simulator. The obtained flight trajectories antenna location, antenna field of view, aircraft model and satellite features have been modelled adopting a commercial of the shelf software package simulator. Link visibility and outage times, as well as aircraft antenna-to-satellite elevation angles and aircrafts bank and pitch angles occurrences statistics, have been computed considering aircraft manoeuvres.

Link availability computation over time has been characterized with different aircraft type airliner and civil utility aircraft.

The link budget analyses have been performed with several kinds of UT antennas taking into account specific installation constraints on the aircraft.

Multidimensional link budgets have been performed by considering the aircraft flying over the ECAC service area and crossing the en route (ENR), traffic manoeuvring (TMN) and oceanic remote polar (ORP) aerospace domains. This entails that aircraft manoeuvring with banking angles in the range $[0 \div 35°]$ have been considered. In particular, link budget results have been obtained over the ECAC area on a grid of $1°$ per $1°$ resolution.

The actual antenna gain value used in link budget analyses has been calculated according to the values of the adopted antenna radiation pattern and taking into account the composition of the:

- The satellite elevation angle associated with the geographical site for the aircraft;
- The banking angle of the aircraft during manoeuvres.

Three cases have been analysed as for the number and position of the antennas on the aircraft:

- One single antenna installed on top of the fuselage.
- Two antennas installed at $45°$ with respect to the zenith directly above the airframe
- Two antennas, one installed on top of the fuselage and the other installed on bottom part (fill in antenna).

In particular, for each case different types of antenna radiation patterns have been considered referring to possible installation solution.

The values show in Table 5 has been obtained considering the difference of signal to noise ratio ($\Delta C/N0$) on four points of the ECAC area and whereas the mobile terminal during both en route (ENR) at the stage of manoeuvres (TMA). The comparison is carried out using the single antenna case as a reference compared to cases with dual antenna (both 45 ° and 180°).

On the basis of the results, the major conclusion of this trade off on the user terminal configuration, is that the following alternative are selected:

- Single antenna installed on top of the fuselage.
- Two antennas at 45° with respect to the zenith above airframe.

Both alternatives are retained to take into account variable installation and operational conditions of the aircraft.

4 Conclusions

This paper seeks to underline the importance of a system design process defined through a well posed approach that allows to realize a preliminary set of systems architectures when the user requirements are not firmly defined. In order to cope with these uncertainties, a systematic process has been defined by ANTARES system team. A subset of five alternative system architectures has been selected which is highly representative and particularly appropriate to evaluate the impact of requirements variability on the system design. The system design has been performed by identifying a wide set of possible Design Options representing different technical choices. Moreover, considering the user requirements, several options have been analyzed.

References

1. Recommendation ITU-R S.726, Maximum permissible level of spurious emissions from very small aperture terminals (VSATs)
2. ETSI EN 301 428, Satellite Earth Stations and Systems (SES); Harmonized EN for Very Small Aperture Terminal (VSAT)
3. Recommendation ITU-R S.580-5, Radiation diagrams for use as design objectives for antennas of earth stations operating with geostationary satellites
4. ITU-R SM.329, Unwanted emissions in the spurious domain
5. ITU-R S.465, Reference earth-station radiation pattern for use in coordination and interference assessment in the frequency range 2 to about 30 GHz
6. ETSI EN 301 473, Satellite Earth Stations and Systems (SES); Aircraft Earth Stations (AES) operating under the Aeronautical Mobile Satellite Service (AMSS)/Mobile Satellite Service (MSS) and/or the Aeronautical Mobile Satellite on Route Service (AMS(R)S)/ Mobile Satellite Service (MSS)
7. EUROCONTROL/FAA Future Communications Study Operational Concepts and Requirements Team - Communications Operating Concept and Requirements for the Future Radio System VERSION 2.0 (COCR V2)

Iris Operations Studies SIRIO–Study for the Satellite Air Ground Interface with Reliable and Interoperable Operations

Manuela Rossi and Laura Anselmi

Telespazio, Via Tiburtina, 965, 00156 Rome, Italy
{manuela.rossi,laura.anselmi}@telespazio.com

Abstract. SIRIO is one of the three Operations Studies included in the framework of the ESA Iris Programme, which aims to design and develop a new Satellite Communication system for Air Traffic Management. SIRIO focus is on the definition of a Service Model and Business Model for the provision of ATS and AOC communication services via satellite. The definition activity is based on the analysis of the ATM service value chain, and takes into account also the impact of the regulatory framework. Revenue Model was designed and together with associated CAPEX and OPEX estimates based on the Reference system architecture coming from the Manufacturer Study constitutes the basis for a Business Case Analysis. Positive Preliminary results coming from the business case and based on different Public/private Partnerships financing schemes are shown.

Keywords: Iris, SIRIO, Satellite Communication System, Air Traffic Management, Service Model.

1 Introduction

SIRIO is one of the three Operations Studies included in the framework of the ESA Iris Programme: the study is lead by Telespazio in collaboration with Hispasat, NATS, EGIS AVIA and Telespazio France.

The objective of the Iris Programme is to design and develop a new Satellite Communication system for Air Traffic Management: such a system consists of a new communication standard and a satellite system infrastructure. The Programme is divided into three Phases:

- Phase I was completed in 2009 and it defined satellite link feasibility and requirements, including top level design options for the new Satellite Communication System;
- Phase II focus on technical activities such as the design and development of the new Satellite Communication standard, and Satellite System ,together with. Service related activities which include the definition of the business case and investigation of future services provision schemes. Phase II.1 (2009-2012) includes detailed definition of all system elements. Three main activities are :

G. Giambene and C. Sacchi (Eds.): PSATS 2011, LNICST 71, pp. 74–83, 2011.

- (1) the design of a dedicated Satellite Communication System (ANTARES phase B study) by Manufacturing Industry,
- (2) three parallel studies (including SIRIO) of the satellite system operations performed by service providers/operators,
- (3) a study of the feasibility of adapting Inmarsat's Swift Broadband system. Phase II.2 (2012-2016) which would deal with the development of the system and the procurement of the subset for validation and complete the Iris Phase II in full alignment with the SESAR Development Phase.

- Phase III foresees ESA technical support to SESAR Service Validation and System Certification activities.

2 SIRIO

SIRIO and the other Operations studies are included in Iris Phase II.1; several activities are currently in progress; and in the following sections, what is available today as preliminary results of the study is presented.

The name SIRIO contains all the main elements of the study: it investigates the use of Satellite for the provision of air-ground communication, with the eye of a Satellite Service Provider and Operator, with the aim to satisfy requisites of reliability and interoperability.

The industrial team was organized in order to cover all areas of activity with the necessary expertise: Telespazio act as coordinator with the intention to characterise the study with a clear operational service view and user/market driven approach in collaboration with Hispasat, which has been involved as Satellite Owner and Satellite Operator; NATS is an Air Navigation Service Provider and therefore brings specific expertise on certification and technical service validation, while EGIS AVIA provides experience on standardisation, certification of aviation communication systems. and Telespazio France.

The SIRIO study is focused on the following objectives which are detailed in dedicated sections:

1. Service model definition: includes the definition of several options for service provision scenarios, taking into consideration also the impact of the regulatory framework
2. Service funding / cost-recovery mechanisms analysis: includes activities related to the identification of a revenue model based on the service model, and which will lead to the definition of the business case.
3. Business Model and Financing schemes identification: this activity includes the proposition options for financing deployment of the Iris "Subset", and of the full operational system architecture,
4. Deployment scenario/timeline investigation: this activity contains all issues concerning timeline of system deployment and certification steps; this means to propose a deployment strategy to reach full operational capability by 2020; and to determine which activities are required for the validation of the service .

2.1 Service Model Definition

The value chain for the provision of satellite-based communication services for aviation is composed by several main stakeholders ranging from the Satellite Owner to the Airlines and Air Navigation Service Provider (ANSP) which are the final users of the services.

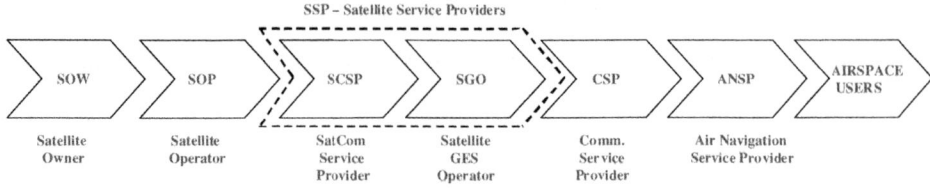

Fig. 1. Satellite-based Communication - Value Chain

The stakeholders identified in the figure above are:

- The Satellite System Owner (SOW) owns the space segment.
- The Satellite Operator (SOP) is in charge of operating the space segment
- Satellite Communication Service Providers (SCSP) are in charge of marketing the satellite capacity.
- The Satellite GES Operator (SGO) is in charge of the provision and operation of Satellite Ground Earth Stations' and Terrestrial Circuits to provide satellite access via terrestrial links
- The Satellite Service Provider (SSP) covers both SCSP and SGO roles, and is in charge of providing the satellite-based communication service to the Communication Service Provider (CSP).
- Communications Services Providers (CSP) is the entity who bundles communication services and represents a single interface for procurement of telecommunication services for ANSPs and Airspace Users.
- Air Navigation Service Providers (ANSPs) and Airspace Users (Airlines) are the end users of the service.

The Service Model definition activity is based on the detailed analysis of the value chain. Starting from the ATM service value chain, different grouping of value-chain roles have been studied, identifying actors and roles to be performed to deliver the service to the end users.

For the Service Model definition a set of topics has been taken into account and a balanced mix of the following key issues seems to be the preconditions to have a viable Service Provision Model:

- Adoption of an ATM-dedicated satellite-based communication standard, widely recognised by Airline companies, airspace users and ANSPs'.
- Focus on the end users needs to make sure that such a service is suitable for aviation safety communications in terms of safety, operations, costs and technical requirements (e.g. coverage, capacity, service interoperability)

- Profitability and bankability: A satisfactory profitability should be guaranteed for the private shareholders, especially in a PPP approach with investments of the private sector
- Market competition: Avoid a monopolistic framework that would have a negative impact on service provision; indeed, the Service should not contain any feature which would restrict competition among operators in the future.

The service model scenarios analyzed in the study, were defined taking into consideration that service provision set up will not limit the number of Ground Earth Stations, but allow incremental deployment of Space and Ground Segment sub-elements, and a competitive framework will be guaranteed through the presence of multiple CSP and/or SSPs.

Next figure shows the service model baseline.

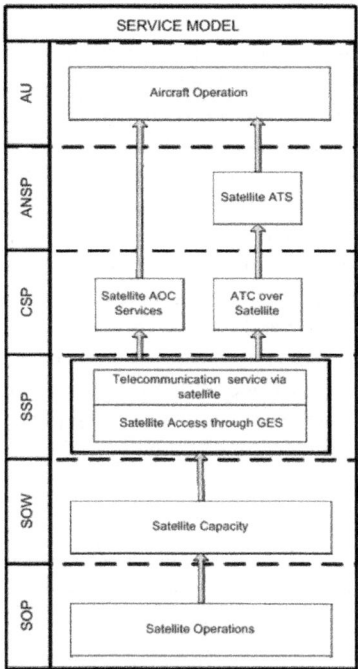

Fig. 2. Service Model for ATS and AOC services provision

It has to be noted that as far as commercial satellite systems are considered, the roles of the Satellite System Owner (SOW) and Satellite Operator (SOP) are usually performed by a single entity. In other types of programmes, where the development of a satellite system is funded (or co-funded) by National Space Agencies for scientific mission purposes, or by Governments for military purposes, the SOW and the SOP can be different. In the framework of the Iris programme, the two roles are considered to be separated because this is what usually happens when the development of a satellite system is funded (or co-funded) by National Space Agencies for scientific mission purposes, or by Governments for military purposes.

It is also assumed that the Satellite Communications Service Provider (SCSP) and the GES Operator (SGO) are performed by a single entity: the Satellite Service Provider (SSP):

Several Scenarios can been investigated starting from the service model presented, depending on different combinations of the following options:

- the satellite system architecture adopted (centralized, decentralized)
- grouping of entities (i.e. if the CSP and the SSP can be grouped together into a single role)
- who is the entity (or entities) which owns and operates the Ground Earth Stations (SSP,ANSP)

It is also essential to consider not only satellite communication but also future A/G Mobil Data-Link communications.

The A/G Mobil Data-Link terrestrial communication infrastructure for ATS services provision could be either owned/operated by ANSP or by any CSP. This rationale applies to the AOC service provision by CSP.

In the design of the service model, also the impact of the regulatory framework and liabilities issues has been considered to define the responsibilities and liabilities of entities involved in the Service Model and also of entities contributing to set-up the provision of the SATCOM services.

Actually, in order to comply with the safety requirements,, at least one actor in the value chain must be certified. according to SES legislation .The certified Service Provider will have to obtain a certificate from EASA, which will state its compliance with the applicable requirements and implementing rules.

Possible scenarios taken into account are single certified Pan-European Service Provider or Multiple certified Service Providers separately for Satcom, Air/Ground com and Air/Air com, as shown in the following table.

Table 1. Certification Models

Certification Model 1		
Entity:	Certified for:	Certification includes:
ANSP	ATS	
CSP	Bundled AMS services Bundled AFS services	Data Link Service SATCOM service
SSP	N/A	
PENSP	PENS services	
Certification Model 2		
Entity:	Certified for:	Certification includes:
ANSP	ATS	
CSP	Bundled AMS services Bundled AFS services	Data Link Service
SSP	SATCOM service	SATCOM service
PENSP	PENS services	

2.2 Service Funding and Cost-Recovery Mechanisms

In order to develop investment profiles for each entity in value chain and to identify associated financial support to develop, procure and operate full system starting from the definition/identification of the ATS/AOC revenue flow, a cost assessment and related sensitivity analysis has to be performed. In particular:

- Conduct a revenue model that take into account the Service Model for AOC and ATS services via satellite, considering that ATS and AOC will have different revenue flows within the entities of the provision chain
- Develop a cost model that takes into account Capital costs (from Phase B studies) and Operating costs (determined by Operator Study) all phased with the implementation timeline
- Conduct a Sensitivity Analysis on main revenue and cost risks, considering in addition that infrastructure cost recovery will be based on revenues from ATS as for AOC revenues will depend on the market penetration levels

In SIRIO the revenue model was conducted taking into account that service for ATS communications shall be provided with an auditable and established subscription and/or communication charges that reflect the cost of service provision to the ATM user. The following assumptions have been applied:

- Mandatory carriage year start: extrapolating the "Implementing Rule on Data Link Services" that can be considered assuming similar obligations (and exemptions)
- Traffic model; starting from the traffic forecast in terms of IFR flight movements for the EUROCONTROL Statistical Reference Area (ESRA) two extrapolation has been carried out (1. until 2016, using the three growth scenarios from the Medium-Term Forecast IFR Flight Movements 2010-2016; 2 until 2030, using the three growth scenarios from the Long-Term Forecast IFR Flight Movements 2008-2030)
- SATCOM equipment ramp-up, based on the assumption that there will be a mandate to equip the aircraft with the new SATCOM system and the start of full operation of the new SATCOM system will be in 2020.

Different charging schemes and related revenue models apply to ATS and AOC, since AOC charge and the associated revenues for the operator are market-driven and therefore the revenue model is designed following a market approach, which includes presence of competition among operators. Moreover, the CSP can sell directly to the end users.

Instead for ATS the ANSP is mandatorily present in the ATS provision value chain since Air Traffic Services are regulated; the revenue model in this case has been designed with the goal to guarantee to the service provider a minimum profitability of its investment.

2.3 Business Case and Financing Schemes

The reference system architecture considered is given by ANTARES and is the basis for a preliminary estimate of the CAPEX and OPEX, which together with the revenue model, allows for the business case analysis.

The Space Segment architecture considered consists of 2 geostationary satellites and 1 spare., while two options for the Ground Segment architecture are considered, one is representative of centralized GS architecture with a single SSP and the other one for a distributed GS architecture, which foresees the presence of 3 SSPs.

For each architecture two different funding scenarios have been investigated: the two scenarios differ in the sharing of CAPEX between public and private actors.

Each scenario considers both AOC and ATS flow. The AOC revenues have been designed following a market approach: from the revenues estimated in the revenue model . Instead the ATS revenues flow has been designed with the goal to guarantee a minimum profitability of the investment (calculated in terms of combination of economic parameters, such as IRR, NPV and pay back period).

In both scenarios, public investments are considered to cover at least the following elements:

• 1st CAPEX outflow
• Non recurring ground segment elements.

Preliminary results of analysis which is still in progress, shows the cash flow trend of the Service Provider, in the scenario characterized by a Centralized Architecture and the larger public financing scenario between the two considered.

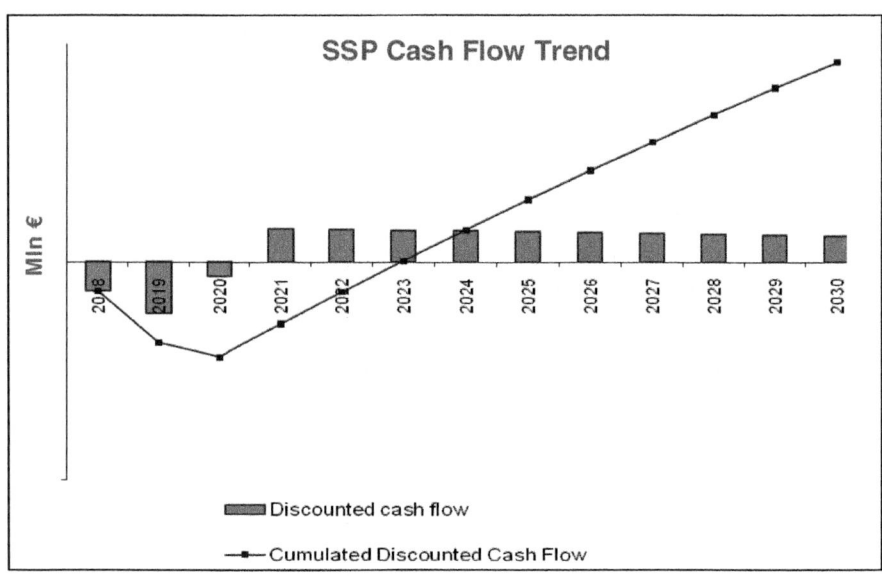

Fig. 3. Business Model Preliminary output- SSP Cash Flow Trend

The outcome of the business case is to determine the ATS charge per flight, in order to get the optimum value which allows by one side for a sustainable business case for the private investors and by the other side a low price for the end users. Current analysis give positive outcomes in terms of payback period for the Service Provider investing in the system and in terms of a fair price for the end users.

The financial viability of the business case relies on guaranteed revenue from ANSPs and airspace users for the service provider, coming from the use of the satellite communication link for ATS, as mentioned in the previous section dedicated to cost-recovery mechanisms and on additional revenues given by the AOC services provision.

The infrastructure required for validation of the pre-operational system is included in the financing scheme and all non-recurring elements of both Ground and Space Segments will be funded by ESA. At the same time, additional elements required to complete the operational infrastructure would be financed through other means/by other entities. At this stage, the business is perceived as quite risky in terms of cost-recovery period by the private sector to allow for a full private investment. However, having the guarantee of an Implementing Rule to be issued would assure a certain level of revenues to be collected by the provision of ATS and would improve the associated business case.

Therefore as already mentioned above, several financial schemes are currently under investigation,. in order to bridge the gap between the technical validation stage and the operational service when a private investor could recover its initial investment.

A possible scheme for sharing the investment and the associated risks, is to include both private and public entities, into what is called as PPP.

Public Private Partnerships (PPP) rely on a series of performance and payment arrangements, which are intended to give a private consortia appropriate incentives to deliver good service to the envisaged final customers. Moreover, the partnership with a qualified private partner can limit public risks while helping to ensure that quality services can be provided quickly and efficiently.

Therefore a PPP-structure can be considered as an attractive way of financing transport infrastructure. Nevertheless, a PPP can only work if certain pre-conditions are met, such as clear commitment and vision from all the involved actors. Moreover, the "project" must be attractive in a competitive market and it must be "bankable," meaning that it must look attractive to bankers and other providers of financing.

Among the different ways to develop and implement a PPP, the attractiveness of providing operational service depends on the following key drivers:

- Pre-requisites for each entity in the value chain
- Clear definition of Roles & Responsibilities in Service Provision
- Certainty / Clarity around risk analysis and allocation
- Efficient Risk Analysis and Allocation
- Financial Analysis: business case and in particular the sensitivity analyses highlighting cut-off conditions

2.4 Timeline for the System Development and Operations

All the aforementioned activities should be synchronised with the following timeline, where regulatory and standardisation activities, Service Provision and Business Model key aspects.

Moreover, the Service development and deployment shall be in accordance with the SESAR Master Plan which aims to have the operational system ready by 2020; key milestones are:

- 2012 ESA member states funding decision for starting Iris Phase II.2
- 2012-2016: Development and deployment of the subset, which is the pre-operational system for verification and validation activities
- 2015/2016 First satellite launch
- 2016 start of system validation activities.; the Satellite Communication System will be validated using the Pre-operational System; the subset will become the first building block of the operational ATM System and all assets funded by ESA will be transferred to the System Owner
- 2016-2010 deployment of the operational system (redundant payload) and certification of the Service Provider
- 2020 start of operations

For the Business Model and Financing activities, end of 2010 / beginning of 2011 is the milestone when the decision on the financial model (i.e.: Public or PPP) and the assets to be financed (i.e.: Iris subset...) has to be taken.

Subsequently, in case of a private investment (e.g.: to cover the satellite platform and launch procurement), a consultation process should be launched in order to look for interested parties and the financial risk for the private investors should be timely mitigated by means of:

- Confidence on the revenues
- Confidence on the timeline to start recovering the investment
- Technological risks

For what concerns regulatory issues, a significant advance on standardisation should be achieved in the next two years aligned with the Communication System Critical Design Review (end of 2012). The development of Implementing Rules and Acceptable Means of Compliance by EASA to be applied for the development and verification of pan European CNS/ATM systems is required even before the System Critical Design Review. i.e. not later than mid- 2012.

3 Conclusions

All the activities detailed in the paper were the preconditions to conduct a reliable business case analysis. What is highlighted in the study is that having a guaranteed return is a pre-requisite for industry to invest.

The expected market to be derived from AOC services as it is currently estimated by SIRIO study is not big enough, and therefore cost-recovery schemes need to assume significant revenues from ATS communications, which materialises if there is a requirement for airspace users to use the satellite communication service.

In any case, with a financing scheme which foresees a large public investment it has been possible to have an acceptable return of the investment in terms also of payback period for the private investors, and to determine a fair price for the end users of the communication services (i.e. ANSPs and Airlines).

This positive results encourage for a deeper analysis, once more a final decision on the system architecture will be taken, allowing for a refinement of the system costs and associated business case and PPP scheme.

References

1. SESAR Consortium, SESAR Definition Phase - Deliverable 5:SESAR Master Plan D5 (April 2008)
2. Filippo Tomasello -Chairman Iris Safety Board.: Aviation safety regulation Key messages Toulouse (January 13, 2010)
3. ESA Iris team, Iris Expert Group Report - Phase 1 (September 10 , 2009)
4. ESA Iris team, Iris Phase 1 - Design summary (December 9, 2009

Optimization of the Headers in DVB-S2 for Conveying IP Packets over GSE

Juan Pedro Mediano Alameda, Patrick Gelard,
Raquel Barco Moreno, and Emmanuel Dubois

CNES (French Space Center)
{juan-pedro.medianoalameda,patrick.gelard,
emmanuel.dubois}@cnes.fr
University of Malaga
rbarco@uma.es

Abstract. Convergence has become a key idea in communications over the past few years and IP has become the key 'convergence layer'. GSE (Generic Stream Encapsulation) enables the carriage of IP packets directly over DVB-S2 networks in a very efficient way, and it is considered to be used in the future access encapsulation. Thinking that these future services will be all-IP, this paper is intended to explain how the header of the BBFRAME (Base-Band Frame) of the DVB-S2 can be optimized until be completely removed in future broadcast satellite networks which will use IP/GSE/DVB-S2 only stack. This reduction is intended to separate the physical layer and link layer functionalities provided both by DVB-S2 and to simplify and ease the understanding of the DVB-S2 protocol.

Keywords: DVB-S2, GSE, BBFRAME, IP, optimization, convergence.

1 Introduction

In the future, networks may be founded on the development of an all-IP communication and service infrastructure that will gradually converge the current Internet, mobile, fixed and broadcasting networks. This IP convergence implies the carriage of different types of traffic such as voice, video, data, and images over a single network.

Satellite networks have been massively used for TV broadcasting and currently also follow this all-IP trend since satellite-delivered IP services are growing in every region of the world, not just those with limited telecom infrastructure. There are several factors involved in the growth of broadband satellite IP services: the performance (many satellite operators and hardware providers have integrated advanced IP networking capabilities into their offerings), the reduced cost (both for equipment and bandwidth), the success of multi-site applications based in open standards, but the arrival of a return link by satellite (DVB-RCS) has been the main factor and has been a milestone in these kind of networks.

DVB-S2 is designed as the successor for the popular DVB-S digital television broadcast standard and was ratified in 2005. It provides several functionalities of

G. Giambene and C. Sacchi (Eds.): PSATS 2011, LNICST 71, pp. 84–93, 2011.

physical layer such as modulation and synchronization and also of link layer such as coding, multiplexing and concatenation. The main objective of this paper is to separate these functionalities and simplify and ease the understanding of the protocol just as providing a first step of a future all internet protocols (user plan, control plan and management plan) over GSE only for the DVB-S2 protocol.

Next sections are organized as follows. In sections 2 and 3 an overview of the DVB-S2 system and GSE encapsulation is presented. Section 4 analyses the reduction of the BBHEADER of the DVB-S2 and in section 5 the reduction of the header in terms of gain efficiency is presented. The final conclusions are drawn in section 6.

2   DVB-S2 Overview

DVB-S2 [2] is a digital satellite transmission system developed by the DVB Project from 2003 as the successor of the world-wide known DVB-S standard. This architecture is designed and optimized for broadcast satellite applications such as digital television, content distribution and data transmission.

DVB-S2 implements the most recent developments in modulation and channel coding, with the use of QPSK, 8-PSK, 16-APSK, 32-APSK and especially, the use of concatenated Bose-Chaudhuri-Hocquenghem (BCH) and Low Density Parity Check (LDPC) codes. DVB-S2 makes use of 28 combinations of modulation format and coding scheme (MODCOD) to deliver performance that approaches the theoretical limit for such systems. Together, these can guarantee a low packet error ratio across a wide range of Signal to Noise plus Interference Ratio (SNIR), from -2.3 dB (QPSK) to 16 dB (32APSK). In addition, DVB-S2 provides functionalities of link layer such as fragmentation, multiplexing, assembling and concatenation and of physical layer as synchronization and modulation. The functional diagram block of DVB-S2 is shown in figure 1.

The processing of packets arriving at the DVB-S2 gateway can be summarized below:

1) Input streams (Generic Streams or Transport Streams) arrive at the Mode Adaptation block and each block input data is sliced into a constant DATAFIELD (DFL) in case of CCM mode or in a variable DATAFIELD in case of VCM/ACM mode. After that, a 80-bits BBHEADER is appended. For the VCM/ACM mode, the maximum length of each DATAFIELD is determined by the field MODCOD between different values (11 values for normal FECFRAME n_{ldpc} = 64800 bits, 10 values for short FECFRAME n_{ldpc} = 16200 bits and 3 values for very short FECFRAME n_{ldpc} = 4096) according to the protection required for each of them.

2) The Stream Adaptation block fills any unused bits in the DATAFIELD with padding (all zero bits) and generates a BBFRAME (Base-band Frame).

3) A block code is applied to each BBFRAME to form the FECFRAME, which can have 4096, 16200 or 64800 bits depending on the field TYPE of the PLHEADER. This code is a combination of BCH binary block code and a Low-Density Parity Check (LDPC) code, adding redundancy from 1/4 to 9/10.

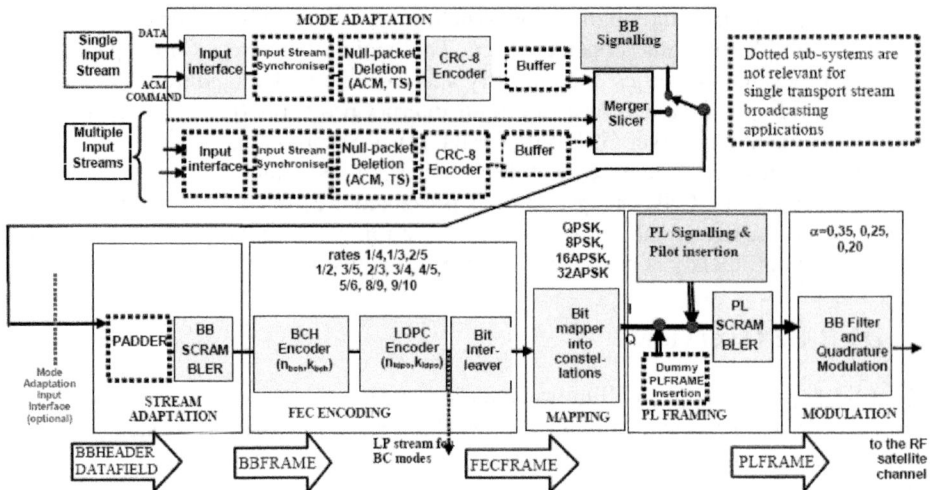

Fig. 1. Functional block diagram of the DVB-S2 system [2]

4) The encoded bits of the FECFRAME are interleaved, mapped to modulation symbols, using either QPSK (parallelism level $\eta_{MOD} = 2$), 8PSK ($\eta_{MOD} = 3$), 16APSK ($\eta_{MOD} = 4$) or 32APSK ($\eta_{MOD} = 5$) to form the XFECFRAME and finally randomized. Then it is sliced into an integer number S of slots of 90 symbols each and embedded in a PLFRAME (Physical Layer Frame), which is prefixed by the SOF (Start of Frame) and the MODCOD. Each PLFRAME length is variable and depends on the chosen constellation. The size of one PLFRAME can be calculated by dividing the total number of bits of the XFECFRAME by the parallelism level (η_{MOD}) and adding the 90 symbols of the PLHEADER.

5) After randomization, the signals shall be square root raised cosine filtered and modulated. The roll-off factor shall be $\alpha = 0,35, 0,25$ or $0,20$, depending on the service requirements.

DVB technology, based on the MPEG-2 system standard, has been increasingly successful in providing IP services and several encapsulation protocols have been proposed, such as MPE (Multi Protocol Encapsulation) [7], ULE (Unidirectional Lightweight Encapsulation) [9] or GSE (Generic Stream Encapsulation) [3], [4]. Although they have been analyzed in the literature [11], [12] and are commonly accepted as the standard ways to carry IP datagrams over existing DVB satellites, Generic Stream Encapsulation (GSE) is known to be a more efficient way to transport IP packets over DVB-S2 and the objective in the next future is to replace the current MPEG2-TS data encapsulation by the GSE direct encapsulation. Figure 2 shows this possible evolution of the protocol stacks from IP/MPE/MPEG2/DVB-S to IP/GSE/DVB-S2.

3 DVB-GSE Overview

DVB Generic Stream Encapsulation (DVB-GSE) comes up to provide a mean of carrying IP based content on DVB-S2 physical layers. Conceptually, it is at the same level in DVB systems as the Transport Stream, offering several functionalities such as concatenation and multiplexing also provided by DVB-S2. All DVB second-generation physical layer standards (e.g. DVB-S2, DVB-T2, etc...) will be "multi-mode", offering the option of using either the traditional MPEG Transport Stream or

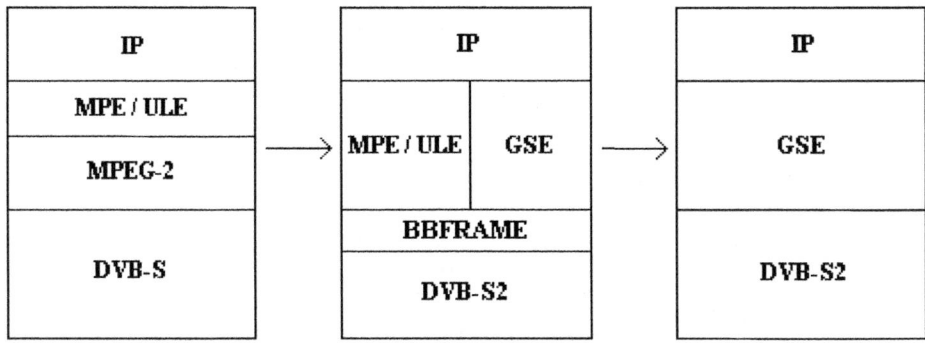

Fig. 2. Evolution of the protocol stacks over DVB

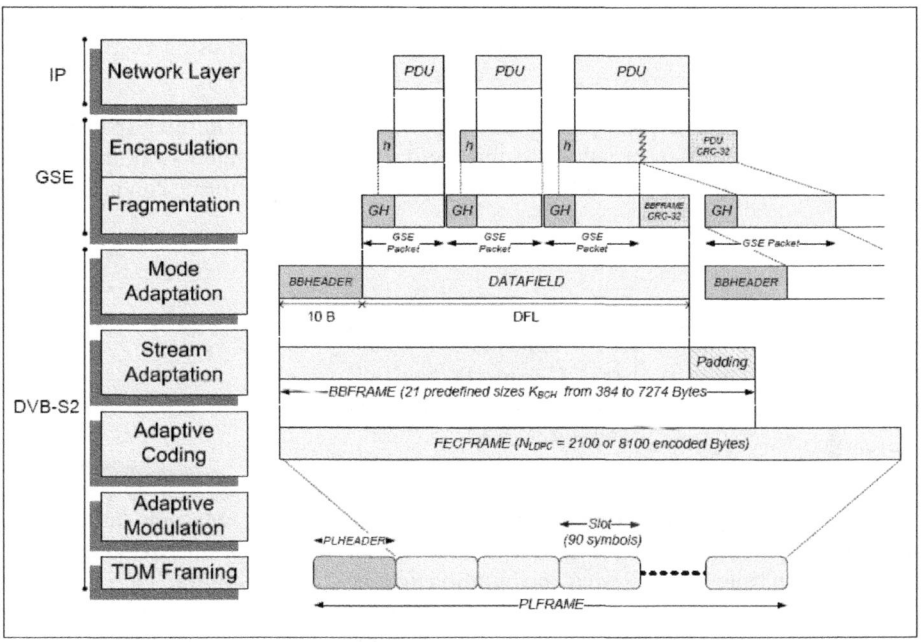

Fig. 3. GSE encapsulation within DVB protocol stack

DVB-GSE. As the network of the future is founded on the development of a converged all-IP communication, the MPEG Transport Stream are going to be replaced likely by the DVB-GSE since the latter is optimized for IP packets, IP-multicast and IPv6 transport (and optionally higher protocols such as MPLS or VLAN). Figure 2 shows a scheme of the evolution protocol stacks over DVB. The tendency is conveying IP packets over GSE (IP/GSE/DVB-S2) for the forward link.

The GSE operation within DVB-S2 protocol stack is depicted in figure 3. An encapsulated PDU (typically IP), prefixed by any optional extension ("h") headers added by the encapsulator, forms the payload of one or more GSE Packets. Each GSE Packet also includes a GSE header ("GH") that contains the length, protocol type and label field (when present). The stream of GSE Packets is placed in the DATAFIELD of a BBFRAME. The sender normally selects the MODCOD (and hence the BBFRAME size) to achieve the QEF (Quasi Error Free) target.

4   DVB-S2 BBHEADER Reduction

As shown in figure 1 and figure 3, after the Mode Adaptation block, a fixed length Base-Band Header (BBHEADER) of 80 bits is inserted in front of the DATA FIELD by the Stream Adaptation block, describing its format. This header adds an overhead to the BBFRAME depending on the LDPC Code and the length of the FECFRAME (normal, short or very short).

All these overhead can be reduced in case of using GSE only. The simplification of the header which is going to be carried out is independent of the protocol encapsulated in GSE (IPv4, IPv6, MPEG2-TS, MPLS...). Next it is presented each field of the BBHEADER, what is used for, its value in case of using GSE and why it does not apply.

1) TS/GS field (2 bits): Transport Stream Input or Generic Stream Input (packetized or continuous). The different values of this field are shown in table 1. With only GSE encapsulation this field should be set to a unique value and therefore can be removed.

Table 1. TS/GS field

TS/GS
11 = Transport
00 = Generic Packetized
01 = Generic Continuous
10 = Reserved

2) SIS/MIS field (1 bit): Single Input Stream (value '1') or Multiple Input Stream (value '0'). The link layer function associated to this field is multiplexing since in DVB-S2 multiple streams may be multiplexed at the transmitter. Generic Stream Encapsulation shall be carried out separately for the incoming data of each generic stream. Each stream is identified by a specific Input Stream Identifier (ISI). Its value is present in the BBHEADER (in the second byte of the MATYPE field) in case of multiple input streams.

The ISI of each (generic or transport) stream in DVB-S2 is signalled in the S2 satellite delivery system descriptor describing the stream. This descriptor is located in the Network Information Table (NIT). This multiplexing function can be resolved by upper layers, and so this field can be removed.

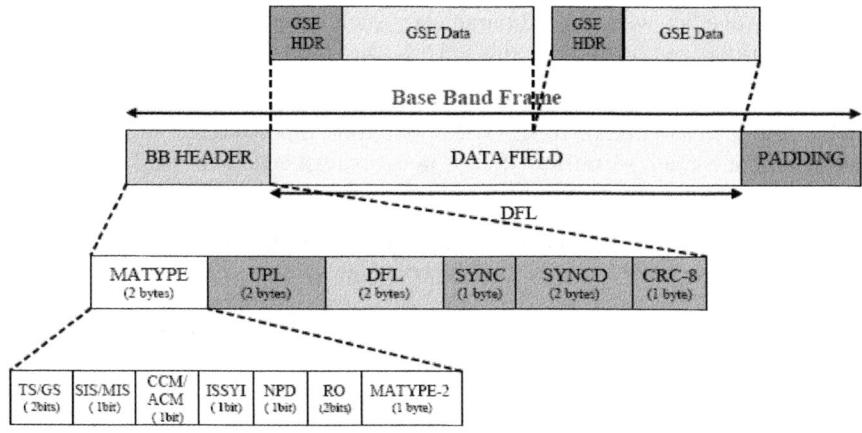

Fig. 4. DVB-S2 BBFRAME format [2]

3) CCM/ACM field (1 bit): Constant Coding and Modulation (CCM, value '1') or Adaptive Coding and Modulation (VCM is signalled as ACM, value '0'). This field could be removed due to the fact that CCM can be considered a particular type of ACM. In fact, the field MODCOD (5 bits) which is part of the PLHEADER of the PLFRAME defines 29 different types of combinations of modulations and coding (28 types and the DUMMY FRAME insertion) and there are 3 possibilities left. We propose to use one of them (29_D, 30_D or 31_D) to signal the CCM mode, or even better, set it by the configuration at the beginning.

4) ISSYI label (1 bit): Input Stream Synchronization Indicator. The function associated with this label allows guaranteeing a constant-bit-rate and is used only for packetized input stream (Transport Stream or Generic Packetized Stream). In case of DVB-S2 encapsulating GSE packets, this field must be always set to 0 (inactive) since GSE is not packetized stream and therefore can be removed. ([2], Annex D.2).

5) NPD field (1 bit): Null-Packet Deletion. The associated function aims at identifying and removing MPEG null-packets because the Transport Stream rules require that the bit rates at the output of the MUX and the input of the DEMUX are constant in time, and the end-to-end delay is also constant. In order to fulfil such requirements in an ACM environment, the null-packet deletion function shall be activated [2]. This field must be set to 0 (inactive) in GSE/DVB-S2 and therefore is not relevant for GSE packets.

6) RO field (2 bits): transmission of the roll-off factor (α). The roll-off factor is a measure of the excess bandwidth of the Base-Band filter (BB Filter) and it varies between 0 (the roll-off zone becomes infinitesimally narrow) and 1 (the non-zero portion of the spectrum is a pure raised cosine). It can take three different values (0.35, 0.25 or 0.20) [2] and it determines the symbol rate.

The roll-off factor does not need to be retransmitted every BBFRAME as once the transmitter and the receiver know the roll-off factor, they can determine the symbol rate. In fact, in the 'Satellite_Delivery_System_Descriptor' [6], the roll-off factor is transmitted together with several parameters such as orbital position or polarization, so this descriptor can be used for this purpose and then these two bits can be removed of the BBHEADER.

7) MATYPE-2 field (8 bits): if SIS/MIS = Multiple Input Stream, this field is used to send the Input Stream identifier (ISI), if not it is reserved. This field can be removed on the same reason than the SIS/MIS field.

8) UPL field (16 bits): User Packet Length in bits, in the range 0 to 65535. The link layer function associated to this field is concatenation. In case of MPEG Transport Stream, this field has to be set to $188x8_D$. For Generic Continuous Streams this field must contain the unique value 0000_{HEX} and can be removed because it is a unique value and this function is already done at GSE level.

9) DFL field (16 bits): Data Field Length in bits. This length can be easily calculated at the physical layer thanks to the 'GSE_Length' field inside the GSE header as shown in figure 5. For a particular MODCOD, the length of the BBFRAME is known and unique. The point here is to calculate how many bits DFL has and how many are padding. In DVB-S2, a Base Band frame always starts with the header of the first GSE packet because there is no fragmentation at the Mode Adaptation level (the fragmentation is just before this, GSE packets are not fragmented between Base Band frames).

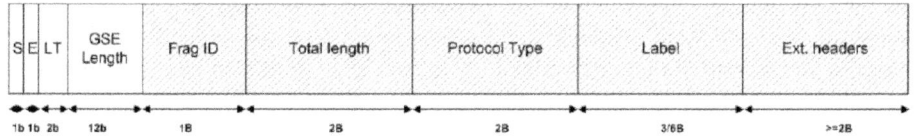

Fig. 5. GSE Header (GH) format [3]

The GSE Length field (fourth field) in the GSE Header indicates the length of the packet after it, so it is easy to calculate the length of the first part of the BBFRAME (16 bits + 'GSE_Length' bits). After that, it should be determined if the next part of the BBFRAME is padding or is another GSE packet (because the padding is provided at GSE level). This is also easy because the padding frame has the first 4 bits of the GSE Header equal to 0000. If the next four bits after the first GSE packet are equal to 0000, the rest of the bits until K_{bch} must be padding and equal to 0. If it is not, it is not a padding frame, and the size will be given by the field 'GSE_Length' as the previous one and so on [3].

10) SYNC field (8 bits): copy of the User packet Sync-byte. In case of MPEG Transport Stream this field has to be set to 47_{HEX}. For Continuous Generic Streams:

SYNC = 00_{HEX} [8]. This field can be removed from the BBHEADER because the synchronization of the packets inside a BBFRAME is already done at the GSE level, since all BBFRAMEs starts with the beginning of a GSE packet.

11) SYNCD field (16 bits): is defined only for packetized Transport and gives the distance in bits from the beginning of the DATA FIELD and the first User Packet (UP) from this frame (first bit of the CRC-8). This field is not defined for Generic Streams and therefore can be removed. Moreover in case of using GSE encapsulation, each PLFRAME must start with the beginning of one GSE packet, so the synchronization at this level is already done and this SYNCD field is redundant.

12) CRC-8 field (8 bits): error detection code applied to the first 72 bits of the BBHEADER. Obviously this field can be removed as the header has been removed.

5   Results in Terms of Efficiency for the Optimization of the BBHEADER

This reduction can be measured in terms of efficiency by dividing DFL (Data Field Length = K_{bch} - 80 bits of the header) by the K_{bch} (length of the BBFRAME) as shown:

$$\eta = \frac{k_{bch} - 80}{k_{bch}} = \frac{DFL}{k_{bch}}$$

The figure 6 shows a graphic with the different efficiencies of the BBFRAME depending on the LDPC code and the type of FECFRAME (very short, short or normal) before the reduction.

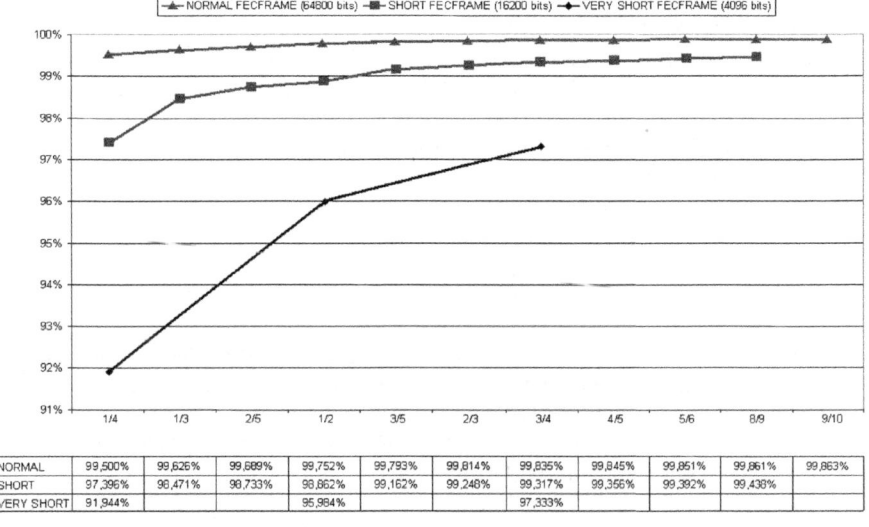

	1/4	1/3	2/5	1/2	3/5	2/3	3/4	4/5	5/6	8/9	9/10
NORMAL	99,500%	99,626%	99,689%	99,752%	99,793%	99,814%	99,835%	99,845%	99,851%	99,861%	99,863%
SHORT	97,396%	98,471%	98,733%	98,862%	99,162%	99,248%	99,317%	99,356%	99,392%	99,438%	
VERY SHORT	91,944%			95,984%			97,333%				

Fig. 6. BBFRAME Efficiency for each LDPC code (DFL/K_{bch})

For the very short FECFRAME (n_{ldpc} = 4096 bits) the efficiency of the BBFRAME varies between 91,9% (code rate 1/4) and 97,3% (code rate 3/4), for the short FECFRAME (n_{ldpc} = 16200 bits), the efficiency of the BBFRAME varies between 97,39% (code rate 1/4) and 99,43% (code rate 8/9) and for the normal FECFRAME (n_{ldpc} = 64800 bits), between 99,5% (code rate 1/4) and 99,8% (code rate 9/10). So the maximum efficiency loss introduced by the BBHEADER is 8,1% (very short FECFRAME with code rate 1/4) and these losses can be reduced to 0% with the solution provided in this paper.

6 Future Work and Conclusions

The reduction of the BBHEADER explained in the paper makes possible to remove double behaviours such as multiplexing or concatenation also provided by GSE encapsulation just as simplify and ease the understanding of the DVB-S2 protocol giving a clear functional separation between functions of physical layer and functions belonging to the link layer. Therefore, DVB-S2 would be in charge of the physical layer functions such as modulation, coding and synchronization, and GSE of the link layer functionalities such as concatenation and multiplexing.

Although the performance is not the goal of this paper, this reduction also achieves an obvious improvement of the processing at the receiver and the increase of the capacity of 80 bits per BBFRAME. This improvement increases the bit-rate between 0,14% and 8,06% depending on the code and the modulation.

A future step is to understand the synchronization at the satellite terminal (RCST) and study the possibility of inserting MPEG2 packets (which carry the Network Clock Reference) inside GSE packets by the scheduler. As it is known, these NCR packets have to be inserted inside certain GSE packets known in advance, and this can create a cross-layer problem which has to be resolved. The addressing and signalling have also to be carefully studied in these systems since currently MPEG2-TS is used instead of GSE to transport the information tables.

The reduction of the GSE header is another point which could be taken into account since, for example, the 'Total_Length' field is only used for providing information to the RCST to keep memory in the reception buffers.

Finally, all these improvements could be studied in standards such as DVB-T2, DVB-C2 and DVB-SH.

References

1. Digital Video Broadcasting (DVB) Home Page, http://www.dvb.org
2. ETSI EN 302 307 V1.2.1 Digital Video Broadcasting (DVB); Second generation framing structure, channel coding and modulation system for Broadcasting, Interactive Services, News Gathering and other broadband satellite applications (DVB-S2) (August 2009)
3. ETSI TS 102 606 V1.1.1 Digital Video Broadcasting (DVB); Generic Stream Encapsulation (GSE) Protocol (October 2007)
4. ETSI TS 102 771 V1.1.1 Digital Video Broadcasting (DVB); Generic Stream Encapsulation (GSE) implementation guidelines (June 2009)

5. ETSI EN 301 790 V1.5.1 Digital Video Broadcasting (DVB); Interaction channel for satellite distribution systems (May 2009)
6. Final Draft ETSI EN 300 468 V1.10.1 Digital Video Broadcasting (DVB); Specification for Service Information (SI) in DVB systems (July 2009)
7. Final Draft ETSI EN 301 192 V1.4.1 Digital Video Broadcasting (DVB); DVB specification for data broadcasting (June 2004)
8. ETSI TS 101 162 V1.2.1 Digital Video Broadcasting (DVB); Allocation of Service Information (SI) and Data Broadcasting Codes for Digital Video Broadcasting (DVB) systems (July 2009)
9. IETF RFC 4326: Unidirectional Lightweight Encapsulation (ULE) for Transmission of IP Datagrams over an MPEG-2 Transport Stream (TS)
10. Information Sciences Institute (University of Southern California), Internet Protocol (IP), RFC 791 (September 1981)
11. Fairhurst, G., Matthews, A.: A comparison of IP transmission using MPE and a new Lightweight Encapsulation. In: Advances in Satellite Communications, IEE Colloquium, London, UK (2003)
12. Hong, T.C., Chee, W.T., Budiarto, R.: A Comparison of IP Datagrams Transmission using MPE and ULE over Mpeg-2/DVB Networks. In: Proceedings Fifth Int'l Conference on Information, Communications, and Signal Processing (ICICS 2005), Bangkok, Thailand (December 6-9, 2005)

A Low Complexity Concealment Algorithm for H.264 Encoded Video over DVB-S2

Susanna Spinsante[1], Claudio Sacchi[2], and Ennio Gambi[1]

[1] DIBET, Universitá Politecnica delle Marche
Via Brecce Bianche 12, I-60131 Ancona, Italy
{s.spinsante,e.gambi}@univpm.it
[2] DISI, Universitá degli Studi di Trento
Via Sommarive 14, I-38123 Trento, Italy
sacchi@disi.unitn.it

Abstract. Due to the high level of compression, H.264/AVC video information is very sensitive to channel errors, and in particular the Variable Length Coding applied at the encoder side, together with Motion Compensation, can amplify the effects of transmission errors during the decoding phase. If, in addition, we consider that in real time applications lost or damaged data cannot be retransmitted, the importance to introduce concealment techniques in order to recover lost or corrupted video data emerges, through the exploitation of spatial and/or temporal correlation among the residual available information. In this paper we present a technique for polygonal edge interpolation, to improve the performance of Intra concealment in the case of video frames rich in details. Experimental results show that the proposed low complexity scheme may provide substantial improvements to the perceived video quality, confirmed by an increase in PSNR values.

Keywords: Satellite video broadcasting, video coding, video transmission, error concealment.

1 Introduction

The amazing spreading of Mobile TV, video-enabled cellular phones, and portable video terminals favored the broad adoption of the H.264/AVC video coding standard [1], mainly thanks to its increased compression performance (up to 60%) with respect to all the previous video compression techniques. For the same reason, the high compression efficiency attainable by H.264/AVC made it possible to deliver High Definition (HD) video over satellite, for example through DVB-S2 [2,3]. Considering the improvements in channel coding (LDPC) and modulation techniques (e.g. QPSK, 8PSK, 16APSK and 32APSK), DVB-S2 standard achieves a capacity increase of 30% compared to its predecessor DVB-S, under the same transmission conditions. The inclusion of H.264/AVC in DVB-S2 specification brings an additional gain of a factor 2 in bit rate at the same perceived video quality, when compared to the previously adopted MPEG-2 video compression. DVB-S2

G. Giambene and C. Sacchi (Eds.): PSATS 2011, LNICST 71, pp. 94–103, 2011.

already provides advanced features and characteristics targeted not only to consumer but also to professional applications, such as the increased number of modulation modes available, the Adaptive Coding and Modulation (ACM) tool, very powerful FEC techniques, new roll-off factor values for a tighter bandwidth shaping. The joint adoption of H.264/MPEG-4 coding further extends the range of supported services, targeted to both average viewers and professionals.

However, as well known, the transmission of compressed video streams over error-prone networks and channels (such as the RF broadcasting channels) is vulnerable to bit errors and packet losses that lead to data corruption and serious video quality deterioration. As a consequence, error concealment algorithms may be implemented at the receiver to enhance the quality of the reconstructed video. Error concealment has the attractive advantage of being encoder-independent: to compensate the effects of transmission errors, and to approximate the missing data, only the residual available correlated information in the received video signal is exploited, without the need of any specific customization at the encoder side. A rich technical literature exists in the field of error concealment techniques (just to mention a few examples, see [4,5,6]); many proposals relate to spatial only, or temporal only solutions, better known as Intra and Inter concealment, respectively, or introduce a mix of different strategies. Other solutions involve the adoption of image processing operators, to be applied to each single frame, usually borrowed from still image compression techniques.

This paper focuses on the development of an interpolation technique for Intra concealment, optimized for the case of polygonal edges, in order to overcome some limitations found in classical Intra approaches ([7,8]). Common techniques, such as bilinear and bi-cubic interpolations, are widely used for their low computational complexity; however they often cause blurring artifacts around the edges, which are annoying to visual perception. The proposed technique improves linear and directional interpolation, to efficiently and accurately recover lost macroblocks (MBs) containing a single, but polygonal, edge. A two-step approach is applied, by which a one-dimensional interpolation follows an edge detection phase. By means of the directional reconstruction, typical smoothing effects due to pixel interpolation are avoided, and image details preserved. Further extensions may include multi directional edges in a single MB that require a more complex analysis of the image information, to select the proper interpolation strategy. A major constraint in the design of such a technique for the recovery of polygonal edges is the limited complexity and computational load required, that are necessary in order to ensure real time applicability in video decoders. Many interpolation algorithms proposed in the literature for image recovery suffer from high complexity and computational requirements, and are not suitable in real time video applications, though able to provide very good performance in image reconstruction. Intra error concealment performance are especially efficient in the following cases: 1) high motion frame areas, for which the temporal recovery is not satisfactory, due to the very low correlation among subsequent scenes; 2) when data losses occur at scene changes in the video sequence; 3) when image variations are present at a local scale (MB level), and the most suitable interpolation algorithm is selected on the basis of the available information. Most of the Intra concealment solutions presented in the literature recovers the missing pixels of a MB by applying interpolation algorithms (such as linear or bilinear ones) on the pixels of the available neighboring and correctly received MBs. This is motivated by the fact that adjacent MBs in the same frame usually show minimal variations in their pixel values. An improvement to interpolation based approaches may be represented by a polygonal technique, which exploits the presence of two dominant directions in image areas rich in details, which usually feature geometric patterns, such as in the case of objects' edges.

The paper is structured as follows: Section 2 will describe the proposed concealment technique; Section 3 will show some simulation results. Finally paper conclusions will be drawn in Section 4.

2   The Proposed Concealment Technique

The human visual system is particularly sensitive to edges' distortions, as those due to blurring or blockiness artifacts that may happen in a coded video frame. As a consequence, a polygonal interpolation process could substantially improve the reconstruction results provided by a linear or bilinear interpolation, by preserving the real edge directions in the image, and avoiding the generation of artificial edges. To such an aim, before moving to the true interpolation operation, it is necessary to perform a local, pixel-based analysis of the stream to decode, in order to preserve the original image structure and obtain an as much faithful as possible reconstruction of the video content. Obviously, a constraint on the computational effort required is always to be considered, especially for real-time applications. Available pixels located at the edges of the missing MB allow recovering information about the lost geometric pattern, and are used to reconstruct the missing pixels by a polygonal interpolation. The solution proposed in the following cannot be applied in any case, but only when a single, polygonal edge in the missing MB is to be recovered. In this sense, the linear interpolation scheme may be seen as a specific case of application of the polygonal interpolation solution. The edge direction detection step necessary to perform polygonal interpolation is executed by applying the well known Sobel operators, with a simple but effective modification with respect to their classical adoption. The Sobel operators S_x and S_y applied to pixels $P(x, y)$ of an image I provide the gradient components G_x and G_y to compute the gradient vector amplitude and direction:

$$|G| = \sqrt{G_x^2 + G_y^2} \ .$$
$$\Theta = tan^{-1} \left(G_y / G_x \right) \ . \tag{1}$$

If the edge located by the Sobel operators along the boundaries of the missing MB has a direction included in $[-5°, +5°]$ horizontally, and $[-85°, +85°]$ vertically, the edge is discarded, as it represents a direction parallel to the MB boundaries, i.e. an edge that does not intercept the missing MB itself. This edge does not

Fig. 1. a) Original video frame, b) Corrupted video frame (the sample missing MB of interest is evidenced)

need to be reconstructed by the interpolation step. In order to better illustrate the proposed scheme, let us consider Figs. 1 a) and b), where a video frame and a corrupted version of it, featuring several missing MBs in a chessboard pattern, are shown. Suppose that the missing MB evidenced in Fig. 1 b) is to be recovered. As shown in the enlarged version of the image detail provided in Fig. 2 a), a correct reconstruction of the missing pixels may be accomplished only by a polygonal interpolation of the lost edge, the geometry of which is determined by the intersection of the two straight lines found by the edge detection algorithm, and the direction of which can be provided by an edge direction detection step. An example is shown in Fig. 2 b).

Given the quantities P_{1a}, Θ_{1a}, P_{1b} and Θ_{1b}, as represented in Fig. 3 a), it is possible to derive the analytical expression of the two lines r_1 and r_2 that divide the missing MB into four areas, as shown in Fig. 3 b). The proposed algorithm associates the missing pixel to one out of the four areas in the MB, according to the intersection points P_{r1}, P_{r2}, P_{r3}, and P_{r4} previously located (Fig. 3 c)). Considering, as an example, the case of a missing pixel in the area #1, only P_{r1}

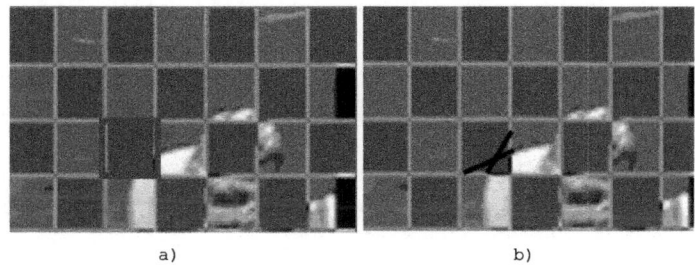

Fig. 2. a) Detail of the missing MB to recover, b) Edge direction detection on the missing MB

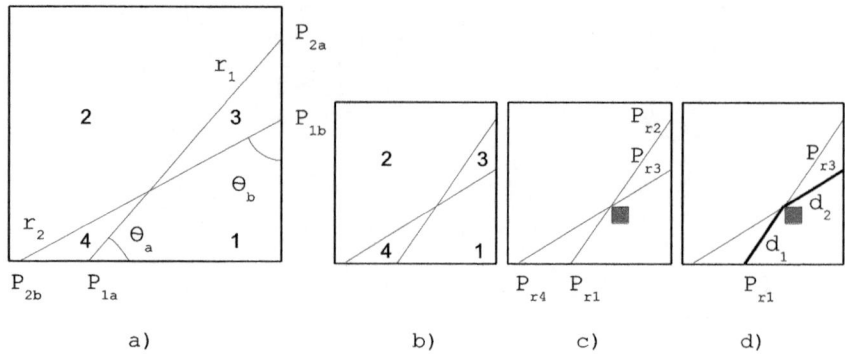

Fig. 3. a) Geometric representation of the polygonal interpolation scheme, b) The four areas located in the missing MB, c) Areas and pixel association, d) Geometric quantities for the polygonal pixel interpolation

and P_{r3} contribute the value of the interpolated pixel, according to their relative distances d_1 and d_2 (Fig. 3 d)) and the general formula (2), that is valid when n different points are used in the reconstruction:

$$p_0 = \frac{\sum_{i=1}^{n} d_i \cdot P_{ri}}{\sum_{i=1}^{n} d_i}. \tag{2}$$

Similar computations are applied when the missing pixels to recover are located in the other areas of the MB subjected to concealment.

3 Simulation Results

The interpolation technique proposed in this paper is integrated in a more structured concealment strategy, according to which a preliminary edge detection and classification step establishes the features of the missing MB to recover. Based on the outcome of this process, the proper Intra concealment technique is selected (bidirectional or polygonal). If the number of edges revealed inside the correctly received MBs adjacent to the missing MB is greater than 2, in order to search for a possible dominant direction to recover, a refinement step is applied. Such a refinement relies on the application of a Directional Entropy (DE) calculation block, to decide if the number of revealed directions (angles) denote a single dominant direction, a limited number of dominant directions, or so many directions that they cannot be accounted for. The DE block implements the following formula:

$$DE = -\sum_{i=1}^{n} p(d_x) \cdot log_2(p(d_x)) \tag{3}$$

where $p(d_x)$ is the probability density function of direction d_x, estimated by enumerating all the edges located by the edge detection step, and falling into 8 groups of directions, from $-\pi/2$ to $\pi/2$, in steps of $\pi/8$. A control parameter β is also defined as:

$$\beta = \frac{-\sum_{i=1}^{n} p(d_x) \cdot log_2(p(d_x))}{Max(DE)} \qquad (4)$$

and compared against a set of thresholds, to differentiate the possible cases. In our experiments we empirically set threshold values at 0.3 and 0.75: by this choice, when $0 < \beta \leq 0.3$, a single directional edge area is located; when $0.3 < \beta < 0.75$, a textural multidirectional edge area is found; when $\beta \geq 0.75$, a complex multidirectional edge area is identified. In the case of complex multidirectional edge areas, the bilinear interpolation could be used, motivated by the multitude of different edges and directions featured by the image. In the textural multidirectional case, quite complicated to cope with, advanced algorithms with increased computational complexity should be conceived.

In order to evaluate the performance of the proposed concealment technique, the operations discussed have been integrated in the JM Reference Software v.14 decoder [9]. Simulations have been performed by encoding a synthetic video sequence obtained by collecting frames extracted by well known test sequences, showing different features and dynamic effects, and then applying a packet loss simulator in order to delete the 1%, 2%, 4%, and 10% amount of the original encoded data. Such values are reasonable, in the case of link degradation due to rain or other atmospheric impairments (like e.g. scintillations). Rain attenuation in high frequency band like Ka-band is a major factor for lowering the link capacity in satellite broadcasting services. The interesting analysis of packet-error-rate (PER) performance of DVB-S2 contained in [10] evidenced that PER curves achieved for all the 28 coded modulation modes of DVB-S2 have the typical waterfall pattern. This means that small SNR variations may result in strongly increased PERs. Each mode is characterized by a SNR threshold guaranteeing error-free transmission (See Tab.1 of [10]). However, the SNR margin allowed to have quasi error-free is very narrow (0.3 dB). A decrease of the SNR of 1 dB below the assigned threshold may involve PERs value larger than 10^{-2}. This is the reason why, error concealment is necessary to guarantee satisfactory QoS (Quality of Service) also in the presence of relevant packet loss. The application of a concealment technique at the decoder may enforce the performance provided by DVB-S2 through its ACM tool, that allows the transmission parameters to be changed on a frame by frame basis, depending on the particular conditions of the delivery path for each individual user. The H.264 video sequence is encoded in the Baseline profile, with a CIF 4:2:0 format, a Search Range of 16 pixels, 1 reference frame, Intra Period = 15, and a dispersed slice group map type. The behavior of the proposed concealment solution is compared to the performance obtainable by applying the default concealment approach included in the original Reference Software, that combines a bilinear interpolation scheme for the spatial concealment of all the type-I frames, and an enhanced boundary matching scheme for the Inter concealment of type-P frames. The applied scheme is selected on the basis of a motion index associated to the motion vectors computed during decoding. The global performances over the entire video sequence are evaluated with respect to the quality of the decoded sequence, described by the Peak Signal to Noise Ratio (PSNR) of the Y component, and the Structural Similarly Information quality metric (SSIM). The former is typically used to estimate the video quality in an objective way; the latter is a numerical tool to estimate subjective quality perceived by the user. A more precise definition for the PSNR would be "similarity to original", as it calculates a quality difference, by comparing each pixel in a distorted (reconstructed) frame to each pixel in the corresponding original frame. As the difference between frames becomes smaller, the PSNR index gets bigger. The PSNR index was calculated by the following formula:

$$PSNR = 10 \cdot log_{10} \frac{255^2 \cdot mn}{\sum_{i,j=1}^{m,n} (x_{ij} - y_{ij})^2} \tag{5}$$

where x and y represent the original and reconstructed frames, respectively, m and n are frame dimensions, i and j denote the pixel position in the frame. Only the luminance component of each pixel was considered for PSNR computation.

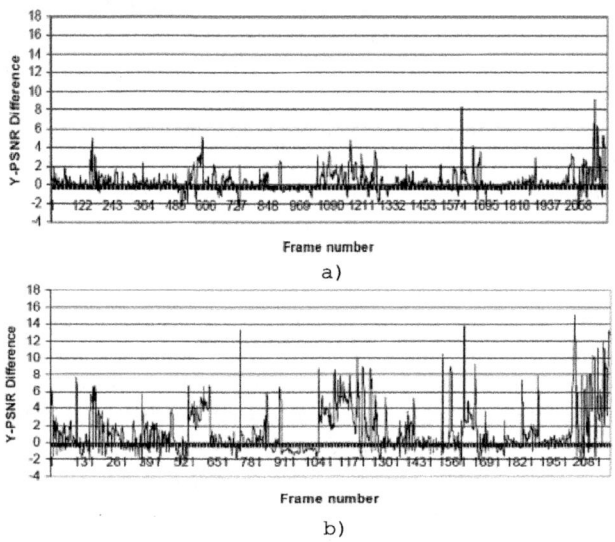

Fig. 4. Y-PSNR gain due to the proposed concealment framework, with respect to the original JM decoder, for a packet loss rate of a) 1% and b) 10%

Fig. 4 shows the performance of the modified decoder, with respect to the original one, expressed as the PSNR gain due to the proposed technique, averaged over 5 decoding iterations, for packet loss ratios of 1% (Fig. 4 a)) and 10% (Fig. 4 b)). On average, the proposed solution is able to improve the quality performance of the decoder in a remarkable way, and the improvement is more evident when the packet loss rate increases. On the other hand, the negative values for the PSNR shown in the graphs correspond to frames for which the proposed technique provides lower PSNR values than the ones obtained by

Table 1. Average Y-PSNR performance for different PER values: comparison among the concealment solutions

% PER	Avg. Y-PSNR (dB) JM Ref. Software	Avg. Y-PSNR (dB) Proposed
1%	36	36.39
2%	33.93	34.75
4%	31.31	32.42
10%	27.35	26.68

Table 2. Average Y-SSIM of the video sequence: comparison among the concealment solutions

Concealment technique	Y-SSIM (dB)
JM Reference	0.974
Proposed	**0.985**
Bilinear	0.975

the application of the original JM Reference Software concealment. A frame-by-frame analysis of the results, together with the visual inspection of these specific frames, reveal that the lower PSNR values are basically due to the Inter concealment strategy, which is not optimized (proper threshold setting is missing), and to noise effects that in some cases distort the edge detection performance. It is obvious that the concealment strategy included within the JM Reference Software is complete either in the Intra and Inter sections; the solution herein tested, on the contrary, has been optimized in its Intra component but not in the Inter one, which is the object of research activities currently ongoing. Table 1 shows how the quality gain provided by the proposed solution, in terms of average Y-PSNR over the whole video sequence, increases as the % PER affecting transmission increases, thus confirming the effectiveness of the concealment technique described in the paper, with respect to the solution included in the JM Reference Software. Results about the SSIM estimation confirm the better perceived video quality of the sequence concealed by means of the proposed technique, with respect to the original JM concealment, as reported in Table 2. A detailed analysis, performed frame by frame over the decoded sequence, shows that the most relevant quality gain is obtained for high motion sub-sequences, and frame areas rich in details, which confirms that the goals of the proposed concealment approach have been fulfilled. In such conditions the concealment process performed through polygonal interpolation provides very satisfactory results, because it is able to recover and preserve the details in the frame that most affect the video quality perceived by the user. An example is shown in Fig. 5, where a detail of the MB recovered by the proposed approach is presented, together with the whole concealed frame. In Fig. 5 a), the evidenced MB shows the edge of the player's shoulder reconstructed by the polygonal interpolation

a)   b)

Fig. 5. Polygonal interpolation based concealment: a) Detail of the reconstructed MB, b) The concealed video frame

Fig. 6. Detail of the proposed scheme performance: a) Corrupted frame, b) JM concealment, c) Bilinear approach, d) Proposed concealment

scheme. The quality of the reconstruction makes the effects of video data losses not perceivable by the user.

A further example of the effective results provided by the proposed concealment, even at the expense of a limited complexity, is provided in Fig. 6, where a sample frame is shown, in which the suggested scheme applies different recovering strategies, according to the different properties exhibited by the frame sub areas at the decoder.

4   Conclusion

In this paper, a novel error concealment algorithm has been proposed for H.264 coded videos in the framework of DVB-S2 satellite TV broadcasting. The proposed algorithm relies on polygonal edge interpolation in order to improve Intraframe concealment performance. Visual improvements in performance in terms of increased video quality are evident, also in the case of high packet error rates that are possible in satellite broadcasting, due to atmospheric and rain attenuations. It is shown that the gain in quality increases as the effects of channel impairments increase, manifested by higher percent values of the packet error rate.

References

1. ITU-T Recommendation H.264 (11/2009), Advanced video coding for generic audiovisual services — ISO/IEC 14496-10:2008, Information technology - Coding of audio-visual objects (MPEG-4) - Part 10: Advanced video coding, Version 13 (November 2009)
2. Digital Video Broadcasting (DVB): Second generation framing structure, channel coding and modulation systems for Broadcasting, Interactive Services, News Gathering and other broadband satellite applications, ETSI draft Specification, EN302 307, V1.1.1 (June 2004)
3. Morello, A., Mignone, V.: DVB-S2: The Second Generation Standard for Satellite Broadband Services. Proc. IEEE 94(1), 210–227 (2006)
4. Yi, J-W., Cheng, E., Yuan, F.: An Improved Spatial Error Concealment Algorithm Based on H.264. In: 3rd International Symposium on Intelligent Information Technology Application, pp. 455–458. (2009)
5. Beesley, S., Armstrong, A., Grecos, C.: An Edge Preserving Spatial Error Concealment Technique for the H.264 Video Coding Standard. In: Ph. D. Research in Microelectronics and Electronics, pp. 113–116 (2006)
6. Nemethova, O., Al-Moghrabi, A., Rupp, M.: Flexible Error Concealment for H.264 Based on Directional Interpolation. In: International Conference on Wireless Networks, Communications and Mobile Computing, pp. 1255–1260 (2005)
7. Weibin, Z., Zhiheng, Z., Zhiliang, X.: Error concealment for video sequences based on Sobel edge detector. In: International Conference on Communications, Circuits and Systems, pp. 723–726 (2008)
8. Senel, H.G.: Gradient Estimation Using Wide Support Operators. IEEE Trans. Image Proc. 18(4), 867–878 (2009)
9. JM Reference Software v.14.0, http://iphome.hhi.de/suehring/tml/
10. Chi, W.S., Seo, K.-D., Lee, I.K., Chang, D.-I.: Joint Source-Channel Coding Scheme for SVC-based DVB-S2 Satellite Broadcasting System. In: Proc. of 2010 Int. Conf. on Consumer Electronics (ICCE 2010), pp. 1–2 (2010)

OAM Technique Based on ULE for DVB-S/S2 Transmission Systems

Raffaello Secchi, Nimbe Ewald, and Gorry Fairhurst

Electronics Research Group (ERG),
School of Engineering, Aberdeen King's college
Fraser Noble Building, AB24 3UE, Aberdeen, UK
{raffaello,nimbe,gorry}@erg.abdn.ac.uk

Abstract. This paper presents a technique to monitor the performance of DVB-S/S2 transmission systems at the Link Layer (LL) level using the Unidirectional Lightweight Encapsulation (ULE) protocol. The technique is based on the use of extension headers and is validated for different traffic patterns. It has been designed to provide an Operation, Administration and Management (OAM) method for IP-based DVB-S/S2 systems using the MPEG-2 Transport Stream. Such methods can be used to identify edge-to-edge performance and connectivity issues in a layer 2 network.

Keywords: DVB-S, DVB-S2, ULE, OAM.

1 Introduction

Digital Video Broadcast for Satellite (DVB-S/S2) [1] standards have been used to successfully deploy standard transmission networks with over a hundred million DVB-S receivers deployed worldwide. DVB-S/S2 not only provides a platform that may be used for traditional television broadcast services, but may also be used to support a growing number of IP-based and interactive services [2,3,4]. The growing use of IP services, both for TV broadcast, augmented services, backhaul/contribution and Internet access is about to revolutionise the satellite service portfolio.

In general, Internet Service Providers (ISPs) and content providers do not manage a DVB-S/S2 satellite platform directly. They typically lease satellite access and bandwidth resources from satellite service providers or third-party operators. This hierarchy is common to many other transmission networks – ISPs often utilise the service offered by transmission operators. Therefore, an ISP that needs to manage the Quality of Service (QoS) guarantees provided to a customer must be able to separately validate the quality of the underlying transmission service.

One method to verify the operation of a service offering is to make performance or connectivity measurements across a portion of the transmission network, sometimes called edge-to-edge measurements between the ingress and egress of a

G. Giambene and C. Sacchi (Eds.): PSATS 2011, LNICST 71, pp. 104–115, 2011.

transmission network segment (such as DVB-S/S2 transmission system). These measurements must necessarily be below the IP or Layer2 Virtual Network Provider (L2VNP - RFC 4664), because the end-to-end network services (e.g. including allocation of IP addresses) are not the responsibility of the transmission operator.

Methods of this form provide an important tool for Operations, Administration, and Management (OAM) in most managed terrestrial networks. Edge-to-edge measurements can help in verifying performance of the transmission service and also are particularly valuable in the location of faults, by indicating whether a reported issue is a result of a transmission fault, network fault, configuration error, etc.

In many cases, current methods are based on proprietary tools that pertain to specific network architectures [5,6]. However, standardisation groups are also realising the importance of having common tools, although the individual methods need to be related to the specifics of the transmission technology (e.g., OAM for MPLS [7], the Metro Ethernet Forum [8] and the Telecommunication Standardisation Sector (ITU-T) [9], IEEE 802.3ah [10] and 802.1ag [11] standards for Ethernet). However, no work has been published to specify OAM methods for DVB-based satellite system, specifically at the link layer.

The common techniques for OAM used on DVB networks rely on monitoring signal level and synchronisation at the physical later and on IP-level techniques for bi-directional services such as netperf or ICMP (ping). DVB systems allow also Transport Stream metrics to be monitored, such as the Continuity Counter, or the Adaptation-Field timestamp. However, these are not easily related to the packet performance experienced by individual edge-to-edge paths through the transmission network.

This paper proposes an OAM technique based on the Ultra-Lightweight Encapsulation (ULE) method. The ULE protocol stack provides a multi-protocol encapsulation for a MPEG-2 TS that supports an extension-header mechanism that can be used to tag any packet sent on a transmission link with additional information – in this case a timestamp option [13] that can be utilised by an OAM framework. We define the use of this timestamp and explain how it may be utilised to monitor the QoS performance of DVB-S/S2 endpoints at the link-layer (e.g., delay, jitter, ordering, and bandwidth consumption). This method is validated using a set of traffic profiles.

The remainder of this paper is as follows: a brief description of ULE and its features is given in Section 2. Section 3 describes the performance monitoring techniques used in this work, while Section 4 describes the testbed used and shows the experimental results. Finally, Sections 5 and 6 provide the discussion, conclusions and future work.

2   Ultra Lightweight Encapsulation

DVB-S/S2 systems adapt variable-sized MPEG-2 Audio/Video payloads to fixed-size 188 B Transport Stream (TS) packets. This may be used to transmit IP datagrams using adaptation protocols such as Multi-Protocol Encapsulation (MPE) [1] or Unidirectional Lightweight Encapsulation (ULE) [12]. MPE specifies a

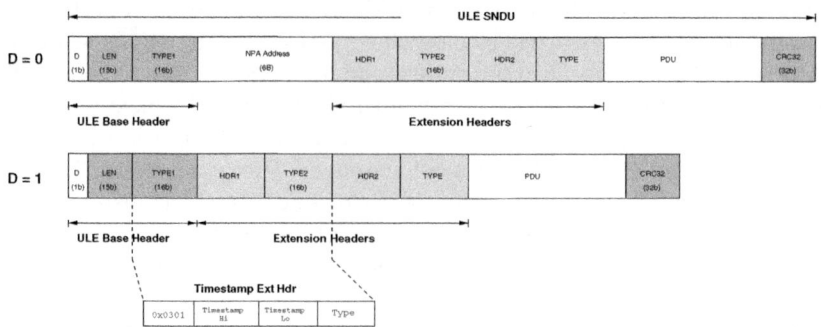

Fig. 1. Format of the service network data unit of ULE

method to adapt variable-size network data packets (typically IP packets) to fixed-size 188 B TS packets. It builds upon the section syntax of the MPEG standards and is widely deployed for IP encapsulation over DVB-S/S2 systems. Its minimum header is 12 B (plus a cyclic redundancy check field – CRC-32) while the maximum payload size is typically 1.5-4 kB (1 kB in some systems, e.g., [14]). Many systems restrict use of MPE to IPv4 traffic, although in principle it could be extended to other payloads using a LLC/SNAP extension or by defining new interpretations of the payload.

ULE adopts an IP-centric encapsulation. It mandates only the functions strictly necessary for transmission over a transport stream (the base header has only three fields header in contrast to eighteen of MPE). This improves encapsulation performance requiring a minimum base header only 4 B long, but also supporting a maximum payload of 32 kB. ULE was defined to be multi-protocol, so it can readily support IPv6, MPLS, IEEE 802.1pQ VLAN tags, etc. ULE also supports an extension header mechanism similar to IEEE 802, and IPv6 that allows optional additional information to be added to a encapsulated packet when needed. This also allows new extensions to be added to the protocol when new applications and needs emerge (e.g., the timestamp option introduced in RFC 5163). This contributes in improving network/receiver-processing efficiency. Several research papers have been published on encapsulation efficiency achieved by MPE and ULE [15,16,17,18].

2.1 ULE Header

The ULE header is shown in Fig.1. The first three fields form the base header, these are: Destination Address Absent flag (D), Length and Type, form the 4 B base header. If the D-bit is 0, the 6 B link-layer receiver address is present in the header at the end of the base header. The 15-bit Length field indicates the size in bytes of the ULE packet from the byte following the base field to the last byte of the CRC-32. The 16-bit Type field indicates the type of network packet using IEEE ether-type convention (e.g., 0x0800 for IPv4 or 0x86DD for IPv6) carried in the ULE payload. Type fields lower than 1536 (decimal) correspond to extension headers.

More specifically, a ULE extension header is identified by a 16-bit Type field, which specifies a header length field (H-LEN) field and a header type field (H-Type) [13]. Optional extension headers are identified by H-LEN values from one to five, while the value zero is reserved for mandatory extension headers. The H-LEN indicates the length in bytes for each optional extension header. Receivers not required to process the extension header can therefore access the next header, skipping as many bytes as indicated in H-LEN.

The receiver is required to extract and process every mandatory extension header present in the ULE packets (and to silently discard payloads with an unknown type). In contrast, a receiver can silently ignore optional extension headers without impairing communication. In other words, the expression optional in the context of extension headers refers to the receivers ability to process or not the extension header, not to the transmitters requirement to insert it. The payload associated is always passed to the higher layer protocol handling, irrespective of whether the option is understood. An optional extension header can therefore be used to carry OAM signalling, which may be utilised by specific terminals.

2.2 The Timestamp Extension Header

RFC 5163 [13] defines three types of ULE extension headers, MPEG-2 TS-Concat, PDU-Concat and Timestamp. The first two are mandatory extension headers designed to carry more than one (TS or IP) packet in a single ULE payload. The Timestamp header is an optional extension header aimed to support monitoring and measurement of an operational ULE link.

The first byte of the base header (TYPE1) is 0x03, indicating a total length of 6 B. The extension header then comprises two fields: the timestamp (divided into an HI and a LO part), and the next type field. The 32-bit timestamp records the number of microseconds past the hour in Universal Time Coordinates (UTC) recording the time of encapsulation. A timing resolution of one microsecond is sufficient for OAM and performance evaluation.

The Timestamp extension header used in our experiments complies with RFC 5163 and is shown in Figure 1. Since we encapsulated IPv4 packets and we did not include any layer-2 address (D=1) or extension headers apart from the timestamp, the ULE overhead per IP packet was 14 B (RFC 4326 identifies cases where this format may be safely used). This is a rather small overhead with respect to the maximum transmission unit (MTU) of Internet packets. Moreover, as shown in the results section, not every packet needs to carry a timestamp to achieve good OAM performance. Accurate link monitoring can be achieved by inserting a timestamp every several tens of kilobytes, significantly reducing the timestamp overhead cost in terms of capacity.

3 Traffic Monitoring

Systems for traffic monitoring are developed to assess performance, detect faults, etc. The level of required accuracy depends on the application. The assumptions here are that the service will be monitored at the receiver. The collected data may be retrieved either by a return channel or collected by some other means, though this is not the subject of the current paper.

This paper considers two performance indexes commonly used for traffic monitoring [3]: one-way delay (OWD) and delay jitter. The timestamps for the i-th packet collected at sender and receiver are respectively indicated by s_i and r_i. The OWD and the delay jitter are defined as:

$$owd_i = r_i - s_i \tag{1}$$
$$jitter_i = |owd_{i+1} - owd_i|$$

These metrics may be used for all packets using a TS or only packets associated with specific flows (e.g., identified by a particular DiffServ code-point when supporting multiple levels of QoS). If ULE encapsulation is performed as soon as packets are delivered to the satellite interface, the OWD takes into account any queueing delays at the terminal. This allows network operators to verify SLAs for delay (either for all traffic or specific high-priority flows). Also, measurements of sender and receiver bit rates can help identify the causes of delay, namely users not respecting the SLA maximum allowed bit rates or ISPs under-provisioning network capacity.

Synchronisation between sender and receiver is necessary to measure the one-way delay. The receiver needs to obtain a reference clock, e.g. using the Global Positioning System (GPS), the Network Time Protocol (NTP), or by reconstructing the clock present in the TS network clock reference (NCR). The reference clock is not required to measure delay jitter, since this only requires compensating the drift between sender and receiver clocks. In both cases, the two endpoints need to agree on the method to insert the timestamps to reconstruct sender bit rate estimations at the receiver.

We implemented our proposed method using a timestamp attached to a ULE-encapsulated packet every time a predetermined amount of data has been sent (partially including the packet that carries the timestamp). This method (byte counting) is considered more accurate [17] than inserting a periodic timestamp every N packets when packets length are subjected to jitter.

Several factors influence the accuracy of measurement. First, since the transmitter and receiver may run several real-time processes at the same time, the timestamp could not necessarily reflect the exact transmission or reception time of a packet whether a process pre-empt the receiving/transmitting routine. This problem is platform-dependent, which is typically offset by employing dedicated hardware.

If ULE timestamps may be used to perform bandwidth measurements, the ULE timestamps may not take into account MPEG overhead. Two methods, packing and padding, can be used to encapsulate ULE packets in MPEG TS packets [16]:

- In padding mode, each packet is encapsulated and put into a TS packet independently from other packets in a flow. If the last fragment of a ULE packet does not fit the TS payload, padding bytes are added to complete the 188 B packet.

– In packing mode, data from the next ULE packet can be used to complete a TS payload. Thus, a ULE packet in packing mode does not have to start at the beginning of a TS packet. If the encapsulator is not able to complete a TS packet because no data is available, it starts a timer. If new data is available within a pre-determined time interval, the encapsulator completes and delivers the packet; otherwise, when a timer expires, the encapsulator releases the packet with padding of the remaining bytes.

In addition to increasing encapsulation efficiency by reducing the average amount of padding per cell, packing mode also increases the estimation accuracy when using ULE timestamps.

Packet loss can be a cause of inaccurate estimations. A receiver error may induce erroneous bandwidth estimation if the ULE packet carrying the timestamp is lost. In many circumstances lost packets can be detected and not included in the statistics even without the presence of a sequence number. For instance, if the sender inserts a timestamp every N packets, the receiver is able to identify a missing timestamp when more than N packets have been received and no timestamps have been detected. In general, the probability to acknowledge an incorrect timestamp sample decreases as N increases. As observed in the next section, collecting a timestamp every 10-20 kB (or approximately every 10 packets) could be a good compromise between sampling rate and robustness of estimation against packet losses.

4   Experimental Results

This section describes our Linux-based testbed used to evaluate the suitability of ULE as an OAM tool. The testbed implemented an IP over DVB-S2 stack using ULE. The TS was received and analysed by the real-time MPEG decoder, dvbsnoop, which was adapted to support ULE.

4.1   Test-Bed Description

The testbed consists of a unidirectional DVB-S2 link between two Linux computers. The sender has a DVB-S2 modulator (Advanced Digital TVB590). An Application Program Interface (API) allowed transmission parameters (frequency, symbol rate and coding rate) to be controlled. The receiver comprises a DVB-S2 card (Hauppauge WinTV) [19] configured to process the transmitter TS. The transmission rate was fixed to 1 Mb/s. In some cases, the bandwidth was varied enforcing the insertion of blocks of null TS packets at regular intervals. This emulated a Constant Rate Allocation (CRA) in a Time Division Multiplexing (TDM) environment where a station receives a periodical transmission opportunity. The NCR was inserted periodically to ensure synchronisation between transmitter and receiver.

A user-space ULE/MPEG encapsulator was developed based on tcpdump packet capturing library. The program captures Ethernet frames and builds ULE/MPEG encapsulated packets in a circular buffer. A concurrent process manages (using a modulator API) the transfer of 4 kB-data blocks from the buffer to the DVB card buffer. The Ethernet Type field is copied into the ULE Type field. If the Type field is not present (the Ethernet Type/Length field is less than 1536), the ULE Type is described by a logical link control/sub-network access protocol (LLC/SNAP) header. In this case, the encapsulator operates in bridging mode [12] copying the entire Ethernet frame into the ULE payload field.

The encapsulator supports both padding and packing encapsulation modes. In our experiments, the packing mode was used to increase efficiency. The timeout duration (packing threshold) was set to 100 ms as recommended in [16] for multimedia traffic. The timestamp is the only extension header inserted in ULE packets.

Figure 2 shows two video traffic profiles from traces used in our tests. Both traces are carried over UDP with an average transmission rate of 64 kb/s. The flow using H.261 encoding carries a TV stream characterised by high bit rate variability (the peak-to-average ratio is about 4.7 over 200 ms). The H.264-encoded stream is a video-chat yahoo trace with more regular bit rate (the peak-to-average is 1.6 over 200 ms). These two different profiles allowed the evaluation of estimator properties for different level of traffic burstiness.

The real-time flows were multiplexed with other Internet traffic, including web browsing and file transfers. The competing traffic was captured on a LAN and replicated on the DVB link multiplexed with the video flow. The encapsulator

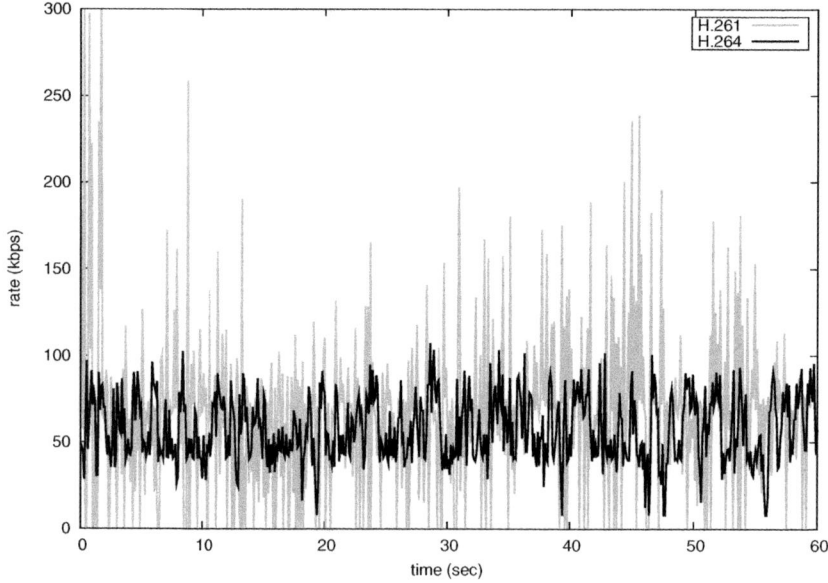

Fig. 2. Traffic profiles used in experiments

inserted the timestamps on video packets only. We tested two cases: 1) the competing traffic has a bit rate comparable to that of the video flow (low load) and 2) the competing traffic presents a bit rate approximately ten times larger (high load) than that of the video flow. Although the overall link capacity is not fully utilised in both scenarios, the competing traffic causes considerable jitter and queueing delays especially in high-load scenario. This is due to the high level of burstiness of HTTP traffic, which manly consists of relatively small and frequent data transfers.

4.2   Results

Figures 3 and 4 display the complement of the cumulative distribution function (CDF) of OWD estimates for the H.261 and H.264 traces, respectively. Each graph reports the CDF for the high and low-load scenarios when the fraction of video packets with timestamp (SR) is 5%, 10% and 100%, e.g., if SR=100%, all video packets contain a timestamps.

In the high-load scenario the competing traffic causes large delays to video packets of both traces. However, increasing the workload impacts differently the two types of traffic. For instance, the proportion of packets exceeding one second OWD is larger in H.264 traces (approximately 24% vs. 18%). On the other hand, the tail of CDF drops faster in H.261 graphs than H.264, which means that H.261 (TV) packets can experience large delays with higher probability.

This discrepancy is mainly due to the different level of burstiness of the two streams. The H.264 flow has a more regular pattern, H.264 packets tend to be more spread, resulting in a higher probability of meeting a burst of cross traffic and, consequently, higher OWD delays. The large rate variations of H.261 flows also result in longer delay if a queue is congested. The timestamp method is able to reveal the OWD behaviour with high accuracy in both scenarios, even when only a small fraction of packets (5%) contain a timestamp. This means that infrequent packet probes are sufficient to discover LL delay anomalies and monitor the queueing behaviour.

Figures 5 and 6 show the complement of jitter CDF for H.261 and H.264 traces, respectively. Each graph reports statistics from the high-load and low-load scenarios in top and bottom diagrams, respectively. Again, the CFD is displayed when 5%, 10% and 100% of packets are sampled. Since jitter measures the variation between consecutive OWD samples, major differences between jitter CDFs in the low-load and high-load scenario are observed when the timestamps are far apart. It is observed that if all the video packets are sampled, the majority of jitter samples lay between 0 and 200 ms and jitter distributions are alike. Conversely, taking a timestamp every 20 packets produces much more visible variations (spreading samples between 200 ms and 1200 ms when the queue is loaded), which allows detecting easily congestion conditions. As a result, a few jitter samples, which represent a low overhead, are sufficient to detect the queue state.

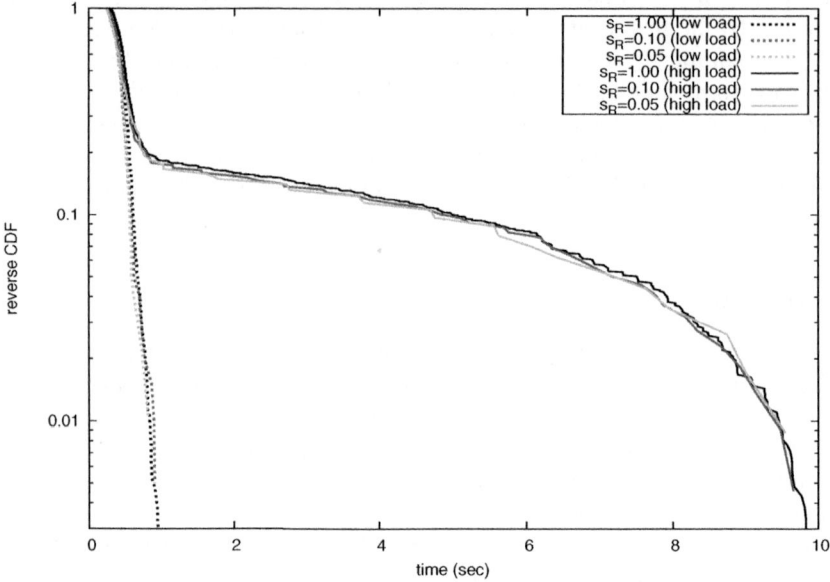

Fig. 3. Reverse CDF of OWD for the H.261 flow

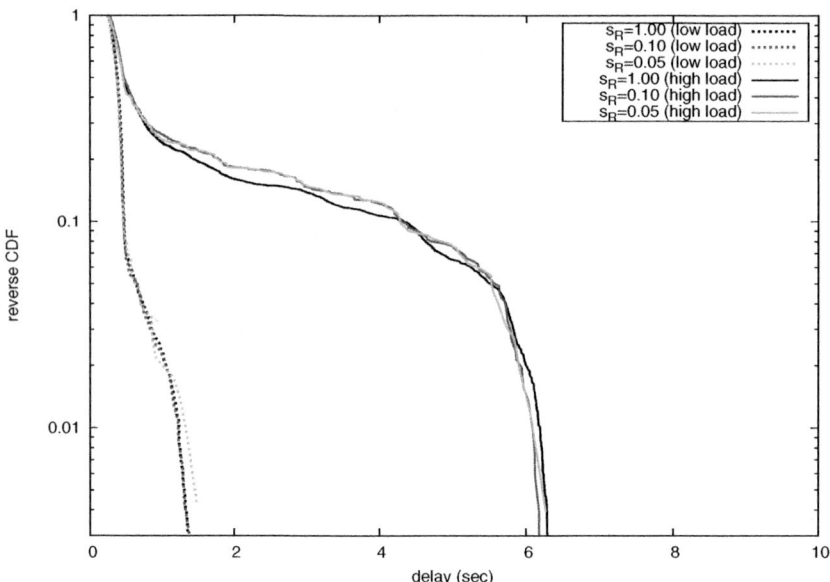

Fig. 4. Reverse CDF of OWD for the H.264 flow

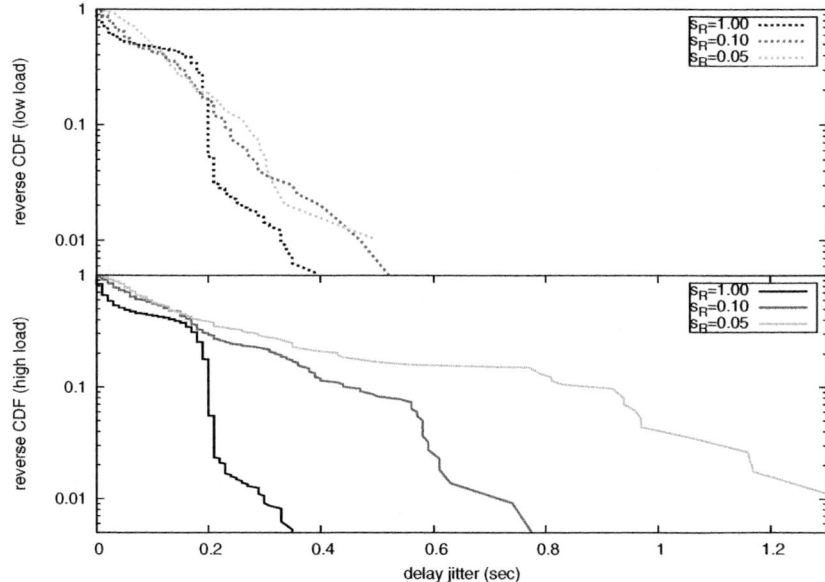

Fig. 5. Reverse CDF of jitter for H.261 traffic

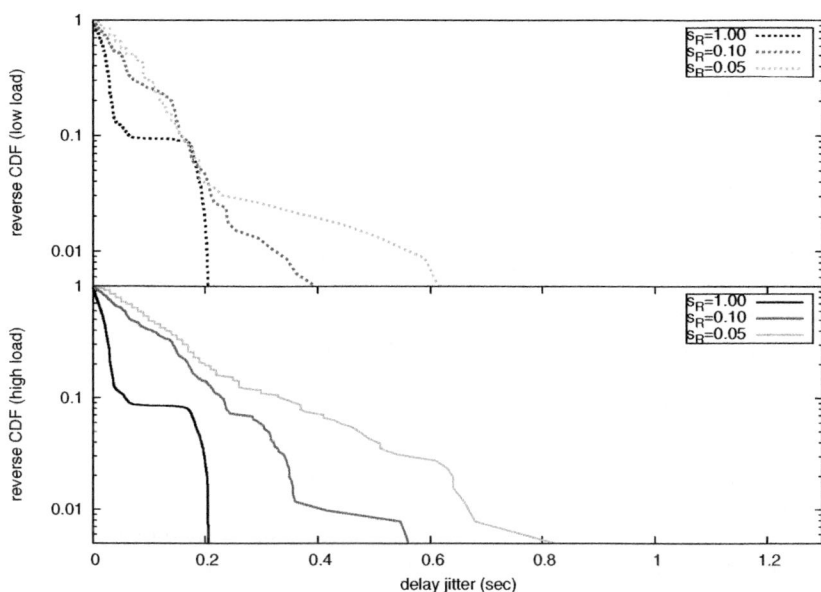

Fig. 6. Reverse CDF of jitter for H.264 traffic

5   Future Directions

The paper focuses on unidirectional streams, and performance measurement at the receiver. Further work may analyse the effectiveness of the method with a range of deployment scenarios and traffic types. In addition, a return link is desirable to transmit performance data to the sender.

When a return link is provided, the architecture could be extended to verify ordering and perform other test by comparison with a log of the sequence of packets transmitted at the sender. It may also be possible to use a mandatory header to carry an echo request formatted as a mandatory TEST extension (e.g., to verify a remote receiver has processed a specific packet, although not the focus of the present paper), allowing pro-active testing of the stream.

This paper describes use of an OAM method for ULE. In addition to ULE, DVB-S2 also supports the Generic Stream, using the GS Encapsulation (GSE) protocol. GSE supports direct transmission of IP traffic over the DVB-S2 baseband frame (i.e., TS packets are not used), but since ULE extension headers were developed to be compatible with GSE, a similar OAM timestamp method could be used with GSE. This method would need to be subtly different in the way timestamps were utilised, and this requires further work.

DVB transmission systems such as DVB-T and DVB-C could also be used with ULE and DVB-T2 and DVB-SH also support GSE. There is therefore potential to develop a common OAM technique that span a range of DVB deployment scenarios.

6   Conclusions and Future Work

This paper presents an implementation of ULE as an OAM tool to monitor DVB-S/S2 transmission systems. We show that the timestamp option can be used effectively for this purpose. This extension header provides timely OAM data with low overhead. It can be used to check compliance with a given service level agreement, to verify system performance and to assist in locating a performance or connectivity issue.

We developed a proof-of-concept testbed to implement this new tool. ULE timestamps allow several system parameters to be monitored such as user bandwidth consumption, one-way delay, delay jitter, and packet losses. In particular, the one-way delay (OWD) is a critical measure for assessing performance for real-time applications. The delay jitter of the timestamps can be used to detect increasing levels of congestion or non-compliance in the delay levels of a SLA. Our measurements confirm that good estimates can be achieved even for low packet sampling rates. However, OWD measurements need the synchronisation of the sending and receiving terminals, which may require separate synchronisation channels. Our experiments suggest that when properly tuning the sampling rate is possible to detect a range of network conditions. In general, a low sampling rate (e.g., one packet every 200 ms) is sufficient for this purpose.

References

1. EN 301 192, Digital Video Broadcasting (DVB); DVB specification for data broadcasting. ETSI Standard (2009)
2. TR 102 033, Digital Video Broadcasting (DVB); Architectural framework for the delivery of DVB-services over IP-based networks. ETSI technical report (2002)
3. TS 102 034, Digital Video Broadcasting (DVB); Transport of MPEG-2 TS Based DVB Services over IP Based Networks. ETSI standard (2008)
4. TS 102 826, Digital Video Broadcasting (DVB); Guidelines for the implementation of DVB-IP Phase 1 specifications. ETSI (2009)
5. Fujitsu white paper, Ethernet Service OAM: Overview, Applications, Deployment, and Issues (2010), http://www.fujitsu.com/downloads/TEL/fnc/whitepapers/EthernetService-OAM.pdf
6. Cisco systems white paper, Overview of Ethernet Operations, Administration, and Management (2010), http://www.cisco.com/en/US/prod/collateral/routers/ps368/prod_white_paper0900aecd804a0266.html
7. Yasukawa, S., Farrel, A., King, D., Nadeau, T.: RFC 4687, Operations and Management (OAM) Requirements for Point-to-Multipoint MPLS Networks. IETF informational standard (2006)
8. Technical Specification MEF 17, Service OAM Requirements and Framework Phase 1, Metro Ethernet Forum (MEF) (2007)
9. ITU-T rec. Y.1731, OAM functions and mechanisms for Ethernet based networks. ITU rec. (2006)
10. IEEE 802.3ah, Ethernet in the First Mile. IEEE standard (2004)
11. IEEE 802.1ag, Connectivity Fault Management. IEEE standard (2007)
12. Fairhurst, G., Collini-Nocker, B.: RFC 4326 Unidirectional Lightweight Encapsulation (ULE) for Transmission of IP Datagrams over an MPEG-2 Transport Stream (TS). IETF standard (2005)
13. Fairhurst, G., Collini-Nocker, B.: RFC 5163 Extension Formats for Unidirectional Lightweight Encapsulation (ULE) and the Generic Stream Encapsulation (GSE). IETF standard (2006)
14. ATSC Doc. A/94, ATSC Standard: ATSC Data Application Reference Model. Advanced Television Systems Committee (2002)
15. Montpetit, M.-J., Fairhurst, G., Clausen, H., Collini-Nocker, B., Linder, H.: RFC 4259 A Framework for Transmission of IP Datagrams over MPEG-2 Networks. IETF informational standard (2005)
16. Collini-Nocker, B., Fairhurst, G.: ULE versus MPE as an IP over DVB Encapsulation. In: HET-NETs 04 Second Intl. Working Conf., pp. Y70/1–P7019 (2004)
17. Fairhurst, G., Mathews, A.: Internet Protocols over Digital Video Broadcasting Media (IP over DVB) Study of Encapsulation and Protocol Performance (2003)
18. Cantillo, J., Collini-Nocker, B., De Bie, U., Del Rio, O., Fairhurst, G., Jahn, A., Rinaldo, R.: GSE: A flexible, yet efficient, encapsulation for IP over DVB-S2 continuous generic streams (2008)
19. WinTV-DVB-s WinTV-Nexus_s, Product Description (15/02/2010), http://www.hauppauge.com/html/dvb_s.htm

TCP Performance in Hybrid Satellite - WiFi Networks for High-Speed Trains

Giovanni Giambene[1], Silvia Marchi[1], and Sastri Kota[2]

[1] CNIT - University of Siena, Via Roma, 56, I-53100 Siena, Italy
giambene@unisi.it
[2] University of Oulu, Rajakylá, 90570 Oulu, Finland

Abstract. Satellite communications have a significant potentiality in the railway scenario, but suffer from strong variations in the received signal power and need suitable solutions to deal with long channel disruptions due to tunnels. The approach envisaged in this paper is based on the adoption of a hybrid network (satellite and terrestrial WiFi coverage) and a *Vertical Handover* (VHO) scheme to switch seamlessly from one segment to another whenever link quality degrades and a new segment is available. Details are provided for the adoption of MIH and MIPv6 in such a context, referring to the BSM standard. Transport layer performance is evaluated considering different TCP versions (e.g., NewReno, BIC, and CUBIC) as well as possible cross-layer approaches to improve the performance in the presence of VHOs. Design criteria are provided taking into account train speed and overlap area size. Finally, interesting TCP performance results are achieved by updating the ssthresh value and limiting the cwnd value after VHOs by means of cross-layer approaches. This work has been carried out within the framework of the ESA SatNEx III project.

Keywords: Satellite Networks, WiFi, Inter-segment (Vertical) Handover, MIH, MIPv6, TCP.

1 Introduction

Railway passengers are expecting reliable broadband communication services available on board and the demand for such services is increasing. In the railway scenario, satellite communications suffer from strong variations in the received signal power due to shadowing and multipath fading. Shadowing due to obstacles (tunnels, buildings, bridges, trees, etc.) causes link unavailability for short or long time intervals. Short-term events (on the order of tens to hundreds of ms) are due to power line structures, trees, buildings, and small obstacles, in general. While, long disruptions (on the order of seconds) can be caused by relatively-long tunnels [1].

The interest of this paper is on the transport layer performance in the presence of channel disruptions due to tunnels. Several approaches are possible, as follows [1]:

G. Giambene and C. Sacchi (Eds.): PSATS 2011, LNICST 71, pp. 116–130, 2011.
© Institute for Computer Sciences, Social Informatics and Telecommunications Engineering 2011

- *Disruption-Tolerant Networking* (DTN)
- *Performance Enhancing Proxy* (PEP)
- Satellite signal relays (gap fillers)
- Hybrid networking with the use of a local wireless coverage.

This paper focuses on the hybrid scenario (satellite and terrestrial WiFi systems) and considers a *Vertical Handover* (VHO) scheme to switch from one segment to another, and different transport-layer schemes, including cross-layer approaches. The study made here based on a hybrid network and a local wireless coverage could also be applied to other environments, such as dense urban areas and railway stations.

We consider a scenario where a Ku-band GEO bent-pipe satellite is used for communication services to the train during the trip. As soon as the train approaches a tunnel, a VHO is performed to switch seamlessly the link to a local WiFi coverage [2]. Within the tunnel, more WiFi *Access Points* (APs) are needed to provide a seamless coverage, thus implying horizontal handover procedures that are considered not to be problematic since they do not entail a *Delay-Bandwidth Product* (DBP) change.

Traditionally, the decision about transferring a connection from one link to another is made by comparing the link quality on the basis of *Received Signal Strength* (RSS), power degradation estimation, and *Bit Error Rate* (BER). Moreover, positioning information could be used in the train scenario to decide about VHO due to the predictability of tunnels positions [3].

2 Survey of TCP Problems Due to VHOs in the Railway Scenario

Several problems affect TCP performance during a VHO procedure [4]; in particular, the sudden DBP variation and a significant change in the *Round-Trip Time* (RTT). Another aspect is that during a VHO there could be a time interval during which the train exchanges data over both satellite and WiFi paths (e.g., ACK and TCP packets propagate quickly via WiFi so that they could anticipate those sent via the satellite link: out-of-sequence problems) according to a 'soft' VHO scheme.

Parameters that are crucial for the TCP performance during a VHO are: the duration of the handover phase (including layer 2 and layer 3 signaling), the overlap area between the satellite coverage and the WiFi one (allowing the execution of the VHO in advance with respect to the physical link disconnection), the average train speed, the buffer sizes (sender, satellite network gateway, and WiFi AP).

We examine below the detailed effects on TCP behavior due to the two possible VHO cases for our scenario [4],[5]: from satellite to WiFi and from WiFi to satellite. We refer here to 'soft' VHO procedures. Moreover, we consider that a mobile collective terminal on the train is the receiver of TCP downstream flows.

2.1   VHO from Satellite to WiFi Network

In this case (train entering the tunnel), we could expect that DBP and RTT have significant reduction in moving towards the WiFi network. The problems that TCP could encounter during this handover process are:

- TCP packets are anticipated (received out of order) due to the fact that when TCP packets are still propagating to the satellite, new TCP packets can be quickly received via WiFi. Then, TCP packets need to be reordered at the receiver before being delivered to upper layers (this might cause a receiver window partly closing).
    - DUPACKs are generated due to the arrival of TCP packets out of order. This may cause useless retransmissions and halving the *congestion window* (cwnd).
- ACK packets are anticipated as soon as the new WiFi path is activated (many ACKs could be anticipated if the handover is performed when the cwnd value is high). In this case, large bursts of data could be injected in the WiFi AP buffer to be sent to the mobile collective terminal on the train, thus entailing the risk of packet losses due to buffer overflow.

2.2   VHO from WiFi to Satellite Network

In this case (train leaving the tunnel), DBP of WiFi is much smaller than that of the satellite network (during the WiFi segment cwnd is temporarily reduced). The problems that TCP encounters during this handover process are as follows:

- Too short RTO at the beginning of the connection via satellite with the risk of timeout expiration and cwnd reset (multiple RTO expirations could lead ssthresh to 1, thus slowing the cwnd increase, restarting from the congestion avoidance phase for TCP NewReno).
- Slow cwnd increase at the beginning of the connection via satellite (if the process starts when TCP is in the congestion avoidance phase) with consequent slow traffic injection increase in the satellite segment.

3   Scenario Description and Handover Procedure

We consider the reference network architecture shown in Figure 1, where a collective terminal on a train receives a downstream Internet traffic via a GEO bent-pipe satellite. When the train approaches a tunnel a new WiFi connection is used by means of a VHO. This can be realized under the assumption that the train collective terminal supports multiple air interfaces (i.e., satellite and IEEE 802.11 interfaces) and that WiFi APs are located in the tunnel. Terrestrial wired links bit-rates have been set to 500 Mbit/s to neglect their impact on congestion events and to emphasize the bottleneck nature of wireless links.

The mobility and related handover events need to be managed at both layer 2 (selection of a new link) and layer 3 (new IP address for the mobile node and

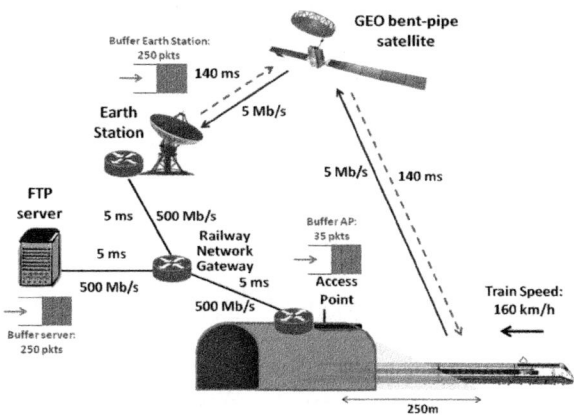

Fig. 1. Reference system architecture

new path for the IP traffic). To support the *Layer 2* (L2) handover procedures we consider that MIH (*Media Independent Handover*, IEEE 802.21 standard) is adopted at layer 2+, referring to the *Broadband Satellite Multimedia* (BSM) standard scenario [6]. With MIH a mobile node discovers a new access link, registers itself to the new network, disconnects from the old network (when some handover criterion is meet), and communicates with the corresponding host via the new access link. As for the IP layer, we adopt MIPv6 that entails an efficient management of IP traffic and a low *Layer 3* (L3) handover latency.

VHO requires some preparatory L2 and L3 signaling exchange to detect the new link and to define automatically the new IP address, *Care of Address* (CoA), by means of *Router Advertisement* (RA) and *Router Solicitation* (RS) messages. Then, as soon as the handover is decided a *Binding Update* L3 message is sent via the new link to the corresponding node (i.e., the sender) to notify about the new IP address of the mobile node. In this work, we consider that the packets are transmitted on the new interface soon after the Binding Update message is sent, without waiting for the Binding Update ACK ('soft' handover). This permits to reduce the VHO latency. The VHO procedure is mobile-initiated.

Let us denote by t_{HO} the total handover latency, that is the time interval from the new link detection instant to the instant when the binding ACK message is received at the mobile node or to the instant when the last TCP packet is received through the 'old' air interface. During t_{HO} there are L2 and L3 signaling messages as well as data packets going through both air interfaces. Note that ACK packets are sent on the new link soon after the Binding Update message is transmitted by the collective terminal. The following Figure 2 describes the (simplified) signaling implemented to support satellite-to-WiFi VHO with MIH; in this case, the most critical issue is the arrival of out-of-sequence ACKs and packets. In particular, Figure 2 describes the case with the maximum t_{HO} value (handover executed -with Binding Update message sent- after the complete reception of a window of data at the satellite).

In order to characterize the overlap coverage area between satellite and WiFi, we write the following constraint:

$$\max\{t_{HO}\} = 2RTD_{WiFi(AP)} + T_{HO\_decision} + \frac{RTD_{e2e\_sat}}{2} + \frac{RTD_{e2e\_WiFi}}{2}$$

$$< \frac{\text{overlap distance (sat} - \text{WiFi)}}{\text{train speed}}$$

$$\Rightarrow \quad \text{overlap distance (sat} - \text{WiFi)} >$$

$$\left(2RTD_{WiFi(AP)} + T_{HO\_decision} + \frac{RTD_{e2e\_sat}}{2} + \frac{RTD_{e2e\_WiFi}}{2}\right) \times \text{train speed}$$

(1)

where $RTD_{WiFi(AP)}$ (on the order of μs) denotes the round-trip propagation delay from the collective terminal to the WiFi AP, $T_{HO\_decision}$ is the VHO decision time (determined as outlined below), $RTD_{e2e\_sat} = 580$ ms represents the end-to-end round-trip propagation delay via satellite, and $RTD_{e2e\_WiFi} = 20$ ms denotes the end-to-end round-trip propagation delay via the WiFi AP.

In our study, referring to the VHO scenario in Figure 2, we reasonably assume that the WiFi AP coverage is able to reach a distance $d_{max} = 250$ m at the entrance (and at the end) of the tunnel. Moreover, the $T_{HO\_decision}$ time can be determined by considering a suitable propagation model and a threshold criterion on the WiFi *Signal-to-Noise-Ratio* (SNR) variation[1]. In particular, we consider the two-ray ground propagation model (path loss exponent is 4) and we adopt the handover decision criterion and related handover distance d_{th} that corresponds to a WiFi SNR increase of 3 dB (threshold) with respect to the SNR at d_{max} (WiFi border of coverage). We obtain the following formula:

$$10 \log_{10}\left(\frac{d_{\max}}{d_{th}}\right)^4 = 3 \text{ dB}$$

(2)

Hence, $d_{max} - d_{th} = 40$ m. Moreover, for a train speed of 160 km/h, we have:

$$T_{HO\_decision} = \frac{d_{\max} - d_{th}}{\text{train speed}} = 0.9 \text{ s}$$

(3)

On the basis of the above values, condition (1) is fulfilled.

Finally, Figure 3 describes the signaling and the maximum t_{HO} time for the WiFi-to-satellite VHO. In this case, the maximum t_{HO} value is complementary to that in (1) with $T_{HO\_decision}$ time depending on the satellite propagation model. The main criticality for TCP during the VHO process is that RTO could expire for those packets that have been sent via WiFi and for which ACKs are sent via satellite.

---

[1] A further element characterizing $T_{HO\_decision}$ is related to the visibility of the satellite in the proximity of the tunnel that depends on the satellite elevation angle, the latitude and the position of the tunnel (and related mountain) with respect to the GEO satellite. The analysis of these aspects is left to a future study.

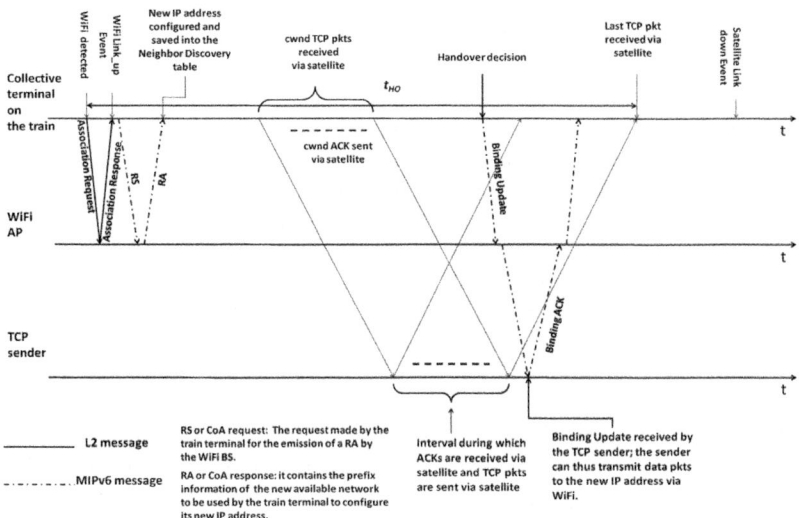

Fig. 2. Satellite-to-WiFi VHO: scenario for maximum t_{HO}

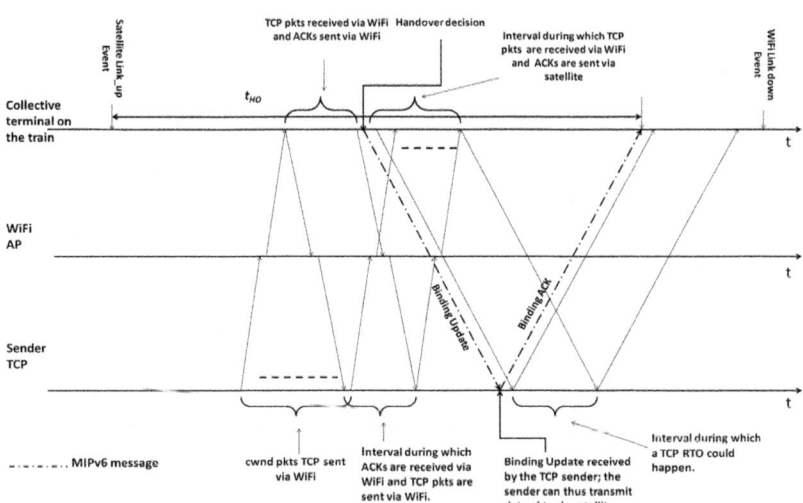

Fig. 3. WiFi-to-satellite VHO: scenario for maximum t_{HO}

4 Techniques for TCP Performance Improvement in VHO

Table 1 provides details on different techniques available in the literature to improve TCP performance in the presence of VHOs from satellite to WiFi networks and viceversa. In particular, we consider both TCP versions that have been proposed for high-speed terrestrial networks and cross-layer schemes that involve L2 and L3. In this paper, we compare two alternative composite methods to support VHOs in the railway tunnel scenario; in both cases we exploit cross-layer signaling that could be supported by the SI-SAP interface of the BSM standard.

- *VHO method #1* – Reordering approach for the satellite-to-WiFi VHO case (see Table 1) and RTT estimation (the timestamp of the Binding Update message is used to update RTO) for the WiFi-to-satellite case;
- *VHO method #2* – Limitation of cwnd on the basis of the WiFi AP buffer size (MAC buffer size limitation technique in Table 1) for the satellite-to-WiFi VHO and ssthresh update for the WiFi-to-satellite VHO by restoring ssthresh to the value assumed before the previous satellite-to-WiFi VHO (this solution is an original proposal that comes from the Hoe approach [12] and requires a cross-layer signaling: the Binding Update message received at the sender triggers the restoring of the ssthresh value).

Moreover, we will investigate TCP-friendliness problems and RTT-fairness[2] issues [14] that arise from concurrent flows experiencing different RTT values (i.e., WiFi and satellite).

5 Results

We have built an ns-2 simulator [15] for the scenario depicted in Figure 1, considering trains that encounter tunnels along their trip with active (persistent) FTP flows. Parameter values are according to Table 2. We have selected the buffer size on the basis of the DBP value (to fill the pipe). The tunnel length is 5.3 km that corresponds to a tunnel duration of about 120 s (we consider here a sufficiently-long tunnel that would create a link disruption in the absence of a local WiFi coverage) for an average train speed of 160 km/h.

We have compared in Figure 4 the behaviors of NewReno, Vegas (as used by SCPS-TP), Westwood+, *Binary Increase Congestion control* (BIC) [7], and

---

[2] For a TCP variant with an average TCP throughput that depends on c/p^d, where c and d are protocol-related constants and p is the packet loss rate, RTT-unfairness [13] is roughly proportional to $(RTT_2/RTT_1)^{d}/(1 - d)$. As d increases, the RTT unfairness increases. For BIC $d = 0.5$ as with NewReno, but c of BIC is much bigger. Note that the occurrence of a tunnel could be macroscopically similar to a packet loss event for cwnd. Hence, the above considerations may also give some insights for our VHO cases.

Table 1. Survey of techniques suitable for TCP performance improvement in VHO

Method	Modifications	Description	Signaling	Satellite-to-WiFi VHO impact	WiFi-to-satellite VHO impact
BIC-TCP [7]	Sender side TCP modification (new congestion control protocol suitable for high-speed networks)	A new cwnd increase function is used with linear, logarithmic, and exponential increase phases. More details are provided in Section 5.	Standard TCP level signaling	No impact	This VHO is managed better since a sudden cwnd increase is supported, thus achieving a better utilization of resources in the new (satellite) network.
CUBIC-TCP [8]	Sender side TCP modification (new congestion control protocol suitable for high-speed networks)	This is similar to BIC, but a cubic cwnd increase law is adopted. More details are provided in Section 5.	Standard TCP level signaling	No impact	This VHO is managed better since a sudden cwnd increase is supported, thus achieving a better utilization of resources in the new (satellite) network.
Reordering approach [9]	Sender and receiver sides TCP modification	This modification allows managing the arrival of out-of-sequence packets due to the VHO from a high RTT network (satellite) to a low RTT one (WiFi). Instead of DUPACKS, ACKs are sent during the interval for which out-of-sequence packets are received (i.e., "HO_mode" time interval), as detailed below. Receiver side "HO_mode": a new header field is used in the ACK to describe the number N of outstanding packets still in the old network. When N reaches 0 the "HO_mode" is over. Sender side "HO_mode": a congestion avoidance phase is performed during the time interval that the receiver needs to acknowledge all the out-of-sequence packets. When an ACK with $N = 0$ is received, TCP returns to its standard behavior.	Cross-layer signaling L3 and L4: when the Binding Update (L3) message is sent, the receiver notifies its TCP layer about the "HO_mode"; moreover, when the sender receives the Binding Update message, it notifies its TCP layer about the "HO_mode".	DUPAKS are not sent so that there is no need to retransmit the delayed packets sent via the old (satellite) network before the VHO decision.	No impact
Adaptive retransmission timer [9]	Sender side TCP modification	When the Binding Update message is received for a VHO procedure, the sender updates the RTO value for the new network on the basis of the *timestamp information* that is included in the Binding Update message (the difference between the Binding Update message arrival instant and the timestamp information is used to update RTO).	Cross-layer signaling L2 and L3	No impact	It is possible to avoid an RTO expiration since RTO is soon updated in the new network. In the standard TCP, the sender updates the RTO value as soon as the ACK is received of the first TCP segment sent via satellite.

Table 1. (*continued*)

Method	Modifications	Description	Signaling	Satellite-to-WiFi VHO impact	WiFi-to-satellite VHO impact
MAC buffer size notification [10]	Sender side TCP modification	The MAC buffer size of the new wireless network reached by VHO (the downlink of the new wireless segment is considered as the downstream bottleneck link) is communicated via the Association Response message by means of a new header field. Then, this information is sent back to the sender by means of the Binding Update message. The MAC buffer size is used to determine an upper bound to the cwnd value at VHO time in the new network.	Cross-layer signaling L2, L3, and L4 by means of the Binding Update message including MAC buffer size information	This method can adapt (reduce) cwnd to the WiFi network, thus avoiding the risk of buffer overflow.	No impact
Forward-RTO (F-RTO) [11]	Sender side TCP modification	F-RTO is executed only after an RTO expiration: the packet generating the timeout is retransmitted, ssthresh is halved, and cwnd is kept unchanged. Then, if the sender receives the ACK of a new packet, two new TCP packets are transmitted and cwnd = ssthresh; otherwise, if a DUPACK is received, the standard RTO protocol is executed (i.e., cwnd = 1 and a slow start phase is performed).	Standard TCP level signaling	No impact	F-RTO can avoid spurious RTO expirations caused by the VHO from a low RTT network to a high RTT network.
Hoe algorithm [12]	Sender side TCP modification	With this algorithm the initial ssthresh is set as half of the DBP estimated on the basis of RTT and the interarrival time between two consecutive ACKs. This algorithm could be modified and performed at VHO time to estimate the characteristics of the new visited network. However, DBP estimation in a new network (visited by VHO) entails a delay due to the arrival of the two first ACKs received via the new network.	Standard TCP level signaling	This algorithm could be used to lower ssthresh in the new visited network (WiFi), thus reducing the risk of buffer overflow and packet losses in the new network.	This algorithm could be used to increase ssthresh in the new visited network (satellite), thus permitting a faster utilization of the satellite bandwidth.

Table 2. Simulation parameters and settings

IEEE 802.11b settings	
Frequency	2.4 GHz
Nominal bandwidth	11 Mbit/s
Transmit Power	15 dBm
Received power threshold (for 250 m radius of coverage)	−74 dBm

System parameters	
TCP packet size (downstream)	1500 bytes
DBP of the satellite link	233 pkts
DBP of the wireless terrestrial link	9 pkts
Sender buffer size = satellite buffer size	250 pkts
WiFi buffer size	35 pkts
Satellite information bit-rate	5 Mbit/s

CUBIC [8] in the presence of VHOs between satellite and WiFi (single flow case). BIC and CUBIC have been proposed for terrestrial high-speed networks with high latency and allow a fast increase of the cwnd value. BIC combines linear, concave, and convex cwnd growth portions based on S_{min} (controlling the TCP-friendliness) and S_{max} (controlling the RTT-fairness), such as: additive increase (linear cwnd increase of S_{max} on RTT basis), binary search (logarithmic cwnd increase: $S_{max}/2$, $S_{max}/4$, ...), max probing (exponential cwnd increase: $2S_{min}$, $4S_{min}$, ...), and again additive increase (linear cwnd increase of S_{max}). All these phases are involved in a WiFi-to-satellite VHO to identify the new maximum cwnd of the satellite network that is much bigger than the WiFi one. In the additive increase phase there is a fast cwnd increase of S_{max} ($\gg 1$) on an RTT basis. This is much faster than the cwnd increase of 1 on an RTT basis for TCP NewReno in the congestion avoidance phase. It is convenient to have $S_{max}/S_{min} = 2^n$ with $n \geq 4$ for a fast cwnd increase. A high S_{max} value allows improving scalability and a low S_{min} improves TCP-friendliness. However, a too high S_{max} could entail a too [7] burst TCP traffic injection with consequent losses (congestion situation). Therefore, we have used [7]: Smin = 0.25 and Smax = 64. With CUBIC, the different phases of cwnd increase are substituted by a cubic law: cwnd $= c\,(t - K)^3 + W_{max}$, where t is the elapsed time from the last window reduction, c ($= 0.4$) is a scaling constant, W_{max} is the maximum cwnd value before the last reduction (e.g., maximum value in the WiFi segment at the WiFi-to-satellite VHO), β ($= 0.8$) is the cwnd reduction factor after a packet loss, and $K = \sqrt[3]{W_{\max}\beta/c}$; S_{max} ($= 160$ packets) now represents the maximum cwnd increment on RTT basis. BIC and CUBIC are implemented as default in several linux versions.

Referring to Figure 4, we can note that both BIC and CUBIC allow the highest cwnd (goodput) values in all conditions and a very fast cwnd increase at the WiFi-to-satellite VHO[3]. Moreover, all TCP versions reduce the cwnd value when a VHO is performed from satellite to WiFi (this change does not necessarily imply a bit-rate reduction because the WiFi link has a much lower RTT than the satellite one). We can see that there is a risk of RTO expiration at the WiFi-to-satellite VHO, as it occurs with Westwood+, thus entailing a cwnd reset to 1 and a slow cwnd increase in the satellite segment (congestion avoidance phase) due to a low value of ssthresh (Westwood+ uses the minimum RTT to estimate ssthresh and, thus, this RTT value is constrained by the WiFi segment in our case).

Figure 5 compares the TCP versions cwnd behaviors (single flow) in a case with random *Packet Error Rate* (PER) of 10^{-3} in the satellite segment. We can note that Westwood+ has a good performance before the second (WiFi-to-satellite) VHO, while its performance is reduced soon after due to a low ssthresh value. Moreover, CUBIC seems to maintain a sufficiently good performance before and after the tunnel.

---

[3] With BIC, the cwnd increase time at the WiFi-to-satellite VHO is about equal to $2\log_2 (S_{\max}/S_{\min}) + 2DBP/S_{\max}$ in RTT unit of time. For NewReno, the same time would be about equal to $2DBP$.

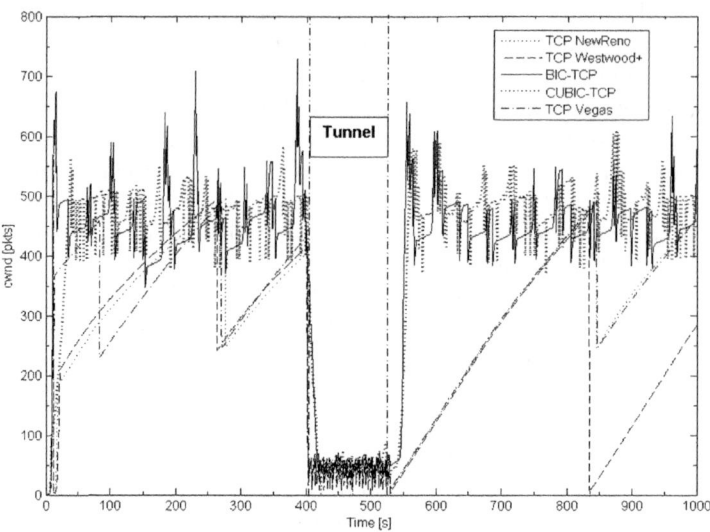

Fig. 4. Comparison of cwnd behaviors with different TCP versions in the presence of satellite-to-WiFi VHO (405 s) and WiFi-to-satellite VHO (525 s); no random errors

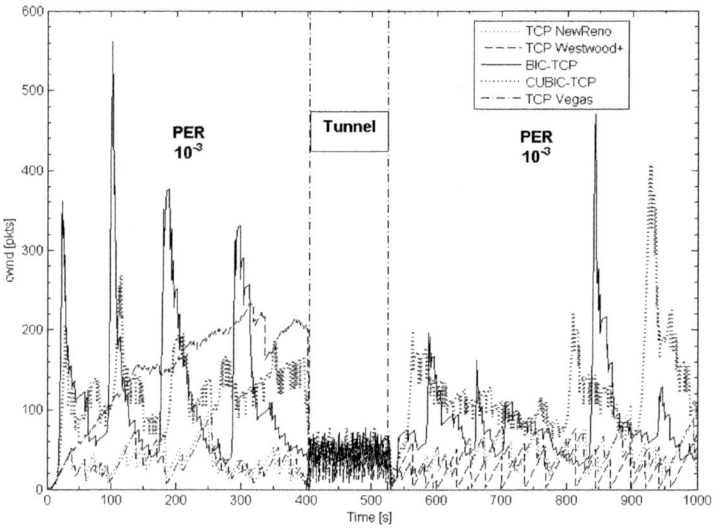

Fig. 5. Comparison of cwnd behaviors with different TCP versions in the presence of satellite-to-WiFi VHO (405 s) and WiFi-to-satellite VHO (525 s) VHO; PER = 10^{-3}

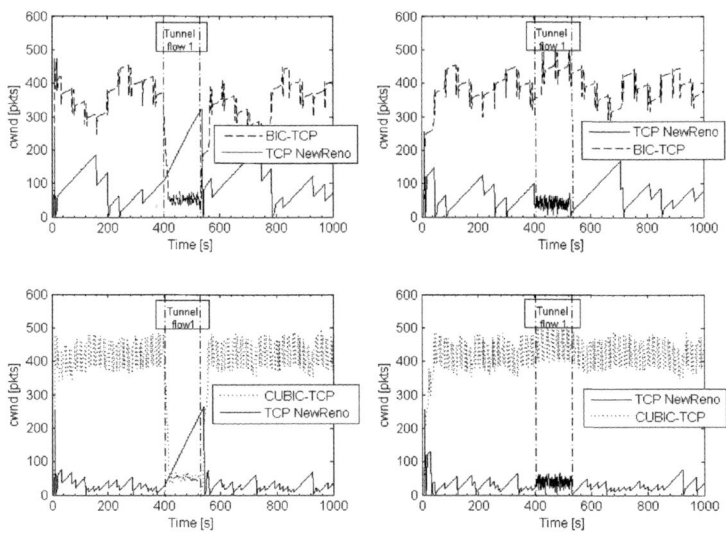

Fig. 6. Cwnd behaviors for two concurrent flows on two trains using different TCP versions (only flow #1 encounters a tunnel); no packet losses are considered

Figure 6 analyzes TCP-friendliness issues in the presence of two concurrent flows using different TCP variants and from two trains[4], so that only flow #1 encounters a tunnel in the time interval under investigation. From these results it is evident that BIC is more friendly than CUBIC with respect to NewReno; this is consistent with the study made in [14] in a scenario with high RTT.

In Figure 7, TCP performance is investigated with VHO methods #1 and #2; we refer here to TCP NewReno as this is the basic scheme (since the buffer size is equal to DBP, the cwnd behavior in the congestion avoidance phase is not linear but curved). We can note that VHO method #2 permits to achieve a faster cwnd increase after the WiFi-to-satellite VHO. Moreover, methods #1 and #2 avoid the occurrence of RTO after the satellite-to-WiFi VHO due to the arrival of packets and ACKs out of sequence (such improvement cannot be appreciated with the cwnd scale in the graph); however, method #1 is unable to avoid an RTO expiration at the the WiFi-to-satellite VHO because the Binding Update message allows measuring the RTT on the basis of the return path, without accounting for the forward path congestion.

Figure 8 shows the instantaneous goodput averaged over 10 repeated runs with different starting phases for two flows on distinct trains. This study permits to appreciate better the performance of method #1 and method #2 with respect to classical NewReno. Note that during the WiFi segment the total goodput is much higher than in the case of the satellite segment, because we have two

---

[4] In the presence of multiple TCP flows (e.g., browsing sessions) for the same train, they experience synchronized losses since they share the use of buffers along the same path.

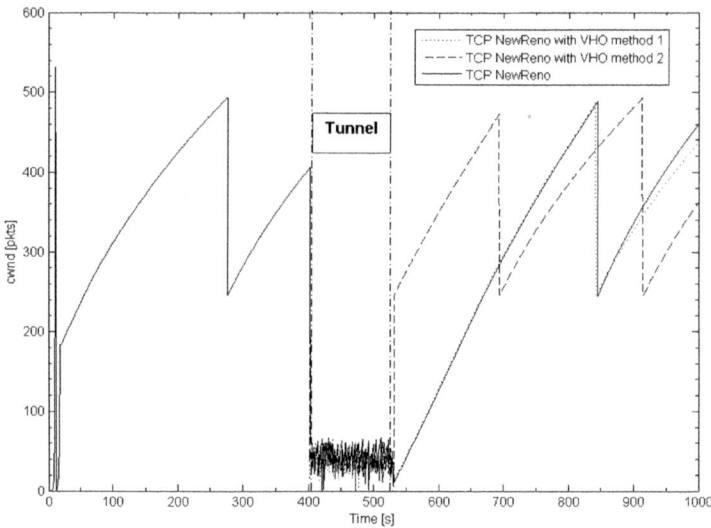

Fig. 7. Comparison of VHO techniques with TCP NewReno and single flow

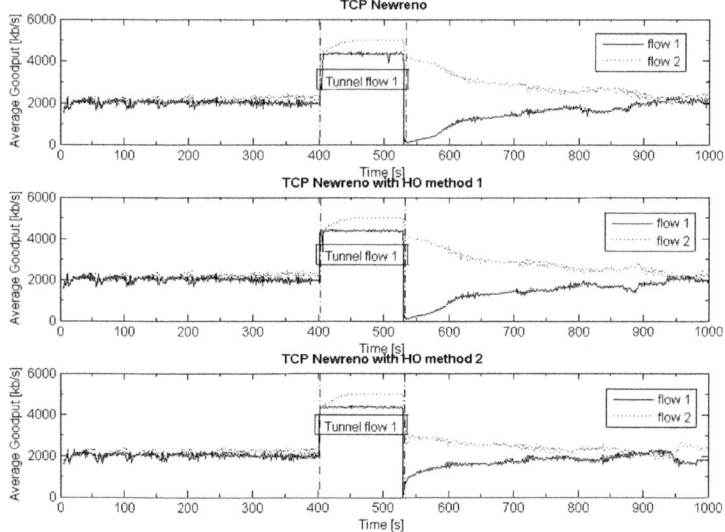

Fig. 8. Mean goodput for two TCP flows (only flow 1 encounters a tunnel)

flows exploiting separate bottleneck links (i.e., WiFi and satellite). These results confirm that method #2 permits to achieve a much better performance at the WiFi-to-satellite VHO.

Finally, Figure 9 compares the two envisaged VHO methods (applied to TCP NewReno), considering two concurrent flows on two trains. Only flow #1

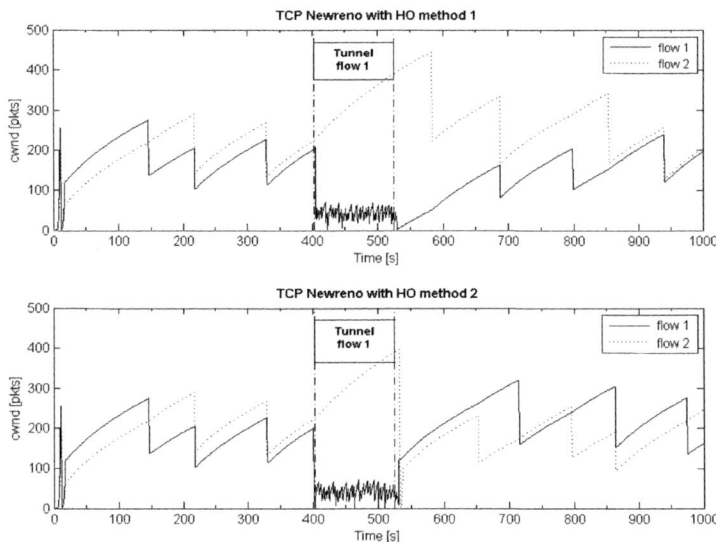

Fig. 9. Cwnd behavior of the two proposed handover methods used with TCP NewReno (only flow #1 encounters a tunnel); no packet errors are considered

encounters a tunnel during the simulation time. We can note that method #2 permits to achieve a better RTT-fairness and TCP-friendliness at the WiFi-to-satellite VHO.

6   Conclusions

This paper has dealt with transport layer performance in the presence of channel disruptions due to tunnels in a GEO satellite railway scenario. The approach considered here is based on a hybrid network with WiFi coverage in the tunnels. Suitable VHO procedures have been described on the basis of MIH and MIPv6 applied to the BSM standard scenario. Moreover, different enhanced TCP versions have been envisaged to increase the speed of traffic injection in the VHO from WiFi to satellite. TCP-friendliness and RTT-fairness issues have been discussed due to the occurrence of VHOs and the presence of flows on distinct trains. Finally, two cross-layer schemes have been investigated (method #2 contains a proposal made in this study) to improve the TCP goodput at the satellite-to-WiFi VHO and at the WiFi-to-satellite VHO. Future work is needed to investigate further TCP-friendliness and RTT-fairness issues considering also other TCP versions (e.g., Scalable TCP), to study the impact of different WiFi AP buffer sizes, and to propose other cross-layer options to improve TCP performance during VHOs. This work has been carried out within the framework of the SatNEx III Network of Experts supported by ESA (ESA Contract RFQ/3-12859/09/NL/CLP).

References

1. ESA Deliverable: Applications of DTN, CoO 1 – Task 3 Applicability of Delay and Disruption Tolerant Networks to Space Systems, SatNEx III (October 1, 2010)
2. Stemm, M., Randy, H.K.: Vertical Handoffs in Wireless Overlay Networks. Mobile Networks and Applications, 335–350 (1998)
3. Zhao, W., Tafazolli, R., Evans, B.G.: Internetwork Handover Performance Analysis in a GSM-Satellite Integrated Mobile Communication System. IEEE Journal on Selected Areas in Communications 15(8), 1657–1671 (1997)
4. Hansmann, W., Frank, M.: On Things to Happen During a TCP Handover. In: Proc. of the 28th Annual IEEE International Conference on Local Computer Networks, LCN 2003 (2003)
5. Luglio, M., Roseti, C., Savone, G., Zampognaro, F.: Cross-Layer Architecture for a Satellite-WiFi Efficient Handover. IEEE Transactions on Vehicular Technology 58(6), 2990–3001 (2009)
6. Hu, Y.F., Berioli, M., Pillai, P., Cruickshank, H., Giambene, G., Kotsopoulos, K., Guo, W., Chan, P.M.L.: Broadband Satellite Multimedia. IET Communications Journal 4(13), 1519–1531 (2010)
7. Xu, L., Harfoush, K., Rhee, I.: Binary Increase Congestion Control (BIC) for Fast Long-Distance Networks. In: IEEE INFOCOM 2004, Hong Kong, China (2004)
8. Ha, S., Rhee, I., Xu, L.: CUBIC: a new TCP-Friendly High-Speed Variant. ACM SIGOPS Operation Systems Review 42(5), 64–74 (2008), http://cse.unl.edu/~xu/paper/ha_ACM_2008.pdf
9. Kim, H., Lee, S.: Improving TCP Performance for Vertical Handover in Heterogeneous Wireless Networks. International Journal of Communication Systems 22, 1001–1021 (2009)
10. Liu, B., Martins, P., Ellatif Samhat, A., Bertin, P.: A Layer 2 Scheme of Inter-RAT Handover Between UMTS and WiMAX. In: IEEE Vehicular Technology Conference, Calgary, Canada (September 2008)
11. RFC 4138: Forward RTO-Recovery (F-RTO): An Algorithm for Detecting Spurious Retransmission Timeouts with TCP and the Stream Control Transmission Protocol (August 2005)
12. Hoe, J.C.: Improving the Start-up Behavior of a Congestion Control Scheme for TCP. In: Proceedings of ACM SIGCOMM 1996, pp. 270–280. Stanford University, California (August 1996)
13. Xu, L., Harfoush, K., Rhee, I.: Binary Increase Congestion Control for Fast, Long Distance Networks, Tech. Report, Computer Science Department, NC State University (2003)
14. Miras, D., Bateman, M., Bhatti, S.: Fairness of High-Speed TCP Stacks. In: Proc. of IEEE AINA 2008, Ginowan, Japan (March 25-28, 2008)
15. NS-2 Network Simulator (Vers. 2.29), http://www.isi.edu/nsnam/ns/ns-build.html

Digital Advanced Rural Testbed for Next Generation Access

Paris Skoutaridis and Graham Peters

Avanti Communications,
74 Rivington Street, EC2A 3AY, London, UK
{paris.skoutaridis,graham.peters}@avantiplc.com

Abstract. This paper provides a first look at the Digital Advanced Rural Testbed (DART) project, which is funded by the Technology Strategy Board under the Network Services Demonstrators Programme. The objective of the project is the design and implementation of a network demonstrator incorporating advanced network infrastructure elements and service enablers. This is done with a view to allowing third parties to experiment with novel business models, applications and services, compliant with the vision of next generation access services.

Keywords: Bandwidth Variation, Content Caching, Demonstrator, Multicast, Next Generation Access (NGA), Quality of Service Variation, Satellite, Technology Strategy Board (TSB), Testbed.

1 Introduction

Over the past couple of decades the internet has played an increasingly integral role to our lives both at a personal and a business level. The resulting increased demand for internet access is also accompanied by a shift to increasingly feature rich content and applications with video content currently being the most bandwidth demanding. The trend for more feature rich, bandwidth demanding content access over the internet has thus far evolved with increasing pace, and is expected to continue to do so. Consequently, Next Generation Access (NGA) services are receiving significant interest in an effort to develop solutions capable of catering for the ever increasing access requirements of both businesses and the public.

The DART project is funded by the UK Technology Strategy Board (TSB) [1] under the Network Services Demonstrator Programme [2]. The project aims to facilitate the experimentation and trialling of business models and capabilities deemed pivotal to the provision of NGA services to remote and rural communities. To that end, DART will offer a test-bed platform incorporating advanced infrastructure and technologies using a mix of satellite, terrestrial wireless and community fibre schemes. The test-bed will be built around a specific set of enabling technologies: real time (or near real time) bandwidth management, quality of service (QoS) management, multicast and content caching. While bandwidth and QoS management are mostly relevant to direct-to-premises end users, the DART network will enable the

G. Giambene and C. Sacchi (Eds.): PSATS 2011, LNICST 71, pp. 131–142, 2011.

trialling of multicasting and content caching to both direct-to-premises end users and remote community central network points (central caches) from where the content can then be redistributed to individual end user's premises.

The DART team comprises five organisations well placed for the design and implementation of the demonstrator, as well as the operation of the demonstrator and provision of support during the experimentation phase of the project. The team is led by Avanti Communications who will make bandwidth available over the recently launched HYLAS1 satellite, and be responsible for the design and implementation of the enablers over the satellite network. Metabroadcast, a London based design and technology company that develops content (video and audio) discovery products using content and social metadata, will utilize their existing software to provide a stream of recommendations for specific content to be cached. The University of Aberdeen will support the recruitment and experimentation with direct-to-premises end users and lead the dissemination activities. The University of Lancaster will provide access to the Wray community network thus enabling experimentation with localised content caching and multicasting. Finally, 21Media, a company specialising in the development of intelligent, platform independent software systems for IPTV, will focus on the exploitation of IPTV usage data over the Wray network and the development of a localized content distribution system.

The remainder of this paper discusses the motivation behind the design and implementation of the DART demonstrator, as well as presenting an overview of the demonstrator itself. The paper is organized as follows. In section 2 we discuss the concept of next generation access and the process that is required for its eventual realization. Section 3 discusses the Network Services Demonstrations Programme of the TSB and how that program aims to address the eventual roll out of NGA services in the UK through stimulating the development and testing of relevant business models and applications. Section 4 presents the DART project in more detail, providing an overview of the key network components and architecture, the specific demonstrator enablers and the way in which it is expected experimentation will be set up and undertaken by third parties. Finally, section 5 concludes the paper with the timeline of DART and the contact details for the project's contact points.

2 Next Generation Access

The ever increasing reliance of modern society on the internet and broadband access is currently putting a strain on existing networks. This problem is compounded by the shift towards more feature-rich, bandwidth demanding applications and services, as well as a shift in consumption patterns and the continuously increasing amount and variety of content available on the internet.

The increased reliance of modern society on broadband internet connections has also led to another phenomenon; that of the digital divide. The fibre networks predominantly used for delivery of broadband connections are largely unavailable in remote and rural areas, meaning broadband access to such locations is extremely limited if not completely unavailable. Consequently, there is a clear distinction between those with and those without access to high speed broadband connections.

Next Generation Access is the term used to collectively describe the services and applications that will be available over the internet in the future, as well as the networks and technologies that will be employed in their provision. As the current trends in services and applications indicate, next generation services and applications will be characterised by significantly increased bandwidth requirements as well as, in many cases, stringent QoS requirements. Given this, next generation networks will be required to cater for the increase in bandwidth requirements as well as to guarantee the provision of suitable QoS for those applications that require it.

At the time of writing next generation access in urban areas of the UK where broadband internet is already attainable is envisaged through the extension of the fibre-optic backbone connections with fibre-to-the-home, instead of the currently employed copper wire or cable connections. Such an upgrade in the last mile access method would significantly increase not only the bandwidth attainable over a broadband connection but also the reliability and quality of the resulting service.

The case is not the same for remote and rural areas where broadband connections are predominantly unavailable. The high cost associated with the deployment of fibre-optic connections removes the economic viability of providing such connections to rural and remote areas of a low population density. Consequently, there is a clear danger that remote and rural areas may be bypassed by future broadband access developments, thus falling even further behind in the broadband revolution compared to high population centres. This would further impact life in such areas both at a private and a business level, and would likely put increasing pressure on the already low populations. Alternative access methods as well as content and application provision paradigms are therefore required in order to ensure that next generation services do not bypass the rural UK.

DART proposes that next generation access to rural areas can only be realised through the combined use of a host of access technologies, in conjunction with intelligent methods of content distribution and application design. In accordance with this proposition, DART proposes the implementation and trialling of specific enabling technologies and network architecture characteristics that have the potential to significantly improve the available offering to the rural community both in terms of service level and quality, as well as in terms of the variety and quantity of the available content and applications.

3 The Network Services Demonstrators Programme

As part of the Digital Testbed Programme, the Network Services Demonstrators program of the UK Technology Strategy Board (TSB) aims at establishing test-bed platforms for the demonstration of network services.

It is the intention of the TSB that demonstrators developed under this programme become national hotspots for experimentation and trials. This should in-turn stimulate development and innovation in fields such as business models, applications and services that rely on advanced network infrastructure and service enablers. Thus, the

Network Services Demonstrators program aims to address the high, unmet demand for advanced platforms over which experimentation can be undertaken in an open and live manner.

The Network Services Demonstrators program comprises two phases that all funded projects are required to follow. Phase 1 involves the definition, design, and implementation of the demonstrator over a period of four to six months. This phase culminates with the generation of a proof of principle business model, application, or service that is enabled by the demonstrator's features.

Phase 2 of the program is the experimentation phase. During this phase the demonstrators will "go live" with access to third parties being provided. Depending on each demonstrator's features, third parties will then be able to experiment on relevant novel business models, applications, and services. The second phase of the Network Demonstrators Programme has a duration of at least one year.

4   The DART Demonstrator

4.1   Objectives and Motivation

The key requirement set by the Network Services Demonstrators Programme on all funded projects is the development of a demonstrator system that enables third parties to trial and experiment with novel and innovative business models, applications and services centred around advanced network infrastructure capabilities and service enablers.

To that end, DART will design, implement and make available for experimentation a demonstrator system that will allow third parties to experiment with specific network infrastructure capabilities and service enablers that have traditionally been either prohibitively expensive to access, or even unattainable altogether.

Through the DART project, experimenters will be provided with access to an advanced Ka band satellite network, thus bypassing the relatively high cost traditionally associated with bandwidth leasing. In addition, the enhanced nature of the design and capabilities of the HYLAS1 network and end user equipment will present experimenters with the opportunity to trial advanced capabilities and service enablers that have thus far been unavailable for testing over satellite networks. These are the bandwidth variation capability, the link QoS variation as well as the ability to experiment with multicasting and caching of content either locally at the end user premises or centrally at a community cache facility.

Through the supported service enablers, DART will be one of the first test beds to allow in depth experimentation with innovative services such as telehealth, telemedicine and precision farming techniques. Experimenters will also have the opportunity to test and develop new business models based on the observation of end user behaviour in rural areas as well as other demographics, or by developing innovative content distribution methods through exploitation of DART's multicast and caching capabilities. In all cases monitoring and feedback of activity over the DART network will be undertaken to allow both experimenters and the DART consortium to analyse the offered services.

Crucially, in addition to the ability to experiment with the above features, DART also presents the unique advantage that these capabilities and service enablers will remain available after the completion of the project for both further experimentation and development as well as for potential commercial exploitation.

4.2 The DART Enablers

Bandwidth Control. Dynamic occasional variation of the link bandwidth is a particularly useful feature that is not currently widely supported by broadband internet networks. Particularly in the case of broadband over satellite, where bandwidth is a premium resource, the ability to dynamically vary the link bandwidth on an occasional basis would not only be extremely useful to both private and business end users but also to network operators.

Dynamic bandwidth control would enable end users to reduce their satellite broadband cost by opting for a lower bandwidth connection for their every day use, while having the option to augment their bandwidth allocation at times that a higher speed connection is required. Examples of such cases include instances when an end user would need to download or upload a large file, attend a videoconference or access other bandwidth demanding content and applications. User-initiated bandwidth variation would require the end user to issue a request to the network, which would respond by increasing the non-contended bandwidth allocated to that link for a predetermined amount of time. In this model there can be several "service levels" defined with respect to the amount of bandwidth increase, each with a predefined cost per unit time.

Dynamic bandwidth variation could also prove attractive to content providers that require high bandwidth, high quality connections for their service. In this scenario a content provider could initiate the bandwidth augmentation of a specific link while their content is being forwarded to the end user. The cost for the temporary bandwidth increase could either be born by the content provider or passed on to the end user via the fee that the content provider charges for their services.

Further to the above, dynamic bandwidth variation would also be a particularly attractive proposition for network operators. Context aware bandwidth variation could be provided by the network operator through the inclusion of appropriate software components in the network management system. Irrespective of the initiation point, dynamic bandwidth variation can potentially increase revenue and reduce network congestion and contention ratios. It also affords the network operator with increased flexibility and adaptability in their bandwidth management processes. Novel and innovative service level agreements and charging methods could also be developed as a result of the ability to dynamically vary the bandwidth allocated to individual links.

Variable Quality of Service. QoS variation is another capability that is currently unavailable in today's broadband internet services. Similarly to bandwidth variation, QoS variation is a capability that has the potential to not only significantly improve the service experience of end users but also to decongest satellite communications networks and improve the overall quality of the provided service. QoS variation is becoming increasingly important in internet networks as a result of the increasingly varied types of services and applications available for consumption, and the significant differences in the QoS requirements. Similarly to bandwidth variation, QoS variation may be provided in a context aware, user initiated or content provider initiated manner.

It is envisaged that users, content providers, and network operators alike would benefit from the dynamic use of QoS variation on individual links. Broadband connections could be allocated default QoS attributes for the majority of the time, guaranteeing acceptable performance during "typical" connection use, with on-demand variation of the connection's QoS attributes taking place when more demanding application or services are accessed by the user. As an example, a lower latency and lower jitter traffic class could temporarily be allocated to a user's connection that engage in a videoconferencing session, or a better protected (e.g. through FEC) traffic class could be allocated to a user that is downloading a large file in order to reduce the download time.

Multicast and Caching. Multicast data distribution is particularly well tailored for satellite communication systems as it takes advantage of the wide area of coverage typical to satellites. Specifically in the case of broadband over satellite multicast can have a significant impact on the bandwidth requirements associated with distribution of common data to a large number of receiver units and, by extension on the number of data distribution sessions that the satellite network can support simultaneously, as it eliminates repetitive transmissions for multiple receivers. Through moderating the bandwidth requirements for data distribution over satellite multicast also improves the scalability of the satellite network to both the receiver population size and the amount of content distributed.

Multicast is equally suitable to both streaming and background traffic classes. Particularly in the case of live content streaming, content providers would benefit significantly from the use of multicast as opposed to distribution of the content over several unicast links. Relevant examples of content would include academic lectures, medical operations, telemedicine where several non collocated experts are required to observe or live sporting events. Similarly, multicast can reduce the bandwidth requirements and overall costs associated with background traffic distribution to several receivers by utilising a single link instead of several unicast links, with the benefit attained being proportional to the content volume and rising sharply with the number of receiving terminals.

When coupled with content caching multicast proves even more beneficial with regards to bandwidth utilisation efficiency. Bandwidth demanding content such as video can be multicast to several caching servers simultaneously thereby significantly reducing the network resources utilisation. The cache servers can then provide end users with access to the cached content for as long as the content is stored. This method of data distribution provides the added advantage that the cached content can be made available to end users for a prolonged time period without the need for additional or prolonged multicast sessions. Furthermore, this is a particularly cost effective and efficient method for making bandwidth demanding content available to communities equipped with a single central cache facility.

4.3 Demonstrator Architecture

The architecture of the DART demonstrator test bed is depicted in figure 1. The DART test bed is composed of the Avanti core network, the recently launched (November 2010) HYLAS1 satellite and two test site categories. The first test site category comprises direct-to-premises users equipped with enhanced two way satellite communication equipment and located in Scotland. The second test site category is a hybrid satellite and cable/wireless mesh network in and around the village of Wray, located in northern England.

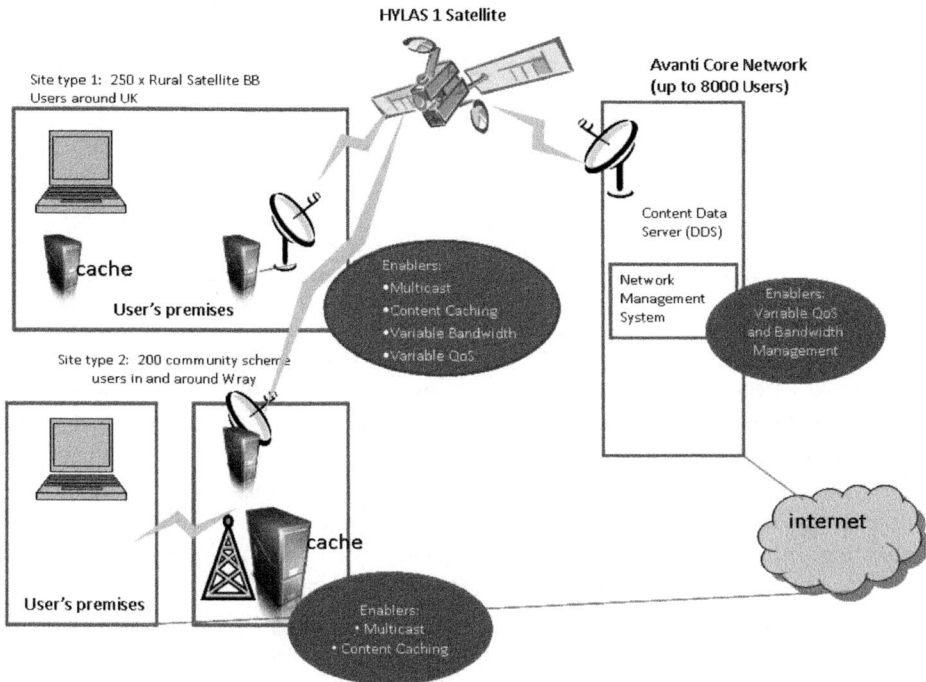

Fig. 1. DART Test Bed Architecture

The end users comprising the first test site category will be primarily recruited from the current Avanti pool of users as well as from the recruitment efforts of the University of Aberdeen. It is estimated that at least 250 existing Avanti end users will participate in testing using DART. The direct-to-premises test site will support trialling and testing of all DART enablers. Multicast and caching services will be tested using an adapted version of the software and hardware developed for an ongoing European Space Agency (ESA) funded project called NXY, whose development schedule is ahead of that for DART. In this scenario, content will be multicast and cached locally on end user equipment and will be accessible by the end users transparently through their browser. A possible realisation of the network components and architecture for multicast and caching is illustrated in figure 2.

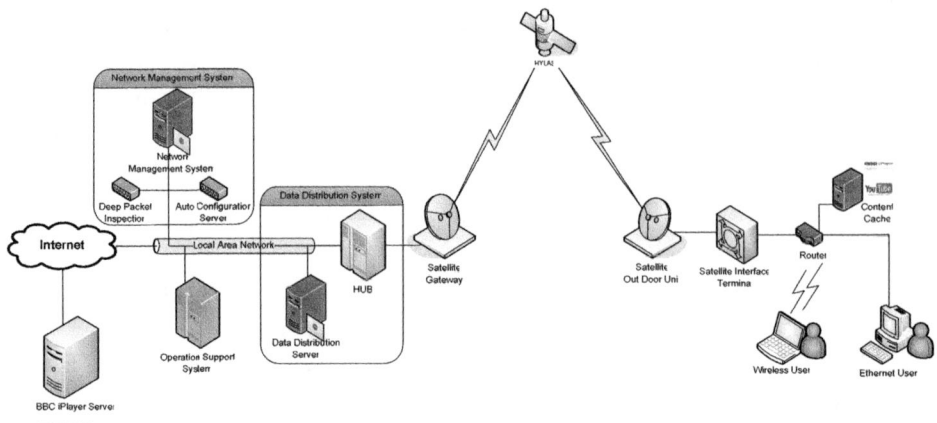

Fig. 2. Multicast and Caching Implementation Architecture

Variable QoS and variable bandwidth testing will also be supported by the registered direct-to-premises end user equipment. Testing of these enablers will be facilitated by the development and distribution to end users of software that enables them to issue bandwidth augmentation requests or changes to the QoS attributes of their links on a short or long term basis. Suitable software for the reception and processing of such requests, as well as the tracking and (possible) billing for the resulting service on the network management side will also be developed and integrated with the existing Avanti network management system.

In addition to the direct-to-premises portion of the network, DART will also provide access to a second site category. This will be the mesh wireless network installed at Wray in 2004 through the National Rural Support Programme (NRSP) [3] and operated since by the University of Lancaster. The Wray network [4] is comprised of a number of mesh wireless routers that connect the village's houses to a central mesh router that provides internet access. Figure 3 illustrates the current architecture of the Wray network. For the purposes of DART, a community satellite caching facility will be added to the network and interfaced with the central mesh router. This facility will allow the caching of content that the community members will then be able to access over the mesh network. The Wray network, augmented with the community satellite caching facility, therefore will provide an excellent opportunity for experimentation and trialling of central caching and redistribution of bandwidth demanding content and the effects that this content delivery method would have on both the backhaul and the last mile networks.

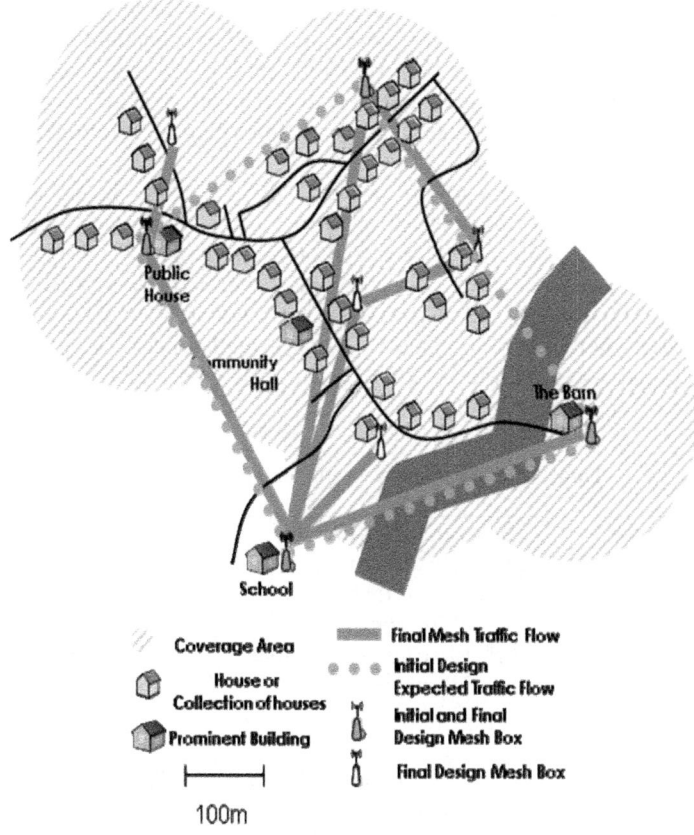

Coverage Area

House or Collection of houses

Prominent Building

100m

Final Mesh Traffic Flow

Initial Design Expected Traffic Flow

Initial and Final Design Mesh Box

Final Design Mesh Box

Fig. 3. Wray Network Architecture

4.4 Accessing the TestBed

The aim of both the TSB and the DART consortium is that accessing the DART demonstrator should be a simple and straightforward process. Given that DART is only one of several demonstrators being developed under the Network Services Demonstrator Programme it is also important that there should be a common process established between the various projects towards registering interest, accessing and using the demonstrators. To that end, a common interface will be developed that will provide experimenters and end users with a means of access to all TSB funded demonstrators.

Access to the DART test bed will therefore require potential experimenters and possibly end users to access the common interface from where they can gather information about the test bed characteristics and capabilities, the trials and experimenters that can be supported over the test bed and any contractual or requirements associated with engaging with DART. The common interface will also allow experimenters to define the key aspects of the trials they intend to perform which will then be notified to the contact person responsible for the setup and management of DART trials.

The next step will involve the exchange of more detailed information between the experimenters and the DART consortium in an effort to scope and define the trials in more detail. The trial would then be set up and "go live" over the DART network. A series of API's will also be created so that experimenters can provide DART with content and associated information such as distribution requirements, charging requirements and so on.

Once a trial is ready to go live over the DART network, the DART consortium would also advertise the upcoming trial to end users registered with the test bed. Advertisement of upcoming or current trials may be done on a suitable part of the common interface, on a suitable web page generated by DART or by direct notification of the DART end users. End users would then be able to view the available content by means of a separate API or through a network portal.

The process of accessing the DART testbed, setting up a trial, and undertaking the trial is summarized in table 1 below.

Table 1. Experiment/trial setup and operation process

Advertisement and Interest Attraction	1.	The testbed activity is generally advertised to all end users so that they are primed for upcoming "offers". End users register general interest	2.	The testbed is promoted to potential experimenters who e.g. content providers, rural service and application providers, network operators and community networks.
	3.	An experimenter finds out about the test bed and becomes interested in potentially using it.	4.	The experimenter approaches TSB or a DART partner for further information.
First Contact and Trial Scoping	5.	Information exchange between experimenter and TSB the experimenter takes place (advice, questions answered etc).	6.	Information exchange between the experimenter and Avanti. (more detailed information about the testbed, information about experimenter's goals).
Trial set up	7.	The experimenter decides to go ahead with the trial	8.	Experimenter agrees to terms and conditions to participate in the trial (probably on-line)
	9.	Experimenter provides more details of the test / trial to be performed. This could be automated (through an API.	10.	Details of the availability of the trial service are advertised to end users. Those interested sign up to T&Cs.

Table 1. (*continued*)

	11. Content provided to the Avanti network through an API along with key information about when to be delivered, charging details, price level etc.	12. End Users see the content as part of a "walled garden". They are free to choose this or free content such as iPlayer.
Trial operation	13. An end User chooses the content they wish to view and agree to pay for the content, in this case either in SD or HD resolution (they could pay more for HD). For non-cached content they would pay using a different charging tariff perhaps.	14. The end user views the material. The service usage is monitored and accounted according to a predefined billing mechanism. The user is charged a low tariff to view videos.
	15. The service tariff is increased after some time to allow understanding of the effect of price elasticity. The tariff change is advised by email to the end user at both the start of the trial and the time of the changeover.	16. Data is collected from the trial, anonymised, and discussions are held with the users on how they value the service and used to assess the commercial exploitation opportunities.

5 Conclusions

Under the TSB Network Services Demonstrators programme, the DART project will address the future provision of NGA services to rural and remote areas through facilitating testing and trialling centred around the key service enablers and advanced network features discussed in section 4. The project will comprise a design and implementation phase and an experimentation phase.

In terms of timeline, phase 1 of the project kicks off on 1st February 2011 with the design activity, followed by implementation, integration and testing and deployment expected to be completed in late July 2011. Phase 2 of the project, the experimentation phase, will kick off in August 2011 with the demonstrator being made available to third party experimenters and will last for 12 months. The project will complete in August 2012 with the final project review.

For further information readers are invited to contact the project manager, Trevor Barker, at Trevor.Barker@avantiplc.com.

References

1. UK Technology Strategy Board (TSB) home page, `http://www.innovateuk.org/`
2. TSB Digital Testbed: Network Services Demonstrators programme description, `http://www.innovateuk.org/_assets/pdf/competition-documents/briefs/digitaltestbednetworkservicescomp.pdf`
3. National Rural Support Programme (NSRP) home page, `http://www.nrsp.org.pk/`
4. Ishmael, J., Bury, S., Pezaros, D., Race, N.: Deploying Rural Community Wireless Mesh Networks. IEEE Internet Computing 12(4), 22–29 (2008)

OLSR-H: A Satellite-Terrestrial Hybrid Broadcasting for OLSR Signaling

Carlos Giraldo Rodriguez[1], Laurent Franck[1],
Cédric Baudoin[2], and André-Luc Beylot[3]

[1] Télécom Bretagne-Institut Télécom/TéSA, Toulouse, France
carlos.giraldo@tesa.prd.fr, laurent.franck@telecom-bretagne.eu
[2] Thales Alenia Space, Toulouse, France
cedric.baudoin@thalesaleniaspace.com
[3] IRIT/INPT-ENSEEIHT, Toulouse, France
andre-luc.beylot@enseeiht.fr

Abstract. Mobile ad hoc networks (MANETs) are proposed for emergency situations because they are self-organized and infrastructure-less. However, the mobility of the nodes and the lack of infrastructure pose challenges to the routing protocols. The signaling of these protocols is affected by the same problems as the data traffic: a multi-hop environment with limited bandwidth, collisions and bit errors. This paper proposes a modification of the Optimized Link State Routing (OLSR) protocol to combine satellite and terrestrial broadcasting for their signaling and evaluates its performance in a forest fire fighting scenario.

Keywords: Routing, ad hoc networking, emergency, satellite.

1 Introduction

Mobile ad hoc networks have awoken the interest of the research community in the last decade. The mobility of the nodes produces frequent topology changes that should be reflected in the routing tables. For that reason, routing protocols exchange network state information to achieve proper routing of data packets to their destinations[1].

Satellites can play an important role in the distribution of routing signaling by offering a broadcast medium ensuring quality of service. This contribution is based on the Optimized Link State Routing OLSR[2] protocol. We chose a link state protocol because the route computation and topology discovery phases are well separated and therefore subject to different optimizations. Finally OLSR is among the popular and standardized routing protocols. OLSR-SAT[3] proposes to modify OLSR to distribute the broadcast signaling messages over satellite. It shows an improvement in packet delivery ratio for data packets traversing several hops.

However, this modification requires all nodes to have a satellite interface. In this paper we relax this constraint by combining satellite broadcast with the default OLSR terrestrial broadcast. This solution is called Optimized Link State Routing-Hybrid (OLSR-H).

G. Giambene and C. Sacchi (Eds.): PSATS 2011, LNICST 71, pp. 143–150, 2011.

The broadcasting procedures of OLSR, OLSR-SAT and OLSR-H are explained in Section 2. A forest fire scenario is described in Section 3 and the simulation results of the protocols in this scenario are presented in Section 4.

2 OLSR Hybrid Broadcast

Proactive routing protocols such as OLSR broadcast network state information in order to support the computation of routing tables. However, broadcasting in a multi-hop wireless network as a MANET is not a trivial task[4].

The main broadcast signaling messages of OLSR are topology control (tc) messages. OLSR broadcasts these messages using multipoint relays (MPRs). The rationale is to reduce signaling overhead by using selective flooding. Each node selects a set of neighbors as multipoint relays. These neighbors will be the only ones allowed to forward broadcast messages from the node. This set of neighbors must be selected so to cover all the 2-hop neighbors.

In OLSR-SAT a geostationary satellite is used to broadcast the tc messages instead of MPRs in OLSR. It mitigates the impairments caused by partition, bit errors and collisions on the wireless medium. It also sets a constant travel delay of ca. 250 ms in place of variable travel delays caused by the multiple hops of the MPR broadcast. However for OLSR-SAT to be operational, all nodes must have a working transmission and reception satellite interface.

A more realistic approach is taken in this paper. A hybrid system combining both MPRs and satellite broadcasting systems is devised as OLSR-H. A node originating a broadcast message always sends it via terrestrial wireless and via satellite

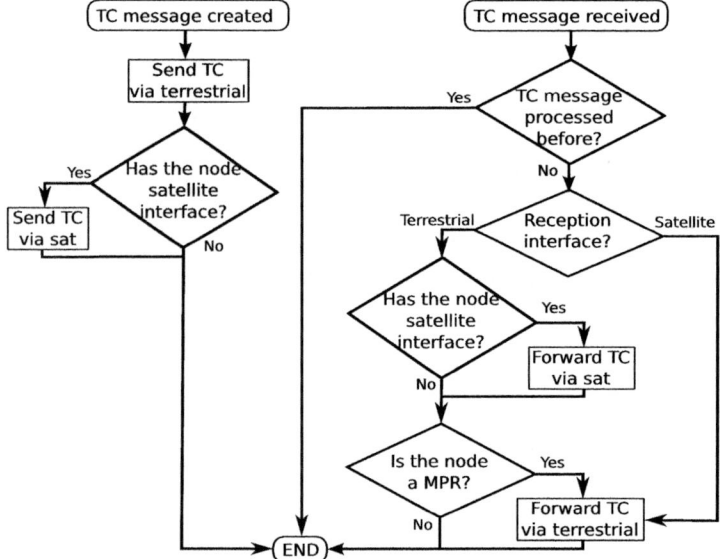

Fig. 1. OLSR-H broadcasting flow chart

if possible. When a node receives a broadcast message it checks if the message was already processed. If not, it continues the process. On the one hand, if the message was received via satellite, it is always forwarded via terrestrial link. On the other hand, if it was received via terrestrial link it is forwarded via the terrestrial network if the node is a MPR and via satellite if the node has a satellite interface. Figure 1 represents a flow chart of OLSR-H broadcasting process.

3   Scenario

The chosen scenario is a forest fire fighting mission. In this context, firemen share information about the fire advance, the topography of the intervention area or their own locations. However a network infrastructure may not exist because of the environment or it may be down because of the fire. The establishment of a MANET among the firemen units offers a primitive data network for these purposes.

The forest fire scenario was described in [3] and implemented using the OMNET++[5] simulator. The same implementation is used in this work (see Figure 2), but not all the nodes feature a satellite interface. Three configurations with 4, 7 and all (28) nodes with a satellite interface are summarized in Table 1. Using command cars as the units carrying satellite dishes is a sensible choice considering energy supply and antenna mounting constraints. Table 2 summarizes the simulation input parameters.

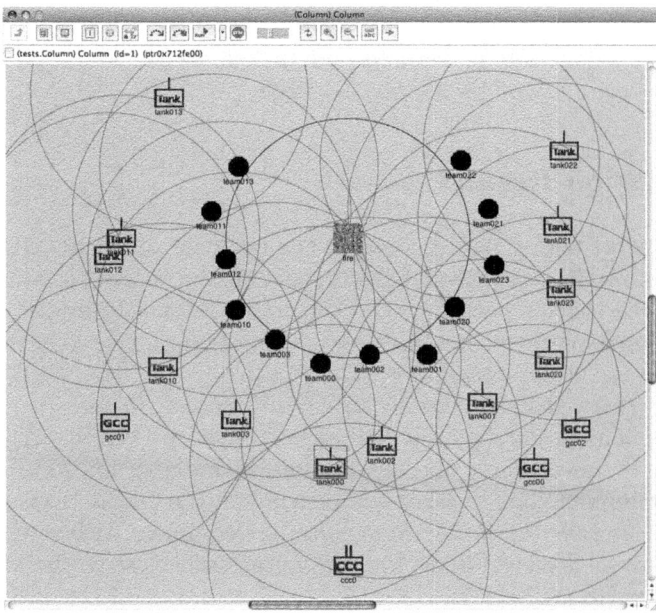

Fig. 2. Snapshot of the forest fire fighting network layout in OMNET++ simulator

Table 1. Forest fire scenario configurations

Name	Broadcast characteristics and nodes with satellite facilities
OLSR	Only tc terrestrial broadcast No node with satellite i/f
OLSR-SAT	Only tc satellite broadcast All nodes with satellite i/f
OLSR-H $r = 4$	Hybrid tc terrestrial and satellite broadcast Column command car + all group command cars (3) with satellite i/f
OLSR-H $r = 7$	Hybrid tc terrestrial and satellite broadcast Column command car + all group command cars (3) + one tanker per group (3) with satellite i/f
OLSR-H $r = n = 28$	Hybrid tc terrestrial and satellite broadcast All nodes with satellite i/f

Table 2. Simulation input parameters

Parameter	Forest Fire Fighting
Attenuation model	Path loss reception
Mobility model	FireMobility
Number of nodes	28
Path loss alpha	3
Transmission power	1 mW
Receiver sensitivity	-90 dBm
Channel model	Rayleigh
Radio technology	802.11g
Radio range (R)	46.25 m
Playground size	$5R \times 5R$
Data packet interval	Exp(100 ms)
Data packet size	1 KB

4   Simulation Results

Simulations were performed to compare the behavior of OLSR, OLSR-SAT and OLSR-H. As in [3], the packet delivery ratio of three data streams was chosen to measure the routing performance. The traffic flows vary in distance (i.e., the number of hops) between the source and destination representing short, medium and long distance communications.

The data packet delivery ratio is shown in Figure 3. As already noticed in [3], the improvement in data packet delivery achieved by OLSR-SAT depends on the distance between source and destination, with higher improvements for long distance communications (>4 hops). Focusing on the OLSR-H results, a similar improvement is achieved if only four nodes are equipped with satellite facilities. The performance achieved in such a configuration is close to the ideal situation where $r = n = 28$.

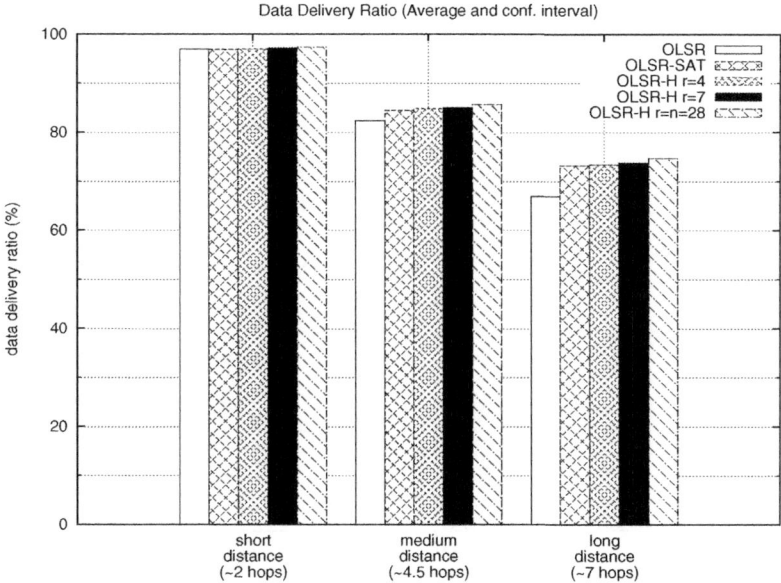

Fig. 3. Data packet delivery ratio comparison among OLSR, OLSR-SAT and OLSR-H

The difference among the three routing protocols is the broadcasting process for the topology control (tc) messages. The delay and the delivery ratio of these messages are therefore analyzed to find the reason of the improvement.

The average and standard deviation for the travel delay of the tc messages are shown in Figure 4. One reason for the data delivery ratio improvement could be the constant tc travel delay of the satellite transmission, allowing the nodes to have a consistent view of the network topology. However, the tc delay of OLSR-H with $r = 4$ is similar to OLSR but the data packet delivery is still improved.

The delivery ratio of tc messages is shown in Figure 5. The tc delivery ratio is improved without a large number of satellite equipped nodes (r). There is a relation between the tc delivery ratio and the quality of the routing decisions and therefore the data packet delivery ratio.

4.1 Detailed Evaluation of the Communication Errors

The tc message delivery ratio was computed at the routing level. The tc message delivery is the relation between the number of tc messages created and the number of tc messages processed. In an ideal situation each tc message created should be processed by all the other nodes ($n - 1 = 27$). The *tc message delivery* is therefore defined as:

$$\text{tc delivery ratio} = \frac{(tc\ msgs\ processed)}{(n - 1) \cdot (tc\ msgs\ created)}$$

An analysis of terrestrial tc transmissions is performed at the link level (that is, WLAN) to classify the source of errors. Only the frames with at least one

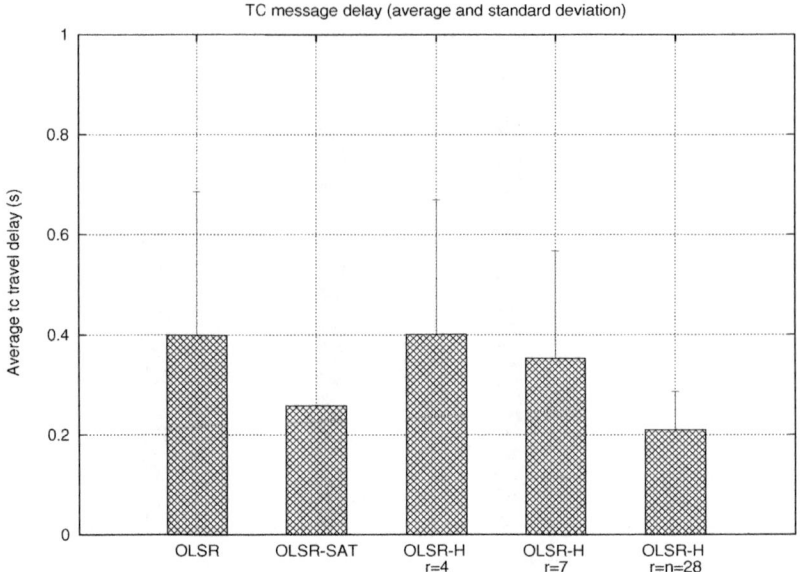

Fig. 4. Comparison of tc message travel delay of the studied routing protocols

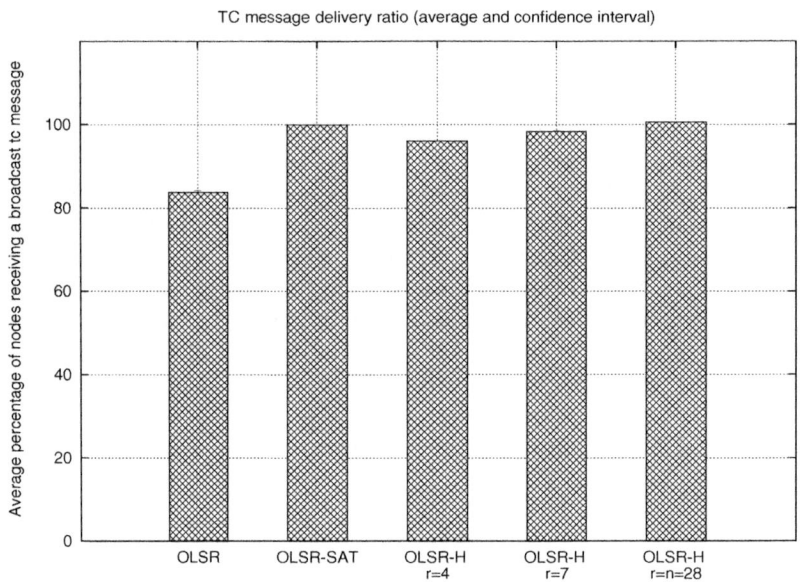

Fig. 5. Comparison of tc message delivery ratio of the studied routing protocols

tc message inside are taken into account. Four counters are set up: *sent frames, correctly received frames, received frames with bit errors* and *received frames with collisions*. In this case the relation between sent frames and received frames depends on the number of neighbors of each frame source.

The results obtained do not depend on the routing protocol: 94% of the frames were correctly received, 4.5% were received with bit errors and 1.5% were received after collisions. These two error indicators are linked to the traffic intensity going through the WLAN medium. It is therefore likely that in real-world conditions, the number of errored frames will be larger. On the other hand, sending tc messages through the satellite link makes it unaffected by the data traffic load.

4.2   Overhead Evaluation

Several frame transmissions are needed to broadcast a single tc message. The number of transmitted frames per tc message created defines the overhead of the broadcast system. Table 3 shows this relation for the terrestrial and the satellite broadcast.

The overhead in the satellite transmissions is due to the satellite propagation delay. While the satellite transmission of a frame is not over, a copy of the same frame could arrive via WLAN to a satellite node. The WLAN frame is then processed and forwarded again via satellite.

The overhead in the WLAN transmission increases with the number of nodes equipped with satellite facilities. The reason is that some nodes have an independent broadcast channel (the satellite) and the MPR algorithm does not take this into account, therefore the MPRs election is not optimized.

Table 3. Broadcasting overhead: transmitted frames per tc message created

	OLSR	OLSR-SAT	OLSR-H(r=4)	OLSR-H(r=7)	OLSR-H(r=n=28)
WLAN	6.51	0	8.41	9.44	14.51
satellite	0	1	2.19	3.36	10.93
total	6.51	1	10.6	12.80	25.44

5   Conclusions

Signaling traffic is a key component of routing. We demonstrate that improving the reliability of signaling communication yields better routing decisions and therefore improves the delivery of data traffic. OLSR-SAT and OLSR-H with $r = n$ are not realistic solutions because they require all nodes to be equipped with satellite terminals. However they yield the maximum improvement we could reach. Then we showed that a similar improvement can be achieved with a lower number of satellite terminals (OLSR-H $r = 4$ out of 28 nodes).

The hybrid broadcast solution produces larger overhead for both terrestrial and satellite systems. Future work should investigate the integration of information about which nodes have satellite facilities in the OLSR MPR election so to mitigate the signaling overhead in hybrid configurations. Also, increasing the traffic load of the network will favour the hybrid broadcast system with respect to OLSR.

Two future directions are worth mentioning. First, OLSR-H impacts two properties of the signaling distribution: the delay required to distribute the signalling and the reliability of the process. The results show that both properties are relevant however reliability seems to be the most significant of the two. This has to be further investigated. Second, while the hybrid signaling distribution scheme was presented in the context of OLSR, it might be used in another context and even help to optimize the route discovery process in protocols such as AODV.

References

1. Biradar, R., Patil, V.C.: Classification and comparison of routing techniques in wireless ad hoc networks. In: International Symposium on Ad Hoc and Ubiquitous Computing 2006, pp. 7–12. IEEE, Los Alamitos (2007)
2. Clausen, T. and Jacquet, P.: IETF RFC-3626: Optimized Link State Routing Protocol (OLSR). The Internet Society http://www.ietf.org/rfc/rfc3626.txt (2003)
3. Giraldo, C., Franck, L., Baudoin, C., Beylot, A.L.: Mobile ad hoc network assisted by satellites. In: 28th AIAA International Communications Satellite Systems Conference, Anaheim (2010)
4. Williams, B., Camp, T.: Comparison of broadcasting techniques for mobile ad hoc networks. In: Proceedings of the 3rd ACM international symposium on Mobile ad hoc networking and computing, pp. 194–205. ACM, New York (2002)
5. Varga, A.: Using the OMNeT++ discrete event simulation system in education. IEEE Transactions on Education 42(4), 11 (1999)

Energy Efficient Cooperative HAP-Terrestrial Communication Systems

Sithamparanathan Kandeepan[1], Tinku Rasheed[1], and Sam Reisenfeld[2]

[1] CREATE-NET Research Centre, Trento, Italy
{kandeepan,tinku}@ieee.org
[2] Macquarie University, Sydney, Australia
samr@science.mq.edu.au

Abstract. In this paper we study the energy efficiency of hybrid high altitude platform (HAP) and terrestrial communication systems in the uplink. The applications of HAPs in the recent years have gained significant interest especially for telecommunications. A cooperative relay based communication strategy on the ground with a single source-relay pair is considered for the hybrid HAP-terrestrial system. We show that cooperation on ground between the terrestrial terminals could improve the energy efficiency in the uplink depending on the temporal behavior of the channels. Thus by having cognitive context aware capabilities in the ground terminals one could exploit the spatial domain to improve the energy efficiency. The energy efficiency can be further improved by means of proper power allocation between the terrestrial source and relay nodes. We consider the decode and forward based cooperative system for our study with a constraint on the overall bit error probability to achieve a predefined quality of service. Results show that considerable gain in the energy can be attained by exploiting the spatial domain by means of cooperation.

Keywords: Hybrid HAP terrestrial systems, high altitude platform, energy efficiency, cooperative communications.

1 Introduction

The rapid growth of bandwidth-hungry telecommunication applications has pushed wireless infrastructure providers and network operators under continuous pressure to exploit the limit of radio spectrum as efficiently as possible. Further, the costs related to rolling out and operation of broadband wireless networks in emerging and Greenfield environments have increased considerably. Notwithstanding, new requirements for flexible network access have emerged within the telecommunications community, spurred by the vision of optimal connectivity, anywhere, anytime. In this context, high-altitude platforms (HAPs) are increasingly being cited as having an important role to play in future systems and applications capable of providing efficient broadband wireless access at lower costs [12]. HAPs as new solutions for delivering wireless broadband, have recently been proposed for the provision of fixed, mobile and broadcast services in the stratosphere at an altitude of 17 km to 22 km [13],[14], [15], [16].

G. Giambene and C. Sacchi (Eds.): PSATS 2011, LNICST 71, pp. 151–164, 2011.
© Institute for Computer Sciences, Social Informatics and Telecommunications Engineering 2011

The HAP systems have features of both terrestrial and satellite communications and thus shares the advantages of both the communication systems. The International Telecommunication Union (ITU) suggests that footprints larger than 100 km radius can be served from HAPs [18]. This would technically allow a single HAP station to effectively replace several terrestrial base stations with a cost-effective and time-efficient deployment and ease of integration to existing networks. The ever increasing cost of infrastructure deployment and the intricacies in site acquisitions and renting have stepped up the complexity in the deployment of next generation broadband wireless networks, particularly in emerging markets. In these areas, HAP systems will be an effective solution to provide the necessary network infrastructure at lower CAPEX and OPEX costs for incumbent and new entrant network operators. Moreover, HAP systems could also be exploited in applications of environment and disaster surveillance to support long-period data transmission and mobile deployments apart from such applications like remote sensing, radio monitoring, weather monitoring etc. These additional applications, along with the savings in physical infrastructure costs could even out the operational expenditures related to terrestrial network deployments. In order to aid the eventual deployment of HAPs, the ITU has allocated spectrum in the 3G band for HAPs, as well as in the mm-wave bands for broadband services at around 48 GHz worldwide and 31/28 GHz for certain Asian countries [17].

The use of HAPs has been proposed in many communication applications [18]. 3G and 4G services are recently considered as principal application of HAPs [17], which offer the possibility of being part of the current 3G networks and services, and also as complementary access solution for 4G networks and its mobile and multimedia services. Multimedia broadcast and multicast services (MBMS) also can be provided by the HAP layer of 3G/4G in order to achieve higher system capacity and lower costs. 3G base stations, integrated on the HAP systems, with wide beamwidth antennas could provide service over large sparsely populated areas. In addition to this, in urban areas, where a higher capacity is needed, smaller cells can be deployed with integrated directional antennas (antenna arrays). All this include exceptional benefits such as offering coverage in a large area, direct propagation paths without obstacles and elimination of the expensive resources spent in ground station installations, maintenance and wire installations.

In the case of a HAP system planned to operate on a given coverage area integrated with a base network already operating as part of the terrestrial network, such hybrid HAP-terrestrial system can be partially or completely deployed as a complementary network, for instance, depending on higher traffic requirements or power/equipment related failures. The notion of HAP platform components deployed as a substitution network will provide for improved balance between the complexity and costs related to operating a large fleet of HAP platforms and the ever increasing traffic requirements in future broadband communication networks. Such consideration will also ease the heavier payload demands and thereby lower the power consumption on the platform improving the overall energy efficiency of the system. Moreover, as the HAPs platform could potentially be linked to multiple ground stations and relay nodes, coordination and cooperation between these stations is an important consideration for reducing the communication energy costs within the system.

In this paper, we consider a cooperative networking scenario based on decode and forward mechanism on the ground between the terrestrial radios for the uplink hybrid HAP-terrestrial system. The terrestrial source node with a bad uplink channel will cooperate with a neighboring terrestrial node in order to transmit its data to the HAP station. The terrestrial node acting as the relay node poses a good uplink channel with the HAP station. For a bit error rate (BER) constraint requirement, we show that depending on the channel gains cooperation can lead to energy efficient communications. We present some numerical results to show the improvement in the energy efficiency. Moreover, we provide an optimization strategy to select the source relay transmission power levels for the cooperative relay link in order to minimize the overall energy consumption due to transmission. The proposed strategy and the related study not only improves the energy efficiency of the overall communication system but also provides seamless connectivity to the terrestrial user when the direct uplink channel is very poor. Having these two potential benefits in mind we present our study in the rest of the paper.

The rest of the paper is organized as follows. In Section 2 we present the hybrid HAP-Terrestrial network model and the corresponding channel models. In Section 3 we present the decode and forward cooperative system followed by the energy efficiency study in Section 4. The optimum power allocation for energy efficiency is presented Section 5. In Section 6 we present numerical results with some concluding remarks in Section 7.

2   The Hybrid HAP-Terrestrial Network Model

In this section, we present the integrated HAP - terrestrial radio network model. We consider an integrated HAP and multiple terrestrial radio system as depicted in Figure-1. The aerial HAP platform is considered to act as a macro cell surrounded by cells served by terrestrial base stations, including the possibility of several ground relay stations. The integrated terrestrial-HAP system enables the network operator to serve high-mobility users characterized with low bit rates and at the same time, serving users with high bit rates with the terrestrial stations. The HAP network will be connected to the terrestrial networks through a series of ground gateway stations. Though, we limit the study to a single HAP system (with a single aerially deployed HAP station) in this paper, it could be conveniently extended for a multiple HAP system (multiple interconnected HAP stations) linked with multiple terrestrial radios. The scenario presented here is motivated by the work conducted in the C2POWER project for power efficient cooperation in multi-radio platforms [21]. In our network model we also assume context awareness meaning that the channel, communications and networking parameters corresponding to the participating nodes are known to each other.

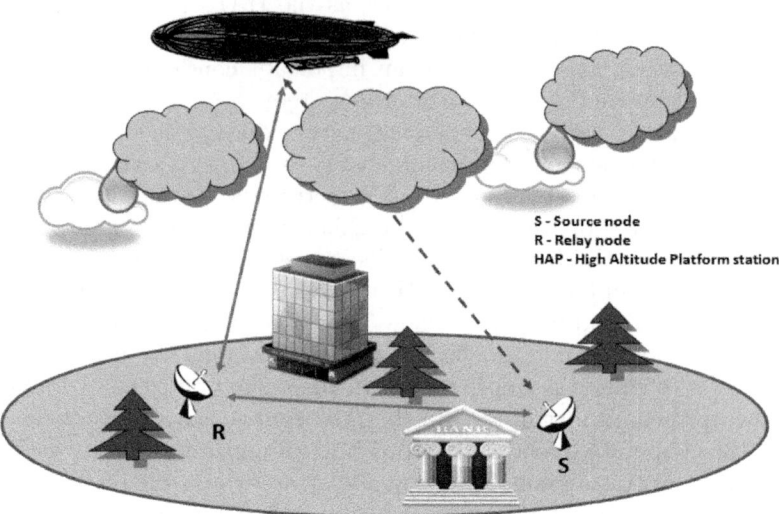

Fig. 1. The hybrid HAP-Terrestrial communication system with cooperative relay link

The cognition can be provided by means of learning or by means of signalling mechanisms, the treatment of such however are not addressed in this paper but simply assume that context aware capabilities are present (which is infact a valid assumption considering the next generation terrestrial systems [21]). The terrestrial nodes are also aware of the neighboring nodes by means of spectrum sensing and cognitive learning capabilities embedded in the radios [9], [10], [11].

2.1   The Channel Models

Theoretical channel models have been proposed in the literature for the uplink terrestrial to HAP communication channel [7]. In our work for the convenience of theoretically studying the energy efficiency we consider a Rayleigh channel model with free space pathloss for the uplink communications having different small scale channel gains with spatial separations. In other words the uplink channel gains for the S $\rightarrow$ HAP and the R $\rightarrow$ HAP can vary due to its spatial separation. The Rayleigh channel model for the uplink clearly describes that no line of sight communication exist due to the position of the mobile terrestrial nodes (in an urban environment) and the stratospheric rain fading conditions. It is also possible to extend our analytical work presented in this paper to a Ricean channel for the uplink however we consider the Rayleigh channel as an example for simplicity.

The channel model between the terrestrial radios is also considered to be a Rayleigh fading channel with a pathloss exponent of α_2, as we present subsequently. Figure-2 depicts the channel models and the related parameters for the hybrid Terrestrial-HAP system, where h_i with $i \in \{1, 2, 3\}$ are the small scale Rayleigh fading channel gains, d_i are the T-R distances, α_i are the pathloss

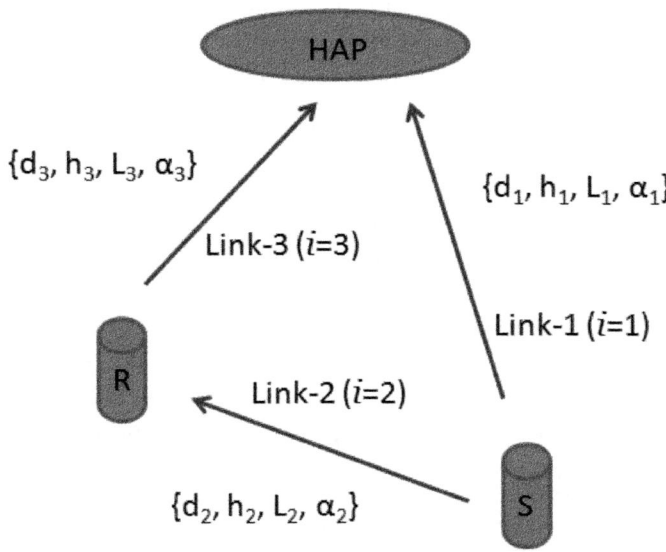

Fig. 2. The channel parameters for the hybrid HAP-Terrestrial system

exponents, and L_i are the mean pathloss for the respective links. The pathloss is given by,

$$L_i(d_i) = L_i(d_0) \left(\frac{d_i}{d_0} \right)^{\alpha_i} \tag{1}$$

where, $L_i(d_0)$ is the pathloss at a reference distance d_0 which is given by the free-space pathloss [8] $L_i(d_0) = (4\pi f_i d_0 / c)^2$, where f_i is the carrier frequency and $c = 3 \times 10^8$ is the speed of light. Moreover, we also define the average power gains of the channels given by $\gamma_i = \frac{1}{t} \int_0^t h_i^2(t) dt$. With the assumption of context awareness, the above mentioned parameters such as the T-R separations d_i, the channel power gains h_i, the pathloss exponent α_i are all known to the terrestrial nodes. Note that since the uplink communication has a free space pathloss model, from the terrestrial node to the HAP, we have $\alpha_1 = \alpha_3 = 2$, where as α_2 would vary with time and the positions of the terrestrial nodes. As mentioned before, the positions and the channel parameters are all assumed to be known to the nodes S and R by means of cognitive learning features in the respective radios.

3   Decode and Forward Cooperative Relay Communications

For the cooperative network model in Figure-2 the received signals for the communications from $S \to HAP$, $S \to R$, and $R \to HAP$ (i.e. for $i = 1, 2$ and 3 respectively) can be expressed in the form of,

$$r_i(t) = \frac{1}{L_i(d_i)} h_i(t) s_i(t) + v_i(t) \tag{2}$$

where, $s_i(t)$ are the transmitted signal components and $v_i(t)$ are the additive Gaussian noise components at the receivers with a double-sided power spectral density of $N_0(i)/2$. Moreover, we consider a BPSK communication system for all the links in our example corresponding to the signal components $s_i(t)$ in equation (2). The bit error probability under Rayleigh fading with additive Gaussian noise channels for all the links is given by [8],

$$\Pi_i = 0.5 \left(1 - \sqrt{\frac{\Gamma_i}{(1 + \Gamma_i)}} \right) \tag{3}$$

where $\Gamma_i = E_b(i)\gamma_i/N_0(i)$ is the average received SNR for the ith link, $E_b(i)$ is the received bit energy given by $E_b(i) = P_t(i)G_i^t G_i^r/[\Delta_i L_i(d_i)]$, $P_t(i)$ is the transmitted signal power, G_i^t and G_i^r are the transmit and receive antenna gains, and Δ_i(bits/s) is the data rate for link i. Note that according to our notations we have $G_1^t = G_2^t, G_2^r = G_3^t, G_1^r = G_3^r$.

Based on the decode and forward strategy for the cooperative relay link the relay node will receive the signal $r_2(t)$ from the source node, detect it, and transmit it to the HAP. The HAP will then receive the signal $r_3(t)$ from the terrestrial relay node and decode it. On the other hand in the direct link transmission the HAP will directly receive the signal $r_1(t)$ from the terrestrial source node. In the subsequent sections we study the energy efficient way for the source node to communicate with the HAP, by choosing between the relay and the direct links, depending on the communication channel conditions to achieve the same bit error probability ξ.

3.1   Overall Bit Error Probability

In this section we present the overall bit error probability for the relay link based on the decode and forward cooperative strategy. If Π_2 and Π_3 are the bit error probabilities respectively of the concatenated two-hop relay link with decode and forward protocol, then the overall bit error probability of the relay link is given by $\overline{\Pi} = \Pi_2(1 - \Pi_3) + \Pi_3(1 - \Pi_2)$. By expanding this expression we have,

$$\overline{\Pi} = \Pi_2 + \Pi_3 - 2\Pi_2\Pi_3 \tag{4}$$

and since the probability values are less than one (small), the last term of the above expression can be ignored, giving us a final expression of,

$$\overline{\Pi} = \Pi_2 + \Pi_3 \tag{5}$$

Then using (3) in (5), we have

$$\overline{\Pi} = 1 - 0.5 \left[\left(\sqrt{\frac{\Gamma_2}{(1 + \Gamma_2)}} \right) + \left(\sqrt{\frac{\Gamma_3}{(1 + \Gamma_3)}} \right) \right] \tag{6}$$

In the following sections we will analyze the energy efficiency associated with the cooperative relay link for a given BER constraint of $\overline{\Pi} = \Pi_1 = \xi$.

4   Energy Efficient Communications with BER Constraint

In this section we study the relative energy efficiency of the relay link and the direct link. For the direct communication link from $S \rightarrow HAP$, with a given BER constraint of ξ, and considering the BER expression in (3), the transmit power requirement $P_t(1)$ at the source node S can be computed as,

$$P_t(1) = \frac{\Delta_1 L_1(d_1) N_0(1)(1 - 2\xi)^2}{G_1^t G_1^r \gamma_1 [1 - (1 - 2\xi)^2]} \tag{7}$$

Likewise, for the relay link from $S \rightarrow R \rightarrow HAP$ the transmit power pair $\{P_t(2), P_t(3)\}$ requirement at the source node and the relay node respectively for the same BER constraint ξ, using the BER expression (6), can be computed as,

$$P_t(3) = \frac{\lambda^2}{G_3^t G_3^r \gamma_3 (1 - \lambda^2)} L_3(d_3) N_0(3) \Delta_3 \tag{8}$$

where λ is a function of $P_t(2)$, and is given by,

$$\lambda = \sqrt{\frac{K_1 P_t(2)}{1 + K_1 P_t(2)}} - 2\xi \tag{9}$$

with $K_1 = \frac{G_2^t G_2^r \gamma_2}{L_2(d_2) N_0(2) \Delta_2}$. From equations (7), (8) and (9) we have the transmit power requirements at the source and relay nodes for the direct and relay transmissions for a given bit error rate constraint. Now, let us define an energy efficiency factor β to compare the energy efficiency between the direct and the relay links, given by,

$$\beta \triangleq \frac{[P_t(2) + P_t(3)]}{P_t(1)} \tag{10}$$

Then, based on the definition of β, one may conclude that direct link is more efficient than the relay link if $\beta > 1$ (or $\beta(\text{dB}) > 0$), or the relay link is more efficient than the direct link if $\beta < 1$ (or $\beta(\text{dB}) < 0$), or both relay and direct links are equally energy efficient if $\beta = 1$ (or $\beta(\text{dB}) = 0$) but considering the overhead due to cooperation the direct link is preferred when $\beta = 1$. Note that in our work we ignore the energy consumptions due to processing at the nodes. As we observe from the above expressions it is very clear that the value of β depends on several parameters such as the pathloss exponent α_2, terrestrial T-R separation d_2, and the small scale channel gains γ_i, etc. With the context aware capabilities at the terrestrial nodes by means of cognitive functionalities the source node could adopt its transmissions between the direct and relay links for improved energy efficiency. In Section 6 we present numerical results for the analysis presented in this section.

5   Optimum Source-Relay Power Allocation for Maximum Energy Efficiency

Suppose the cooperative relay link exhibits better efficiency compared to the direct uplink, in other words when $\beta < 1$ (or $\beta(dB) < 0$), then the source and the relay node transmit power levels need to be carefully chosen in order to minimize the total energy utilization for the transmission. It is observed that the efficiency factor β exhibits a convex property which could be then minimized to have better energy efficiency. From the numerical results presented in the next section we observe the convex property of β clearly. Suppose $X = P_t(2) + P_t(3)$, then the optimal source-relay transmit power levels are given by,

$$\{\hat{P}_t(2), \hat{P}_t(3)\} = \arg\min_X\{\beta\} \tag{11}$$

or equivalently, given by,

$$\{\hat{P}_t(2), \hat{P}_t(3)\} = \arg\min_X\{P_t(2) + P_t(3)\} \tag{12}$$

The optimum transmit power values for the source and relay nodes are then given by taking the first derivative of X with respect to $P_t(2)$ and equating it to zero. The first derivative therefore is given by,

$$\frac{dX}{dP_t(2)} = 1 + \frac{dP_t(3)}{dP_t(2)} \tag{13}$$

Then, by using equations (8) and (9), we arrive at the following expression for $\frac{dX}{dP_t(2)}$, given by,

$$\frac{dX}{dP_t(2)} = 1 + \frac{K_0 K_1^{\frac{1}{2}}\lambda}{P_t^{\frac{1}{2}}(2)(1 + K_1 P_t^{\frac{3}{2}}(2))(1 - \lambda^2)^2} \tag{14}$$

where, $K_0 = \frac{L_3(d_3)N_0(3)\Delta_3}{G_3^t G_3^r \gamma_3}$. The optimum transmit power at the source node is then given by solving the following equation,

$$\left(\frac{dX}{dP_t(2)}\right)_{P_t(2)=\hat{P}_t(2)} = 1 + \frac{K_0 K_1^{\frac{1}{2}}\lambda}{\hat{P}_t^{\frac{1}{2}}(2)(1 + K_1 \hat{P}_t^{\frac{3}{2}}(2))(1 - \lambda^2)^2} = 0 \tag{15}$$

Solving the above equation for $P_t(2)$ is not trivial and hence we use the Gradient Descent method to find $\hat{P}_t(2)$. The gradient descent algorithm for some $\epsilon > 0$ to compute the optimum source power is then given by,

$$\hat{P}_{t_{n+1}}(2) = \hat{P}_{t_n}(2) - \epsilon\left(1 + \frac{K_0 K_1^{\frac{1}{2}}\lambda_n}{\hat{P}_{t_n}^{\frac{1}{2}}(2)(1 + K_1 \hat{P}_{t_n}^{\frac{3}{2}}(2))(1 - \lambda_n^2)^2}\right) \tag{16}$$

where, $\lambda_n = \sqrt{\frac{K_1 P_{t_n}(2)}{1 + K_1 P_{t_n}(2)}} - 2\xi$. The above iteration converges to $\hat{P}_{t_{n+1}}(2) \rightarrow \hat{P}_t(2)$ for sufficiently large n depending on the value of ϵ. The corresponding optimum relay transmit power can then be computed by using equations (8) and (9), which is given by,

$$\hat{P}_t(3) = \frac{\hat{\lambda}^2}{G_3^t G_3^r \gamma_3(1 - \hat{\lambda}^2)}L_3(d_3)N_0(3)\Delta_3 \tag{17}$$

where $\hat{\lambda}$ is given by,

$$\hat{\lambda} = \sqrt{\frac{K_1 \hat{P}_t(2)}{1 + K_1 \hat{P}_t(2)}} - 2\xi \tag{18}$$

Therefor, by using equations (16) and (17), one could compute the optimum source-relay transmit power levels for minimum energy consumption in the co-operative relay link.

6   Numerical Results

In this section we present numerical results to analyze the relative energy efficiency of cooperative relay link with the direct link. First, we consider the analytical results obtained in Section 4. The following simulation parameters are considered for our analysis;

Link-1: From S to HAP (Direct Uplink)
$\Delta_1 = 100e3(\text{bps}), \gamma_1 = 5e-1, d_1 = 20.0025e3(\text{m}), f_1 = 3.5e9(\text{Hz}), \alpha_1 = 2,$
$G_1^t = 30(\text{dB}), G_1^r = 25(\text{dB}).$

Link-2: From S to R (Relay Terrestrial Link)
$\Delta_2 = 100e3(\text{bps}), \gamma_2 = 20, d_2 = 2.5e3(\text{m}), f_2 = 3.5e9(\text{Hz}), \alpha_2 = 2.3,$
$G_2^t = 30(\text{dB}), G_2^r = 30(\text{dB}).$

Link-3: From R to HAP (Relay Uplink)
$\Delta_3 = 100e3(\text{bps}), \gamma_3 = 1, d_3 = 20.0035e3(\text{m}), f_3 = 3.5e9(\text{Hz}), \alpha_3 = 2,$
$G_3^t = 30(\text{dB}), G_3^r = 25(\text{dB}).$

Common parameters
$\xi = 1e^{-3}, N_0(i) = -140(\text{dBw/Hz}).$

Note that some of the parameters above are varied in the results presented subsequently, in this case the varied parameters are explicitly mentioned appropriately.

Figure-3 depicts the power requirements for the direct and relay links for the above given parameters. In the figure we have plotted the required direct uplink power for a BER of $1e^{-3}$ together with the terrestrial relay-node power in dBw, the total relay link transmit power (sum of link-2 and link-3) in dBw, and the energy efficiency factor $\beta(\text{dB})$. From the figure we observe that when the total relay-link power exceeds the direct uplink transmit power then the direct transmission becomes energy efficient and when the total relay-link power is lower than the direct uplink transmit power then the relay-link transmission becomes energy efficient. This is also observed by looking at the β curve in Figure-3, that is the relay link becomes efficient when $\beta(\text{dB}) < 0$ and the direct uplink becomes efficient when $\beta(\text{dB}) > 0$. Furthermore, we observe that the total transmission power for the relay link (or equivalently the energy efficiency factor

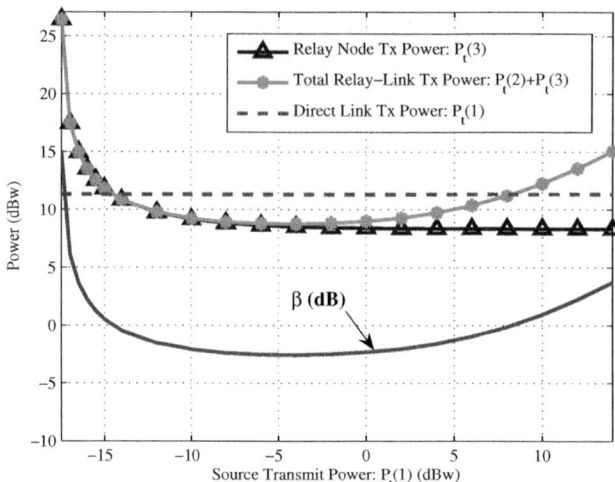

Fig. 3. Comparing the direct and relay links for energy efficiency, for a BER constraint of $\xi = 1e^{-3}$

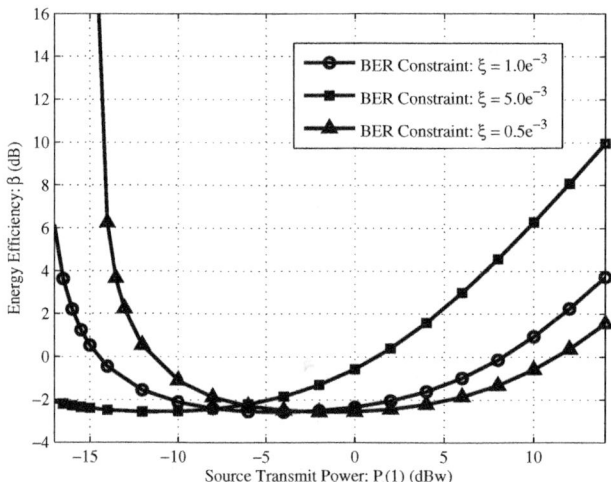

Fig. 4. The energy efficiency factor comparing the direct and relay links, for varying BER constraints

β) exhibits a convex property in the domain of the source transmit power $P_t(1)$. This is a desirable feature to optimize the power consumption by choosing the appropriate source-relay transmit power levels if the relay link is chosen to be the energy efficient link, as discussed in Section 5.

Figure-4, Figure-5 and Figure-6 show the variations on the energy efficiency factor β for varying BER constraint ξ, the direct uplink fading gain γ_1, and the terrestrial channel gain γ_2. As observed in Figure-4, when the BER constraint

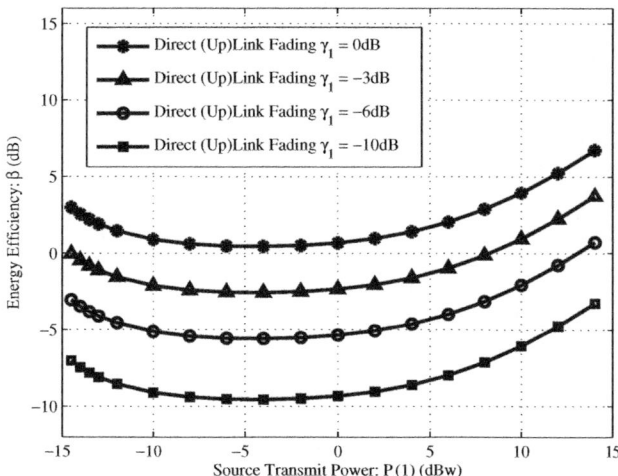

Fig. 5. The energy efficiency factor comparing the direct and relay links, for varying direct up-link channel gain γ_1

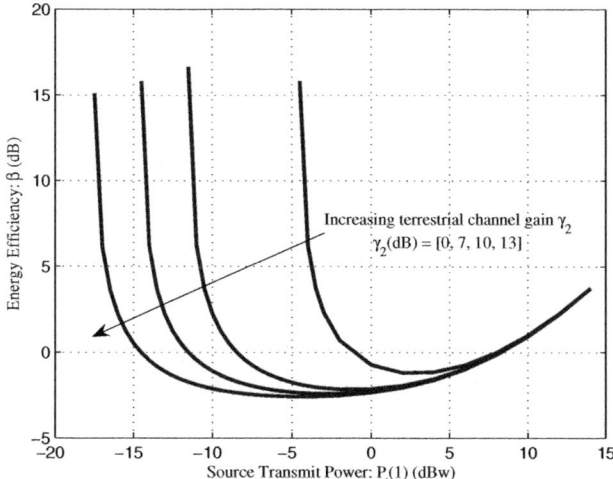

Fig. 6. The energy efficiency factor comparing the direct and relay links, for varying terrestrial link channel gain γ_2

ξ reduces, the source transmit power requirement also reduces for the values of $\beta(\text{dB}) < 0$. In Figure-5, we observe that the range of source transmit power levels such that $\beta(\text{dB}) < 0$ increases when the direct uplink gain reduces (i.e. when fading gets severe), this is because when γ_2 reduces the cooperative relay link becomes more and more energy efficient compared to the direct uplink. In Figure-6, we observe that the source transmit power requirement reduces for the case of $\beta(\text{dB}) < 0$ when the terrestrial channel gain γ_2 increases, as expected.

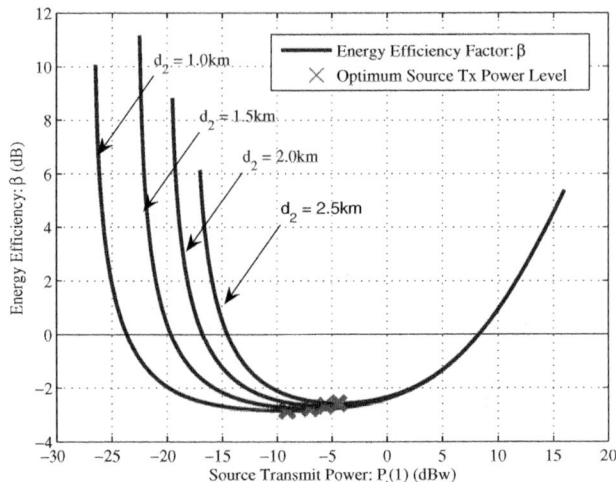

Fig. 7. Optimum source power for the cooperative relay link, for varying terrestrial T-R distance d_2

Finally, we present the results for the optimum power allocation between the source and the relay nodes when $\beta(\text{dB}) < 0$. Figure-7 depicts the variations in β for varying T-R separation d_2 between the terrestrial nodes. In the figure we also depict the optimum source power selection for minimizing the total energy consumption in the cooperative relay link as described in Section-5. As observed in the figure, the optimum source power level increases with increasing d_2 as expected for the same channel condition γ_2.

7 Conclusion

In this paper, we consider energy efficiency of a cooperative HAP and terrestrial communication system. For Rayleigh fading channels, we analyze the energy efficiency for the cooperative relay link and the direct uplink. Our analysis shows that for poor direct uplink channels the cooperative relay link provides better energy efficiency. Numerical results were presented to show the comparative energy efficiency between the cooperative relay link and the direct uplink for various network and channel parameters. We had also presented an optimization framework to select the optimum source and relay transmit power levels for the cooperative relay link based on a Gradient Descent iterative method for the Rayleigh fading channels. Our future work in this area include similar analysis with Ricean and other fading channels for the uplink and also consider cooperation between multiple HAP stations communicating and coordinating through inter-platform links.

Acknowledgment. The authors would like to thank Dr. Francesco Rossetto and Dr. Sandro Scalise from the German Aerospace Centre (DLR), Germany for their valuable comments on the manuscript.

References

1. Djuknic, G.M., Freidenfelds, J., Okunev, Y.: Estabilishing wireless communications services via high altitude platforms: A concept whose time has come? IEEE Communications Magazine, 128–135 (September 1997)
2. AragonZavala, A., Cuevas-Ruz, J.L., Delgado-Penin, J.A.: High-Altitude Platforms for Wireless Communications. John Wiley and Sons, Chichester (2008)
3. Karapantazis, S., Pavlidou, F.: Broadband Communications Via High-Altitude Platforms: A Survey. IEEE Communications Surveys, First Quarter 7(1) (2005)
4. Grace, D., et al.: LMDS from High Altitude Aeronautical Platforms. In: Proceedings of Global Telecommun. Conf. 1999 (GLOBECOM 1999), Rio de Janeiro, Brazil, December 5-9, vol. 5, pp. 2625–2629 (1999)
5. Grace, D., Tozer, T.C., Daly, N.E.: Communications from High Altitude Platforms - A Complementary or Disruptive Technology? IEE Colloquium New Access Network Technologies (October 31, 2000)
6. Grace, D., et al.: Providing Multimedia Communications Services from High Altitude Platforms. Int'l. J. Satellite Commun., 19(6), 559–580 (2001)
7. Dovis, F., Fantini, R., Mondin, M., Savi, P.: Small-Scale Fading for High-Altitude Platform (HAP) Propagation Channels. IEEE journals on Selected Areas in Communications 20(3), 641–647 (2002)
8. Rappaport, T.S.: Wireless Communications: Principles and Practice. Prentice-Hall, Englewood Cliffs (1996)
9. Mitola, J., Maguire, G.Q.: Cognitive Radio: Making Software Radios More Personal. IEEE Personal Communications 6(4), 13–18 (1999)
10. Haykin, S.: Cognitive Radio: Brain-Empowered Wireless Communication. IEEE Journal on Selected Areas in Communications 23(2), 201–220 (2005)
11. Kandeepan, S., De Nardis, L., Di Benedetto, M.G., Guidotti, A., Corazza, G.E.: Cognitive Satellite Terrestrial Radios. In: IEEE Globecom 2010, Florida (2010)
12. Karapantazis, S., Pavlidou, F.-N.: Broadband Communications via High Altitude Platforms (HAPs) - A Survey. IEEE Commun. Surveys and Tutorials, 1st Qtr (2005), http://www.comsoc.org/livepubs/surveys
13. Tozer, T.C., Grace, D.: High-altitude platforms for wireless communications. IEE Electronics and Communications Engineering Journal 13, 127–137 (2001)
14. Grace, D., Thornton, J., Konefal, T., Spillard, C., Tozer, T.C.: Broadband communications from high altitude platforms - The HeliNet solution. In: Wireless Personal Mobile Conference, Aalborg, Denmark (2001)
15. Mohammed, A., Arnon, S., Grace, D., Mondin, M., Miura, R.: Advanced communications techniques and applications for high-altitude platforms. Editorial for a special issue in EURASIP Journal on Wireless Communications and Networking (2008)
16. Mohammed, A., Yang, Z.: Broadband communications and applications from High Altitude Platforms. International Journal of Recent Trends in Engineering 1(3) (May 2009)

17. Fourth Generation Wireless Communications from High Altitude Platforms, Vodafone Market Report (May 2010)
18. Gerla, M., et al.: Communications via High Altitude Platforms: Technologies and Trials. International Journal of Wireless Information Networks 13(1) (January 2006)
19. Lähdekorpi, P., Isotalo, T., Kylä-Liuhala, K., Lempiäinen, J.: Replacing Terrestrial UMTS Coverage by HAP in Disaster Scenarios. In: European Wireless Conference (April 2010)
20. Araniti, G., Iera, A., Molinaro, A.: The Role of HAPs in Supporting Multimedia Broadcast and Multicast Services in Terrestrial-Satellite Integrated Systems. Wireless Pers. Commun., 32(3-4), 195–213 (2005)
21. URL of C2POWER Project, http://www.ict-c2power.eu
22. Holis, J., Pechac, P.: Penetration Loss Measurement and Modeling for HAP Mobile Systems in Urban Environment. EURASIP Journal on Wireless Communications and Networking 2008 (2008)

Eutelsat Experiences on Remote Terminals for High Speed Mobility in Ku and Ka Band

Eros Feltrin, Jean-Marie Freixe, and Elisabeth Weller

Eutelsat SA – 70 rue Balard, 75015 Paris, France
{efeltrin,jfreixe,eweller}@eutelsat.fr

Abstract. Eutelsat plays key role in the provision of mobile services over geosynchronous satellites. Starting with narrowband services for land transports, through broadband maritime service until the most recent IP and multimedia services on board high speed means of transport (trains and airplanes), Eutelsat is present in this market for several years, as services provider or just supplying capacity. This paper gives a technical overview, from the Eutelsat's point of view, of the evolution the mobility via satellite is having. The references are to the existing services in Ku and those planned to be deployed in Ka band.

Keywords: earth stations, antenna pattern, waveform, VSAT, handover.

1 Introduction

For several years the Eutelsat's services portfolio includes mobility, which is today improving from narrow-band towards broadband IP based applications. Furthermore, the new commercial satellites working in Ka band, as the Eutelsat's Ka-Sat, and the introduction of more efficient waveforms leaves hope for a further development of the mobility-via-satellite business.

Nevertheless, the high level of innovation introduced by these new generation satellites makes the exploitation of the Ka band a real technological challenge. In fact, if on the one hand the multi-spot coverage (see Fig.1) permits to dramatically increase the global available capacity and reduce the customer economical charges, on the other hand the ground segment attains a level of complexity never reached before in others satellites services.

Such a complexity concerns the fact that the entire satellite service is provided from a ground infrastructure extended to the Mediterranean Basin comprising eight gateways linked to an optical fibre ring and managing the entire IP services. In the case of mobile services these gateways must be able to manage the spot-beam handover (HO). Additionally, a high level of complexity concerns also the end user mobile terminals that must be able to execute automatically the different steps of the HO process, similarly to mobile terrestrial networks based on standards as the WiMAX [1]. The effect of the HO process on the perceived quality of the service must be negligible.

G. Giambene and C. Sacchi (Eds.): PSATS 2011, LNICST 71, pp. 165–174, 2011.
© Institute for Computer Sciences, Social Informatics and Telecommunications Engineering 2011

Furthermore, at the end user side, one of the most crucial points remains the implementation of reliable antennas capable to apply the polarization or frequency switching required by the HO and matching the constraints coming from the different mobile environments. If in Ku band, the target reliability and performances have been achieved for maritime applications and high speed mobile applications – mainly avionic and railway applications [1 - 3] – it is not yet the case for Ka band which still requires investment for development.

This paper introduces a brief historic of the experiences done by Eutelsat with the Ku band mobile broadband services, hinting at some of the major problems faced in the development phase. Then, the evolutions attended with the capacity coming with Ka-Sat (launched on 26th December 2010 and scheduled to be operational on Q2 2011) are described: the new technologies available for the production of very low profile antennas, the integration of the mobile antennas as elements of a global ground infrastructure, the complexity introduced by the multi-spot coverage in all the layers of a satellite communication system, including of course the radiofrequency devices.

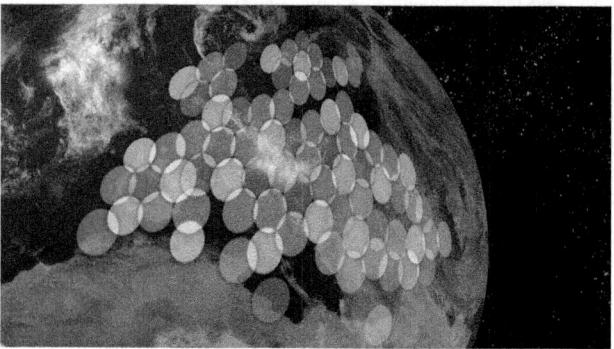

Fig. 1. Example of multi spot coverage in Ka band

2   VSAT Services in Ku Band

2.1 An Overview

Even if the terms very small aperture terminals (VSAT) has been introduced in order to indicate relatively little transmitting and receiving earth stations, today "VSAT" is frequently associated to IP services which end user ground segment is small and cheap. The increased power of the most recent satellite and the more efficient modulation and channel access techniques contributed in about one decade to reduce the cost and size of the terminals, at the point that also the meaning of the term "small" has been reviewed: if on 2009 a Ku band VSAT was generally a 1.8m antenna, today the most deployed dishes are around 90 cm, which is by now considered too big and anti-aesthetic. Additionally, the spreading of VSATs into much extended networks targeting at markets differentiated from the mere rural segment, led to even more dramatic reduction in the price.

Soon, the 21st Century saw the deployment of mobile broadband service in addition to the fix ones. If the narrow-band mobility was already a reality in Ku band, the Euteltracs service being one of the most popular services (with its 36000 remotes deployed throughout Europe), the need for broadband and voice services on-board vessels soon became evident. Even if the Ku band did not propose a world-wide coverage (which on the contrary is proposed by Inmarsat), such a limitation did not impacted the deployment of services in maritime fleets with vessels placed into restricted areas, as the Mediterranean basin, very well covered by geosynchronous satellites.

The civil maritime antennas inherited the technology, as usual, from the military world. Therefore, several manufacturers invested in the development of theirs own maritime stabilized steering antennas. The size of the dishes evolved in the same way as for the fix VSAT: today the catalogs for maritime antennas include also 60cm dishes.

Further progresses, in both the antenna and modems design, have been done in order to extend the satellite broadband services on board trains, airplanes and land transport.

Fig. 2. The low profile satellite antenna on board the French high speed train

Eutelsat participated to the initial projects for the provision of satellite connectivity on board aircrafts (the *Connection by Boeing* service run in Europe on Eutelsat capacity). But the most recent conquest is the deployment of a network providing multimedia services and Internet access on board the high speed trains of the French East line for the railway operator SNCF: the project was initially known as *Connexion TGV* and today has been renamed as *Box TGV*. This kind of applications introduces heavy constraints in terms of robustness under mechanical solicitations (mainly shocks and vibrations), occupied volume and aerodynamic impact. The equipment design must keep into account the applicable standards and the used materials must respect the safety normative.

For *Connexion TGV*, spread spectrum channel access techniques are used in oder to reduce the off-axis interferences generally associated to small antennas. The low profile steering antennas adopted have been designed and manufactured by *OrTeS*, a joint venture between *Teleinformatica e Sistemi* and, *Orbit* who cooperated with Eutelsat through all the phases of the project: from the approval-of-concept to the deployment [1-3]. Fig. 2 shows an high speed train equipped with a OrTeS antenna. Both modems and antennas have been approved for the railway usage, being compliant with the applicable railway standards.

A similar system has been adopted on board business jets for the provision of IP connectivity. In this case the antennas adopted have further reduced size: the application of spreading techniques is then mandatory.

2.2 Return on Experiences

Connexion TGV permitted to all partners involved in the project to mature an important technical background. First of all, from the point of view of the equipment: a fundamental rule is that even if a device complies with the railway standard ,this does not mean that it will work well on board the train because the countless unknown variables. The big headache came from the electrical power supply: electrical buffers and stabilizers are mandatory, but not sufficient, for the mitigation of the voltage interruptions. Particular care must be dedicated to the electrical grounding and, more in general, to the connectivity, which impacts directly the reliability of the installation.

After almost two years from the first two installations, we have the first feedbacks concerning the reliability of the solution. If on one hand the *mean time between failures* (MTBF) required by the customer is respected, unfortunately the technology is suffering a little-bit its youngness, even if on the over 50 terminals installed only a couple of them have been replaced in short time due to manufacturing defects. In some cases, malfunctioning appears and disappears without apparent reasons and corrective actions are difficult to be undertaken without an efficient logs reporting. This is becoming today a big deal: the devices adopted on board the means of transport are so complex that a detailed monitoring and alarms reporting is fundamental. An efficient reporting would help in debugging the failures but also in preventing major problems. On the other hand, there is a risk that such a detailed errors reporting became so articulated that it would be exploitable only by adepts.

Another important point concerns the radome. No environment can prove better then the railway one how important is the mechanical robustness of the radome. This element takes the important role of protecting from impacts the underneath elements but, in case of very hard shocks causing the detachment of some elements of the antenna, the radome prevent the ejection of broken pieces that could be transformed into real projectiles.

Fig. 3 and Fig. 4 clarify this concept: the first shows the deformation of the radome after a very violent impact. Fortunately the fragments of damaged reflector have not been ejected because the radome was not pierced assuring the safety for the antenna surroundings. Fig. 4 shows the spectacular effect of a rail ballast stone that hammered into a radome: the antenna was working perfectly but for safety reasons the radome has been replaced.

 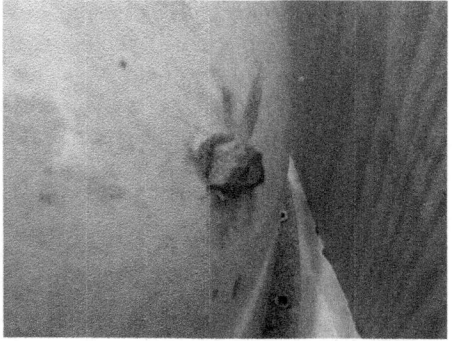

Fig. 3. A radome issued from a very violent impact   **Fig. 4.** A stone of the rail ballast hammered into a radome

These examples prove the importance of having a robust radome, which anyway must be transparent for the electromagnetic waves: R&D activities on this domain are solicited. Additionally, since the radome is the part of the antenna the most exposed to the impacts end than most subjected to replacement, a design permitting to reduce its cost is the welcome. It must be also considered that reducing the height of the antenna would permit to reduce the probability of impact with external corps: in that sense the flat antennas based on phased array techniques have an unmistakable advantage (unfortunately it is not yet proven if they are the best trade-off between performances and size).

3   Ka Band

The usage of small mobile satellite antennas has a direct consequence on the signal spectral efficiency. In fact, reduced gains impose to opt for more powerful (but less efficient) coding schemes, whereas the relatively large beam lobes increases the interferences which can be limited only inserting some kind of spectrum spreading (or modulations with lower cardinality). Then, the cost per transmitted information unit is higher.

The usage of the Ka band represents a milestone in that sense. This sentence is justified mainly by the fact that the cost of the capacity is dramatically reduced. A rough comparison can be done considering the *capacity serving a certain geographical area*. The capacity offered in one Ka-Sat spot beam with 250 km diameter is 237 MHz in both directions. The satellite covers the entire Europe with 82 spots, which corresponds to about 20 GHz per direction. A standard bent pipe Ku band satellite provides few tens 36 MHz transponders over an European coverage. Therefore, the ratio between the amount of Ka and Ku band capacities explains why the Ka band is cheaper.

Ka-Sat major drawback is in the complexity of the ground segment: in the case of mobile services the cellular coverage offered by the multi spot configuration imposes the usage of HO enabled systems, similarly to the most common terrestrial wireless communication devices: for details we refer to [3]. For the scope of this paper is enough remembering that the mobile antennas must be able to swap polarization and/or central frequency on the base of a HO command sent by the base band system, i.e. the modem. On the contrary, such a behavior is not required in terrestrial wireless where the HO does not involve physical reconfigurations in the antenna: these changes on a satellite earth station dramatically increase the hardware complexity.

3.1 Ka Band Antennas Specifications

In most part of the cases, we have already mentioned, mobile satellite transmissions require small antennas. Anyway, the adjective "small" takes several meaning and, commonly, a good trade-off among volume occupation, weight and performances is required. Anyway there are cases where the room available for the antennas is extremely reduced: it is the case of the double deckers high speed trains – where the train gauge profile admit a few centimeters margin for the installation of objects on the roof – and of the airplanes – due to evident aerodynamic reasons.

Furthermore, we cannot neglect that for several customers "small" means "good-looking": it is not unusual to find customers deciding for a small satellite dish, independently on the performances it can achieve, just because it is fine.

All these reasons led Eutelsat to issue an RFI addressed to the antenna manufacturers and asking for the feasibility of very low profile antennas in Ka band. This RFI has been based on a link budget computed for a Digital Video Broadcasting via satellite (DVB-S2) [6] forward link (FW) at a symbol rate of 30 Msym/s, LDPC code rate 2/5.

For the return channel (RC) different configurations based on the DVB with interactive channel (DVB-RCS) standard [7] have been considered, with a single reference code rate, i.e. turbo 1/3. This estimation has been done by Eutelsat with the only aim of having a reference for the identification of the performances required to the ground segment, and does not take into account the possibility of applying adaptive codes and modulations techniques in both FW and RC.

Tables 1 and 2 report the link budget results for the FW and the RC. Even if the computation has been done considering the worst case in terms of both gateway and remote terminals coverage, it proves that the main limitation comes from the *quality factor* of the antenna, i.e. the G/T ratio. Ideally this parameter should be of the order of 14 dB/K, but analysis has been done on the extremes of a physically reasonable interval of values, i.e. 11 – 14 dB/K. In fact, we are aware that the upper value of G/T can be reached with reflector antennas similar to that already used by Eutelsat on board high speed trains [2][3], but the drawback is the size (0.48×1 m) and weight that limit the usage only on board few types of trains.

Table 1. FW link budget results

	Mobile antenna (G/T= 11dB/K)	Mobile antenna (G/T= 14dB/K)
Clear sky margin	3.7 dB	5.6 dB
Yearly availbility	99.7 %	99.83 %

Table 2. RC link budget results

	256 kbit/s	512 kbit/s	1024 kbit/s
Clear sky margin		7.5 dB	
Yearly availbility		99.9%	
BW occupation	500 kHz	1 MHz	2 MHz
Eirp	~35.5dBW	~38.5dBW	~41.5dBW

The singular result of this RFI has been that into a limited cylindrical volume comprised between 800×800×250 mm and 1200×1200×80 mm, the only technology permitting to reach a G/T maximum of 13 dB/K (variable with the elevation position), is based on a phased array with hybrid positioning system (electronic steering in elevation and mechanical azimuth rotation). Obviously, the G/T would further reduce if the volume constraint was more stringent: in order to close the link, less efficient encoding schemes, even down to 1/3 or 1/4 must be envisaged.

Phased array with hybrid positioning is a well known technology already commercialized in Ku band (Cohbam and Starling give two examples). Its implementation in Ka band would only be a geometrical scaling issue. Conversely, the implementation of the circular polarization switch subsystem required for the spot-beam handover, requires a dedicated design that has not been foreseen for the Ku band antennas (where the polarization is mainly linear). Today, the solutions envisaged to enable the polarization switch feature are based on a dielectric polarizer mechanically rotated.

The link budget for the transmission from the remotes shows that the required mobile antenna *equivalent isotropic radiated power* (e.i.r.p.), given by the product $G_T \times P_T$ between the antenna gain and the transmitted power, can reach 41 dBW. Generally speaking, the e.i.r.p. is not a real issue, because it depends on the power P_T available on the radio amplifier. The real problem comes from the fact that small antennas have a small G_T and high side-lobes generating *adjacent satellite interferences* (ASI) which is stronger as more as more as P_T is high. The ASI effect is regulated by standards as better discussed in the following.

Several techniques are used to mitigate the ASI: on one side lower rate coding schemes permits to reduce the e.i.r.p. thanks to a higher protection. on the other side spreading techniques, even if it does not have a direct impact on the link budget, permit to reduce the power density of the transmitted signal. These techniques are well known: for example Eutelsat has adopted a system using a direct sequence spreading scheme with very high *spreading factors* [2] whereas DVB-RCS+M standard [7] foresees spreading based on bursts repetition. Further proprietary techniques (derived from *code division multiple access* - CDMA) have been adopted by others satellite systems on which a direct sequence spreading is combined with the *time division multiple access* (TDMA).

3.2 Interference Limitations

One of the most important points arisen during recent experiences is the compatibility of the antennas with the European ETSI [5] or the American FCC [6] and the Eutelsat standards [4] (as well the equivalent standards issued by other satellite operators) for the earth stations. The compatibility with these standards permits on the one hand to reduce the interferences generated by the emitting antennas on the adjacent satellites (through the control of the off-axis e.i.r.p. density) and on the cross polarized transponders, and on the other hand to reduce the interfering signal density received by the ground stations.

Unfortunately, there is a widespread propensity at neglecting the importance of these standards on the base of several justifications (usually not technical) and always overlooking the effects of these assumptions, in particular when these effects do not appear immediately. In fact, it is very common that a service is not apparently impacted by the choice of one or a set of dishes rather than another (frequently the choice is done only on the base of cost assumptions); but the same service are degraded when the number of remote terminals grows up, the cross transponder changes its load configuration or, more generally, when the loading of the transponders reach the full load conditions.

For the mobile antennas, such a problem is even more accentuated, because the operating conditions of the remote earth stations change with their geographical position or with the used satellite (when satellite roaming is foreseen). As example, we mention the case of a maritime system who worked (apparently) regularly in a well defined geographical area but once moved into another zone under the same satellite coverage the interferences received by the antenna (due to a unconventionally large receiving pattern) from another satellite were so high that the modem was not any more able to lock the signal.

The Eutelsat standard, formally known as EESS 502 or standard M [4], is periodically revised because, differently from the ETSI or the FCC one, it accounts also the standard operational conditions of each Eutelsat satellite. In some cases the standard accounts also the frequency coordination agreements among the different satellite operators.

Even if the standard M seems too stringent to some antenna manufacturers, it does not aim at obstructing the deployment of antennas. Frequently, Eutelsat comes to arrangements leading to relaxing the theoretical constraints in order to match the manufacturing physical limitations with an acceptable impact on the interference effects. As an example, Eutelsat assumptions for the HotBird™ 6 Ka band edge coverage (i.e. G/T=10dB/K) the same e.i.r.p. off-axis density per 40 kHz proposed by the ETSI standard[1], i.e.

$$
\begin{cases}
19 - 25\log\phi - 10\log N \ \ \text{dBW} & \text{for} \ \ 1.8° < \phi \le 7.0° \\
-2 - \log N \ \ \text{dBW} & \text{for} \ \ 7.0° < \phi \le 9.2° \\
22 - 25\log\phi - 10\log N \ \ \text{dBW} & \text{for} \ \ 9.2° < \phi \le 48° \\
-10 - 10\log N \ \text{dBW} & \text{for} \ \ \ \ \ \ \ \ \ \phi > 48°
\end{cases}
\tag{1}
$$

This mask takes into account the possibility of having an adjacent satellite with characteristics similar to HotBird™ 6.

---

[1] Without going into further details, there is a difference between the masks defined in the standard M and the EN 301 459 standard. In particular the standard M replaces the 1.8° angle with $1 \le \alpha < 2$. For the sake of simplicity we maintain here the same notation as ETSI.

Anyway, for the Ka-Sat, which edge coverage G/T=18 dB/K, assuming the hypothesis of having a similar adjacent satellite the mask should be dropped down of 8 dB (because the higher sensibility of the satellite). Since the constraints imposed by a so defined off-axis mask are physically too stringent for most part of the Ka antennas, the standard has been relaxed of 4 dB and the new mask is given by

$$\begin{cases} 15 - 25\log\phi - 10\log N \;\; \text{dBW} & \text{for} \;\; 1.8° < \phi \le 7.0° \\ -6 - \log N \;\; \text{dBW} & \text{for} \;\; 7.0° < \phi \le 9.2° \\ 18 - 25\log\phi - 10\log N \;\; \text{dBW} & \text{for} \;\; 9.2° < \phi \le 48° \\ -14 - 10\log N \; \text{dBW} & \text{for} \;\;\;\;\;\; \phi > 48° \end{cases} \tag{2}$$

Similarly, passing from HotBird™ 6 to Ka-Sat, the cross polar discrimination limitations have been relaxed, in order to keep into account the effects of using the circular polarization, instead of linear, in Ka-Sat.

For the mobile systems, the standard M will be reviewed in the next future in order to take into account further parameters and dynamic behaviors as for example the tracking algorithm performances and the corresponding pointing accuracy or the up-link power control techniques adopted.

Particular care will be taken in considering the applicable operational conditions that could permit to approve systems otherwise unacceptable. This is the case of mobile antennas having an excellent behavior on the azimuth cut (but a bad elevation pattern, which is common in low profiles antennas) that are deployed into restricted geographical area on transports assuring small variations of the installation plane. In this case the antenna skew sweeps a very narrow range and the azimuth pattern – or patterns measured on few degrees inclination angles – can be taken into account.

Even if today these parameters have not been yet included in the standard, Eutelsat will proceed in deep analysis of the antenna characteristics, before approval: for this reason the manufacturer shall be invited to details several key parameters of the antenna (e.g. patterns measured on particular cuts, details of the tracking algorithm, etc).

4 Conclusions and Further Activities

The request for IP connectivity and multimedia services on board transports is today growing up. Several examples of commercial applications can be mentioned: this paper refers to the services deployed in Ku band by Eutelsat and highlights the technical problems Eutelsat faced in the design and implementation of a railway satellite service for which a return on experience is reported.

The launch of Ka-Sat opens further perspectives for the development of new services that take advantage from the dramatically lower cost of the capacity. In the framework of the mobility, some drawbacks come from the higher complexity required to the remote mobile terminals. This subject has been already extensively analyzed in the past [1]: together with a description of the technologies available for the implementation of a Ka band antennas, this paper adds some details concerning the standards to be applied to the earth stations in order to reduce the interferences on the satellite.

Today Eutelsat co-operates with an antenna manufacturer for the implementation of a prototype antenna in Ka band to be used as approval-of-concept for the mobile services on Ka-Sat. In parallel, even if Ka band offers very good opportunities for the technical and commercial development of the mobile services, Eutelsat keep always in its portfolio solutions in Ku band that offer the incontestable advantage of a long experience and of a technology that is ready to use. Thus, others Eutelsat activities in this domain concern the extension of the existing Ku band mobile services to other trains, aircraft and vessels fleets.

References

1. Feltrin, E., Weller, E., Bellaveglia, G., Lo Forti, R., Marcellini, L.: New frontiers and technologies for the mobile satellite interactive services. In: Proceedings for 16th Ka and Broadband Conference, Milan (October 2010)
2. Feltrin, E., Weller, E.: A satellite-based infrastructure providing broadband IP services on board high speed trains. In: Sithamparanathan, K., Marchese, M. (eds.) PSATS 2009. LNICST, vol. 15, pp. 143–152. Springer, Heidelberg (2009)
3. Feltrin, E., Weller, E., Bellaveglia, G., Lo Forti, R., Marcellini, L.: Application of a spread spectrum satellite system for broadband IP services in high speed trains. In: Proceedings for the 14th Ka and Broadband conference, Matera (September 2008)
4. Eutelsat EESS 502 Standard – Standard M – Earth Station minimum and operational requirements
5. ETSI EN 301 459 v1.4.1 – Satellite Earth Stations and Systems (SES); Hormonized EN for Satellite Interactive Terminals (SIT) and Satellite User Terminals (SUT) transmitting towards satellites in geostationary orbit in the 29.5 GHz to 30.0 GHz frequency bands covering essential requirements under article 3.2 of the R&TTE Directive (June 2007)
6. FCC § 25.138, Blanket Licensing provisions of GSO FSS Earth Stations in the 18.3–18.8 GHz (space-to-Earth), 19.7–20.2 GHz (space-to-Earth), 28.35–28.6 GHz (Earth-to-space), and 29.25–30.0 GHz (Earth-to-space) bands
7. ETSI EN 302 307 Digital Video Broadcasting (DVB); Second generation framing structure, channel coding and modulation systems for Broadcasting, Interactive Services, News Gathering and other broadband satellite applications
8. ETSI EN 301 790, DVB- Interaction Channel for Satellite Distribution Systems

SatNEx Phase III – Satellite Communication Network of Experts

Erich Lutz[1], Barry Evans[2], Alessandro Vanelli-Coralli[3],
Tomaso De Cola[1], and Giovanni Giambene[4]

[1] German Aerospace Center DLR,
Oberpfaffenhofen, Germany
[2] University of Surrey, Guildford, UK
[3] University of Bologna, Italy
[4] University of Siena, Italy
{Erich.Lutz,Tomaso.DeCola}@DLR.de, b.evans@surrey.ac.uk,
avanelli@deis.unibo.it, giambene@unisi.it

Abstract. SatNEx is a European Network of Experts for satellite communications, coordinated by the German Aerospace Center DLR. The first two phases of SatNEx were funded by the EU from 2004 to 2009. The third phase, SatNEx-III, comprises 17 partners and is funded by ESA from 2010 to 2013. A core team consisting of DLR, University of Surrey, and University of Bologna is coordinating the SatNEx-III research activities. Specific research tasks are contracted to partners in the frame of annual "Call-off Orders".

Keywords: Satellite communications, satellite systems, satellite technology.

1 Introduction

The primary goal of SatNEx under European Union (EU) sponsorship was to achieve long-lasting integration of the European research in satellite communications and to develop a common base of knowledge. SatNEx was formed in 2004 by 22 partner research organisations and universities with the support of the EU FP6 programme as one of several new Networks of Excellence (NoE). In 2006, funding was renewed by the EU to continue with SatNEx-II up to 2009.

Apart from establishing a critical mass of research effort in satellite communications in Europe, SatNex has established a series of annual summer schools for new researchers, providing a comprehensive programme of advanced technical and scientific lectures which cover specific areas in satellite communications. Also, SatNEx has sponsored and coordinated several satellite conferences, has produced many journal and conference papers as well as four books, and has provided inputs to standards bodies and participated in forming standards.

G. Giambene and C. Sacchi (Eds.): PSATS 2011, LNICST 71, pp. 175–185, 2011.

Now, the third stage of SatNEx, SatNEx-III, continues with the support of ESA with a more focused approach. SatNEx III is comprised of 17 European research organisations and universities. The German Aerospace Center DLR with its Institute of Communications and Navigation leads the team with close support by the core team partners University of Surrey, Centre for Communication Systems Research, and University of Bologna.

1.1 Network Partners

The SatNEx-III network comprises the following partners:

- DLR, Institute of Communications and Navigation (Co-ordinator)
- University of Surrey, Centre for Communication Systems Research (UniS)
- University of Bologna, Department of Electronics, Computer Science, and Systems (UoB)
- Aristotle University of Thessaloniki, Department of Electrical and Computer Engineering (Auth)
- University of Bradford, Communication Systems Engineering Research Group (BRU)
- Consorzio Nazionale Interuniversitario per le Telecomunicazioni (CNIT)
- Italian National Research Council CNR, Istit. di Scienza e Tecnologie dell'Inform. "A. Faedo" (CNR-ISTI)
- Centre Tecnologic de Telecomunicacions de Catalunya (CTTC)
- National Observatory of Athens Institute for Space Applications and Remote Sensing (ISARS)
- Office National d'Études et de Recherches Aérospatiales (ONERA)
- University of Salzburg, Department of Computer Sciences (SBG)
- TéSA Association, Institut Supérieur de l'Aéronautique et de l'Espace (TeSA)
- Graz University of Technology, Institute of Communication Networks and SatComs (TUG)
- Universitat Autonoma de Barcelona, Dept. Telecomunicació i Enginyeria de Sistemes (UAB)
- University of Aberdeen, King's College, Department of Engineering (UoA)
- Universita Degli Studi di Roma "Tor Vergata", Department of Electronics (UROMA2)
- Universidade de Vigo, Department of Signal Theory and Communications (UVI).

Fig. 1 shows the geographical distribution of the SatNEx-III partners.

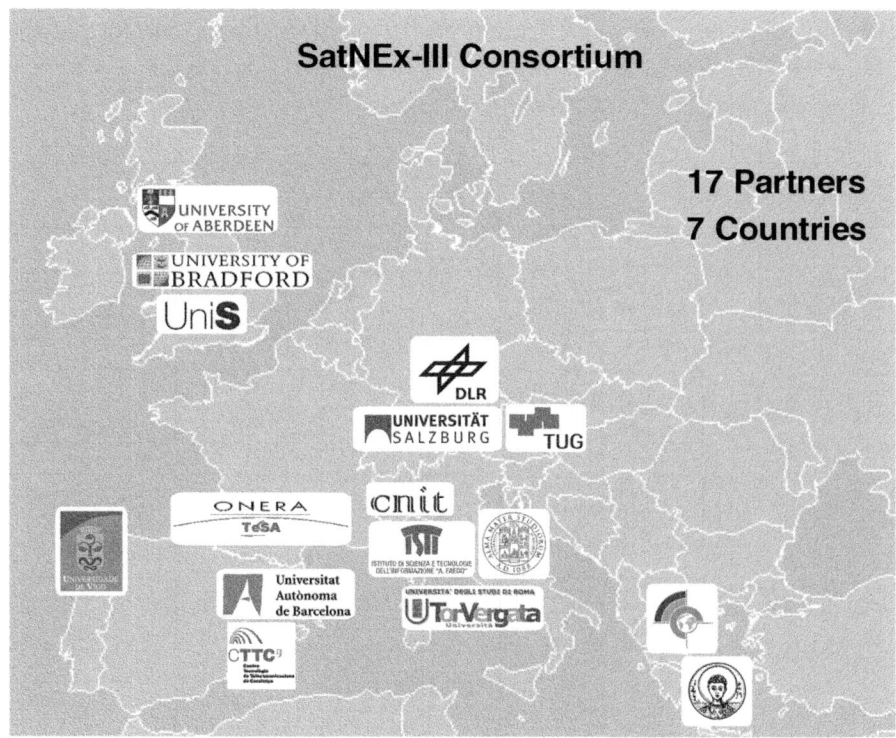

Fig. 1. Geographical distribution of SatNEx-III partners

1.2   General Tasks of SatNEx-III

In general, SatNEx-III will:

- play an important role in exploring new satcom techniques supporting ESA in selecting the right avenues for R&D work plans (TRP, ARTES 1, ARTES 5)
- support ESA in ad-hoc technical actions related to satcom standards
- identify promising terrestrial technology spin-in into space
- play a pivotal role in forming young professionals for satcom.

Throughout the three-years duration of SatNEx-III, **horizontal activities** will be pursued in the following areas:

- Long-term development of satcom visions and systems
- Development of physical and access layer technologies
- Development of networking technologies and protocols
- Satcom training activities and dissemination of SatNEx-III results.

In addition, ESA will activate each year a number of **specific advanced research areas**. In the first project year, SatNEx-III has been charged with the following tasks:

- Concept development for a Terabit/s satellite system
- Exploration of hybrid space/ground signal processing techniques

- Investigation of new protocols for disruption tolerant satellite communications.

Fig. 2 shows the structure of the SatNEx-III tasks.

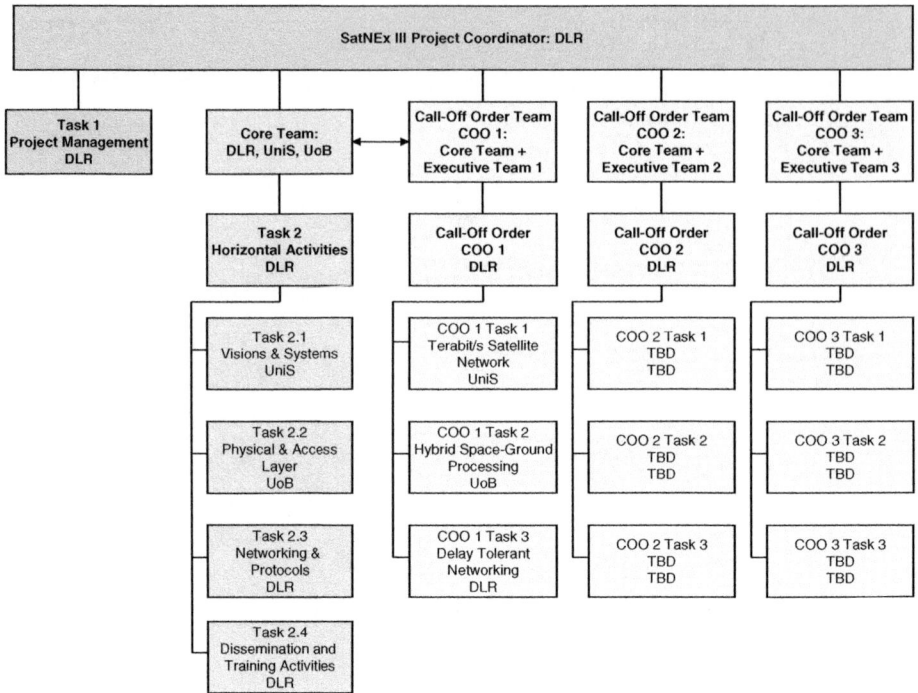

Fig. 2. Structure of SatNEx-III tasks

2   SatNEx-III Horizontal Activities

The horizontal activities last throughout the three-years duration of SatNEx-III, and are performed by the core team consisting of the German Aerospace Center DLR / Institute of Communications and Navigation, University of Surrey / Centre for Communication Systems Research, and University of Bologna. A brief description of the horizontal activities is as follows:

2.1   Long-Term Development of Satcom Visions and Systems
(lead: University of Surrey)

- Identify new satcom applications and related requirements
- Devise new system architectures which can satisfy the new applications and requirements
- Investigate the best options for integrating satellite networks with terrestrial systems

- Identify new research areas in view of the identified satcom applications and requirements
- Exchange and consolidate the results with ESA and other key players (e.g. satellite operators and industry).

First Results

The future vision of communications is one of 7 Trillion wireless devices serving 7 Billion people with pervasive machine to machine communications as part of an Internet of Things. As well as connections of large numbers of devices users will expect ubiquitous service delivery with consistent user experience. Data is doubling every year and video content is becoming pervasive in all services which is leading to demands for massive increases in bit rates and hence bandwidth. Current estimates for domestic users indicate downloads of around 20Mb/s by 2016 (Analysys Mason Report 2010) which would easily transform to the EU target of 30Mb/s by 2020. Other users such as SME's and institutional users will be higher. The increase in video content will also increase the return link rates and this is likely to move towards more symmetry with new services. By 2020 services such as social TV involving multiview, HDTV and 3DTV are likely to be wide spread and business users will be using immersive presence systems for virtual meetings requiring 100's of Mb/s. Even today we see the first remote holographic experiences of art galleries and museums and we can envisage this moving to shared experiences between social groups which will increase the bandwidth demands in both directions. Hence the future is a mixed environment between large numbers of low rate M2M traffic and video dominated high bit rate broadband services.

For satellite systems as with any radio system, spectrum is a major issue and its constraints are already forcing us to be smarter in the use of this resource. Sharing of spectrum between systems and mitigation of interference as well as the use of cognitive radio for more opportunistic use are already investigated. Energy conservation (Green Systems) is another major driver on an end to end basis, which will dictate future systems design with systems becoming always available rather than always on.

Future networks will no longer be designed on a vertical service basis but with virtualisation will be flexible and enable separate virtual networks to be set up from a common infrastructure for different services. They will also be self organising from a system level based on power, spectrum, energy, QoS and cost. Such scenarios bring with challenges of resilience, security and network management. One of the key features for satellite is the movement to internet down load replacing traditional broadcasting.

From the above we see key issues for satellites as;

- Higher capacities either in one large GEO or smaller clusters
- Smarter Payloads to provide connectivity
- Smarter use of the spectrum-sharing and interference mitigation
- Dealing with very large numbers of sensors
- Integrated and hybrid systems to be part of flexible networks
- Delivering video content in non broadcast modes.

2.2 Development of Physical and Access Layer Technologies
(lead: University of Bologna)

- Report on terrestrial trends and possible reuse of emerging terrestrial wireless physical and access layers standards and solutions in satellite networks
- Identify new research areas and assess their suitability to the satellite scenarios identified in the visions and systems task
- Identify physical and access layers standardisation areas of interest to satcoms.

2.3 Development of Networking Technologies and Protocols
(lead: DLR)

- Report on terrestrial trends and possible reuse of emerging terrestrial networking and standard solutions in satellite networks
- Identify new research areas and assess their suitability to the satellite scenarios identified in the visions and systems task.
- Identify networking and protocol standardisation areas of interest for satcoms.

First Results

The study has been primarily concentrated on the standardisation efforts made by IETF (Internet Engineering Task Force), IRTF (Internet Research Task Force), and CCSDS (Consultative Committee for Space Data Systems) about transport layer enhancement and in general networking issues. In particular, the aim of this investigation has been to explore some new areas of research that could be of some applicability in satellite networks, though being conceived for wireless or terrestrial communications.

What has been observed is that transport layer issues cannot be limited to the design of new transport protocols, but also and more importantly to transport layer solutions supporting QoS in multimedia applications. This could be the case of RSVP (Resource reSerVation Protocol) extensions for supporting triple-play applications as to concerns the resource allocation along the routing path. Other aspects related to the QoS management are linked to the assignment of dedicated DSCP codes for admitted capacity over Differv-enabled networks. All these elements are of some importance in integrated terrestrial-satellite networks, when QoS requirements have to be targeted by the service provider to offer the user a satisfactory QoS level.

Another topic being addressed concerns the interconnecting issues between satellite and terrestrial domains, in case NATs, firewalls or in general security countermeasures are applied to filter and shape the incoming/outgoing traffic. In this perspective, the findings of the midcom working group within IETF are certainly of some interest, as they draw the challenges and the possible solutions that can be deployed to allow middlebox communication even in presence of filtering and shaping traffic policies.

Finally, an additional topic considered for a possible application in satellite network regards the synchronisation in IP-based networks. It was noted that this aspect is quite important in environments where disruptions or delay spikes may alter the behaviour of applications. In this respect, the activities of the tictoc working group part of IETF have been considered for possible application to mobile satellite networks. In particular, the expected extensions to the Network Time Protocol (NTP) could result helpful to ensure a more robust synchronisation even in presence of link interruption.

As far as deep space communications are concerned, some investigations about interoperability issues with terrestrial links have been carried out, by taking as reference the standardisation work carried out within the Consultative Committee for Space Data Systems (CCSDS).

2.4 Satcom Training Activities and Dissemination of SatNEx-III Results
(lead: DLR)

The purpose of this task is to support the networking of people within SatNEx. In particular, the tasks are to

- Support exchanges of researchers between partners
- Ensure dissemination of results to ESA, delegations, industry and satellite operators
- Maintain the SatNEx web site
- Secure representation of SatNEx-III in major satellite conferences
- Raise funds for maintaining the current SatNEx summer school initiative for PhD students
- Give annual lectures at ESTEC on subjects of common interest to ESA and SatNEx.

One of the main activities in this area will be the organisation of the next SatNEx summer school which is scheduled for Sep. 5 – 9, 2011 in Siena. For more information see the announcement to be published in the Internet. This summer school will maintain the series of summer schools organised in the frame of the SatNEx project.

3 SatNEx-III Specific Research 2010 (Call-off Order 1)

Three specific advanced research topics, activated by ESA, are investigated every year by specific task teams. In the first project year, SatNEx-III has been charged with the following tasks:

3.1 Concept Development for a Terabit/s Satellite System

lead: University of Surrey; partners: ISARS, ONERA, TeSA, UAB
In this task, the team investigates a number of critical technical system aspects related to the development of a Terabit/s satellite system. More specifically, we will:

- define a reference system architecture and related system assumptions, with particular emphasis on frequency bands for the user and feeder links, and number of beams required
- investigate approaches to achieve flexible resource allocation over the coverage region

- investigate physical and access layer techniques able to efficiently cope with unbalanced traffic distribution
- investigate countermeasures at system/network/user terminal level to cope with the residual ground beams displacement
- assess the fading impact for both the feeder link and the user link and define fading countermeasures
- investigate smart feeder link concepts.

First Results

A baseline European system for a generation after next satellite has been determined for the Terabit/s satellite driven by the capability to accommodate a Terabit/s of traffic. The constraints of spectrum at Ka band (2.5GHz up and 3.4GHz down) and Q/V (4GHz up and 5GHz down) as well as the cost of the user terminal have resulted in a system of 19 gateways operating at Q/V plus 175 (or 88 dual polarisation) beams using 3 colour reuse for the user coverage. The system uses DVB-S2-RCS(NG) air interface and ACM on the Ka band links to combat rain fading. On the gateway links in order to achieve a 99.99% availability a smart gateway system is used connecting all the gateways via an optical network and switching the traffic and load balancing between the gateways to achieve performance.

Assuming that the ACM and smart GES system provide the availabilities a baseline link budget has indicated that the capacity can be achieved providing that the satellite antenna C/I exceeds circa 20dB. Current work concentrates on more realistic modelling of the latter with real traffic within the region. The return link is the limiting performance and several configurations of polarisation reuse have been investigated in an attempt to optimise the C/I which is still marginal. Improvements in the latter using managed return link transmissions are being investigated. The smart GES scheme has been modelled for various numbers of GES and shown to provide the necessary availabilities and we are now investigating the payload routing complexity and the practical switching issues between the GES.

The large numbers of beams required to meet the capacity, result in severe requirements on the payload in terms of the numbers of amplifiers required. Payload power estimates for satellites around 2020 are around 20kW and are limited by the solar panels packaging into the limited launcher fairing. Assuming a single feed per beam around 200 amplifiers rapidly exceeds these limits. The use of dual polarisation halves the number of amplifiers and has been adopted and we are investigating the use of Beam Hopping on the forward link to reduce the numbers still further but there is clearly a trade off in the user terminal complexity and in modifications to the air interface. Another implication of large numbers of smaller beams is the stability of the satellite and the resultant beam movements. This is being modelled and mitigation schemes being proposed.

The final issue being investigated is the performance of the ACM in the Ka band. Here with larger fading, a wider Mod/cod range is desirable and extension to 64 APSK with improvements in the SNR estimation and demodulation as well as improvements in the signalling of the mod/cod changes. See reference 1 for further details.

3.2   Exploration of Hybrid Space/Ground Signal Processing Techniques

lead: University of Bologna; partners: CTTC, TUG, UniS, UVI

In this task the team investigates the potential applicability of hybrid space ground processing techniques with signal digitalisation on-board the satellite. In particular, the following tasks are being performed:

- define a reference system architecture and related system assumptions
- review of literature in the field of data compression, beam forming, Multi User (MU) Multiple Input Multiple Output (MIMO) algorithms with apportionment of tasks between the space and the ground segment
- pre-select candidate architectures and algorithms for hybrid space ground processing relevant to the selected reference system
- perform a detailed trade-off among different options for the hybrid space ground processing in terms of performance and complexity
- make recommendations for further R&D activities in this field.

First Results

In the frame of this task different issues related to signal processing have been investigated.

As a consequence of the increase in the number of spot beams in modern satellite systems, more and more flexible beam management systems are required, leading to more complex beamforming networks on board the satellite. As an alternative, ground-based beamforming (GBBF) techniques are based on the exchange of radiating element signals between satellite and gateway. At the cost of a high feeder link bandwidth demand, more sophisticated and power consuming techniques can be implemented. Flexibility is preserved, and changes in shape, traffic and pointing direction can be accommodated. Finally, hybrid schemes consist in partitioning the beamforming process between the satellite (Coarse Beamforming) and the gateway, with the objective of reducing the feed signal space to a subspace, thus decreasing the required feeder link bandwidth.

The deployment of larger number of beams with high frequency reuse increases the total system throughput, however, increasing the interference among users at the same frequency. Multi user interference mitigation techniques can be considered for controlling the corresponding performance degradation.

Multi user interference mitigation techniques can be fully implemented at the gateway (GW), so that the impact on the complexity of the user terminals (UT) is limited. In the forward link (FL), it consists in the use of *precoding* techniques. The general idea is to combine at the GW (i.e. the transmitter side) the different input signals associated with each user in such a way that the interference levels as seen at each UT (i.e. the receiver side) are controlled and minimized. In the return link (RL), the interference between users can be mitigated by considering *multi user detection* (MUD) techniques at the GW (i.e. at the receiver side). In the sense of information theory, these scenarios are respectively referred to as the multiple input multiple output (MIMO) broadcast channel (BC) and multiple-access channel (MAC).

3.3 Investigation of New Protocols for Delay Tolerant Satellite Communications

lead: DLR; partners: CNIT, CNR-ISTI, SBG, UniS, UoA, UoB, UROMA2

In this task the team investigates the applicability of new DTN protocols to satellite communication systems. In particular, the following tasks are being performed:

- review the state-of-the-art in Delay and Disruption Tolerant Networks (DTN) and current level of definitions, standardisation, and demonstration achieved
- investigate the potential applicability and advantages/drawbacks of DTN to:
  - o mobile satellite telecom networks to efficiently cope with possible link interruptions
  - o broadband satellite telecom networks to efficiently cope with transmission delays, avoiding the use of Proximity Enhancement Proxies (PEP)
  - o deep space communications

  and consider aspect such as the impact on applications support, overall data transfer data
- efficiency and reliability, as well as delay
- collect DTN emulation software modules able to demonstrate the performance in selected application cases
- make recommendations for further R&D activities in this field.

First Results
The study being performed within this task aimed at showing the usefulness of the Delay/Disruption Tolerant Network architecture concepts for satellite and space communications. As to the latter, it has been intensively demonstrated that DTN will be of primary importance in the future deep-space missions, to support manned missions and scientific experimentations, even in case of sporadic network infrastructure interruptions and in presence of very large propagation delays, in the order of several minutes. On the other hand, the use of DTN idea for satellite communication has been only partially applied to satellite communications, since this environment experiences only in part the challenges that have actually fostered the research in the DTN scientific community. In fact, satellite communications show limited propagation delays (wrt to deep space environments) and link interruptions stem from user mobility in harsh environments or from adverse weather conditions, both challenges usually counteracted with solutions implemented at the physical layer. However, this study pointed out that the availability of a DTN-enabled architecture shows important benefits also in this environment, owing to the recovery and storage capabilities, which allow achieving high system performance (e.g., throughput and data delivery delay). In particular, it was observed that the DTN concept turn out to be helpful in case of fixed geostationary satellite scenarios, where the DAMA access scheme or weather adverse conditions (playing some role especially in Ka and EHF frequency bands) may introduce additional delays or unexpected link interruptions. Besides, an even greater advantage is registered in LEO satellite scenarios, where the satellite acts as data-mule. In this case, the storage capabilities of the DTN architecture along with the proactive fragmentation option

implemented within the Bundle Protocol (the protocol defined for the DTN architecture) allow optimising the contact durations, thus resulting in the maximum resource usage.

Finally, the benefits of DTN architecture in the aforementioned scenarios is also being investigated from an implementation point of view, by means of dedicated testbeds implementing the Bundle Protocols and reproducing the peculiarities of fixed/mobile satellite scenarios or deep-space environment. This study is the last part of this task and is still ongoing.

4 Conclusions

At the time of writing, work on the call-off order 1 is going on, and possible subjects for the next call-off order are being discussed. The SatNEx-III project will continue to work on timely research topics related to future satellite communications and will bring together European research organisations in this important field.

Reference

1. Evans, B.G., Thompson, P.T.: Key issues and technologies for a Terabit/s satellite. In: 28th AIAA International Communications Satellite Systems Conference (ICSSC 2010), Anaheim USA, paper number AIAA-2010-8713 (2010)

DTN for LEO Satellite Communications

Carlo Caini and Rosario Firrincieli

DEIS/ARCES, University of Bologna, Bologna, Italy
{carlo.caini,rosario.firrincieli}@unibo.it

Abstract. Satellite communications are an interesting and promising application field for Delay/Disruption Tolerant Networking (DTN). Although primarily conceived for deep space communications and sensor networks, it was immediately recognized that DTN was applicable to satellite environments, in particular to cope with the intermittent channels typical of LEO (Low Earth Orbit) constellation satellite systems. The aim of this paper is to assess the advantages of DTN when applied to LEO satellites. Qualitative assessments are supported in selected cases by preliminary results obtained on a testbed based on GNU/Linux machines. In particular, two application scenarios have been considered, both using a single LEO satellite. In the former, we have one LEO satellite for Earth observation, connected to its gateway stations only at intermittent scheduled intervals due to its orbital motion. The latter is one LEO satellite acting as a "data mule" between a terrestrial sensor network and a remote satellite gateway station, which are never in the satellite coverage area at the same time. The results show the feasibility and the advantages of DTN in LEO satellite communications.

Keywords: DTN, LEO satellites, Satellite communications, Challenged networks, DTN2, ION.

1 Introduction

Delay/Disruption Tolerant Networking (DTN) aims to provide interoperable communications in "challenged networks", i.e. those networks where one or more of the usual assumptions implicit in the use of the TCP/IP stack (short delays, negligible PER, existence of a continuous path between source and destination), no longer hold true. Such networks include deep space communications, a large variety of terrestrial and maritime sensor networks, satellite and airborne communications e.g. Unmanned Aerial Vehicles (UAVs) and many other in both the civil and military fields [1]-[7].

Concerning satellite communications, DTN represents an interesting alternative to the use of PEPs (Performance Enhancing Proxies) in GEO (Geosynchronous Earth Orbit) satellite systems, as shown in [8], but its use is particularly appealing in LEO (Low Earth Orbit) systems, because of DTN ability to cope with intermittent channels, disruption and lack of end-to-end connectivity, typical of both single LEO satellites and incomplete constellations [6], [9]. Hence, this paper aims to evaluate the

G. Giambene and C. Sacchi (Eds.): PSATS 2011, LNICST 71, pp. 186–198, 2011.
© Institute for Computer Sciences, Social Informatics and Telecommunications Engineering 2011

advantages of DTN when applied to LEO satellites, supporting our assessments with results and logs obtained on a real testbed, i.e. on DTN implementations running on GNU/Linux machines.

In the tests we considered two possible applications, both using a single LEO satellite. In the first, a LEO satellite for Earth observation is connected to its gateway stations only at intermittent scheduled intervals. In the second, a single LEO satellite acts as "data mule" between a terrestrial sensor network and a remote control centre. In both cases, to cope with intermittent channel availability (and also the lack of a continuous path in the latter), file transfers using TCP/IP stack would require manual intervention. By contrast, as shown in the paper, file transfers can be performed automatically by DTN even in these challenging scenarios.

2   DTN Outline

2.1   Origin and Motivation

DTN was first conceived to address space communications impairments, as it was glaringly obvious that the usual TCP/IP stack, alone, could not cope [1]. Later, DTN scope was enlarged to cover all challenged networks, whether spatial or terrestrial. To this end, in 2002 the IRTF Delay Tolerant Networks Research Group (DTNRG) was established to promote DTN. As the new architecture must tolerate not only long delays, but also link disruptions, the DTN acronym is often expanded as Delay/Disruption Tolerant Networking. The interested reader is referred to [2] for an informative study of TCP limits in challenged networks, and to [3] and [4] for an exhaustive survey of DTN development. Tutorials and other references can be found on the DTNRG website [5], which is the major source of DTN documentation and software. Although the DTN architecture based on the introduction of the Bundle protocol, described in [6], and [7], is not the sole possible option, it is the most common and we will refer to it in this paper.

2.2   DTN Bundle Protocol Architecture

In order to support communication in challenged environments, the Bundle protocol DTN architecture [6], [7] is based on a new layer, located between Transport and Application, called "Bundle layer". The related protocol (the Bundle protocol) can interface with various transport protocols (including TCP [10] and UDP, but also with new protocols, like Licklider [11], [12] or Saratoga [13]), through "convergence layer adapters". In this new architecture (see Fig. 1), transport protocol end-to-end features are confined to homogeneous network segments (A, B and C), while end-to-end data transfer across the heterogeneous network is provided by the bundle layer; large data packets called "bundles" are exchanged between DTN nodes through a store-and-forward relay. The main innovations of DTN architecture are summarized below.

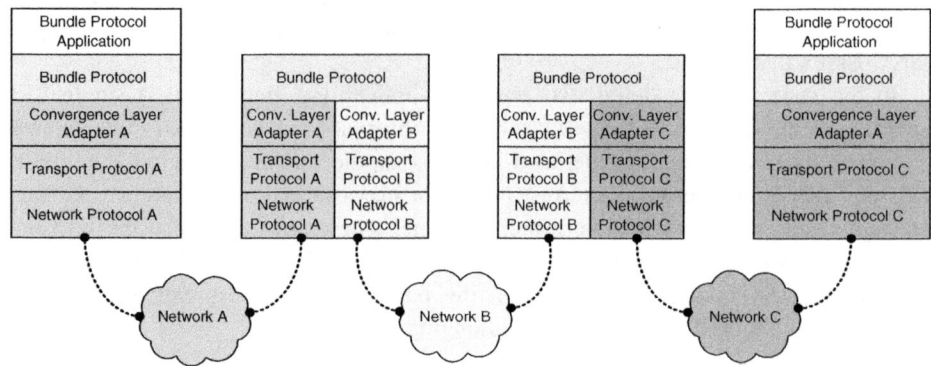

Fig. 1. DTN architecture and protocol stack

2.3 DTN Overlay

First, although TCP/IP protocols are not necessarily replaced, their role is changed. By installing the Bundle protocol on end-points and some intermediate nodes, (e.g. on satellite gateways), the end-to-end path is divided into multiple DTN hops. On each DTN hop a different protocol stack can be used, or, when the same stack is retained, which is the most common case, just different protocols, like TCP, UDP, or different versions of the same protocol (e.g. different TCP variants). Readers familiar with satellite PEPs [14], [15] can easily realize that the DTN multi hop architecture can be seen as a generalization of the TCP splitting concept. In particular, both DTN and TCP splitting PEPs allow the use of specific protocols (or specific versions of the same protocol) on the satellite segment. In such a way, the same advantages of TCP splitting PEPs can be achieved, in terms of goodput, also by DTN [8]. However, it must be stressed that while in the DTN architecture the "splitting" is a direct consequence of the new protocol stack, in PEPs it implies a severe violation of the end-to-end TCP semantics. Concerning security, TCP splitting is incompatible with IPsec, while the DTN architecture has the advantage of a greater flexibility (both end-to-end and hop-by-hop security can be provided). On the other hand, by contrast to PEPs, the DTN architecture is not transparent to end nodes.

2.4 Storage at Intermediate Nodes

The second difference between DTN and customary TCP/IP network is related to information storage. In standard networks, because of usual assumptions of continuous connectivity and short delays, information is supposed to be stored only at end nodes. By contrast, in DTN networks, where the usual assumptions do not hold anymore, information (i.e. data bundles) must be stored at intermediate DTN nodes for long period of times and, when the custody option [6], [7] is enabled, only on persistent memory (e.g. on local hard disks). This feature actually differentiates DTN architecture from usual PEPs, as it makes DTN much more robust against disruptions, disconnections, and temporary node failures [16].

2.5 Bundle Fragmentation

An interesting feature of DTN Bundle protocol is the possibility of fragmenting bundles. Fragmentation can be performed a priori (proactive fragmentation) or a posteriori (reactive fragmentation). The former has been conceived to cope with intermittent periodic connectivity, where there may be a stringent constraint on the maximum amount of data that can be transferred (contact volume) on a DTN hop at each availability time window (contact time). It allows large bundles to be divided "a priori" into multiple fragments compatible with the contact volume. This feature could be useful in single LEO satellite systems, where the contact volume is known in advance. By contrast, reactive fragmentation works a posteriori, triggered by a relatively long disruption. It could be advantageous in satellite communications (both GEO and LEO) with mobile terminals, when obstacles (buildings, tunnels, etc.) may disrupt the satellite signal reception.

3 LEO Satellite Communications and DTN

LEO satellites are characterized by low orbits with a reduced distance from the Earth's surface (160 – 2000 km). Compared to GEO systems they offer the obvious advantage of reduced attenuation loss and a shorter propagation delay. On the other hand, to an observer on the Earth's surface they do not appear fixed in the sky, but fast and constantly moving; for example, at an altitude of 520 km the revolution period necessary to counteract the Earth's gravity is about ninety minutes. As a result, a single satellite can only provide intermittent connectivity with a fixed ground station, while continuous connectivity can be provided only by constellations of several tens of satellites, like those used in Iridium [17] or Globalstar [18], the two most well-known commercial systems. Because of their different implications, we will treat single and multiple satellite coverage separately.

3.1 Single Satellite Coverage

In the case of a single satellite we further distinguish between two possible applications. The first, a data transfer from a LEO satellite to a remote control centre; the second, a "data mule" data collection from a sensor network.

3.2 Earth Observation Scenario

Due to their low orbits, LEO satellites pass over a fixed ground station for short intervals (some minutes) many times a day, thus providing scheduled intermittent connectivity. In this scenario we consider data transmission from a LEO satellite to a terrestrial destination. For instance, a LEO satellite devoted to Earth observation which has to transfer large image files to a remote operation control centre. Here, we have to cope with intermittent scheduled end-to-end connectivity because of satellite motion. The short transmission time window and the possible limited channel bandwidth pose limits to "contact volume", i.e. the total amount of data that can be transferred at each link availability interval. Image files larger than contact volume cannot be transferred during a single pass, and require to be divided into multiple segments for transmission during consecutive passes. In this case DTN could benefit from the "proactive fragmentation" feature of the Bundle protocol. Alternatively, if this feature is not available in the implementation in use (or to avoid the complexity that can derive from the concurrent use of fragmentation and Bundle protocol security extensions), it is possible to use a DTN application, like DTNperf, which can segment a file into a series of bundles of the desired dimension. Further details will be provided in the numerical results section.

3.3 Data Mule Scenario

Here, we consider a source and a sink both located on Earth and connected through two ground stations and a LEO satellite. The two ground stations are a long distance apart, so are never concurrently in line of sight from the satellite. Consequently, there is not a continuous path between them and the LEO sat must act as a data mule. For example, imagine a remote sensor network connected to its control centre via satellite. Data must be first collected at a central node of the sensor network, with DTN and satellite capabilities (the first ground station); the data are then to be transferred to the second ground station. The LEO satellite is alternately in line of sight from one or other ground station, and data transfer can only performed by storing data on the satellite, which must therefore have adequate storage capacity. Note that as this scenario is the most challenging, it is also the most favorable to DTN. The total absence of end-to-end connectivity prevents the establishment of TCP (or TCP-like) connections, while it is perfectly suited to the DTN "store-and-forward" approach.

3.4 Multi Satellite Coverage

Unlike GEO systems, where one satellite can offer continuous coverage of a large area, with LEO systems continuous connectivity requires the deployment of a constellation of satellites (50-70). First generation LEO systems like Globalstar and Iridium, designed in the early '90s and still in use, were primarily designed to provide voice communication and can offer only secondary data capabilities at low bit rates (max 128 kbit/s for the Iridium system). They are soon to be replaced by second-generation systems designed mainly for data communications and Internet access. The present Iridium system, for example, will be replaced by "Iridium Next", which will make use of the same orbits and number of satellites, but its enhanced payloads will be able to offer data communications at various rates up to 8 Mbit/s. Deployment of this second generation will require the launch of 66 active satellites (plus some spares), and is expected to take a couple of years.

Until a LEO constellation is fully deployed, it is difficult to make use of the satellites already in orbit, because gaps in the, moving, coverage area cause intermittent connectivity. DTN could cope with this problem, thus enabling the first satellites deployed to enter into operation, with obvious economic advantage. For example, incomplete constellations could be used for file transfers and non real-time data exchange, thanks to the DTN ability to function despite intermittent connectivity and disruption. Moreover, even after complete constellation deployment, DTN could still offer significant advantages. It could, for example, counteract link disruptions frequently met when using mobile terminals, or remedy the possible temporary lack in free channels during handovers between satellites (in LEO system handover are necessary even for fixed terminal due to the satellite motion).

DTN use in LEO constellations will be the object of future research, and will therefore not further treated in the numerical result sections.

4   DTN Implementations and Tools

4.1   DTN2: Bundle Protocol Reference Implementation

DTN2 is the Delay Tolerant Networking reference implementation. In addition to the reference Bundle protocol implementation, the DTN2 package also contains some DTN basic applications (DTNping, DTNsend, etc.) and DTNperf_2, the DTN evaluation tools used in our experiments and described below. The DTN2 goal is twofold: "to clearly embody the components of DTN architecture, while also providing a robust and flexible software framework for experimentation, extension, and real-world deployment" [5]. In other words, DTN2 aims to be suitable for both study and real use. It runs on Linux (x64 and x86) and other platforms as well. DTN2 can be downloaded from Source Forge (see [5]). The latest release is 2.7. Installation is complex, but configuration is relatively simple, being based on one configuration file for each DTN node. To enable DTN capabilities, it is enough to launch DTN2 as a daemon, which can be done at boot time. Once launched, all users can easily start DTN applications, like DTNperf, on top of it.

4.2   ION: NASA Bundle Protocol Implementation

ION (Interplanetary Overlay Network) is an implementation of the Bundle protocol developed by NASA JPL (Jet Propulsion Laboratory), with the contributions of Ohio and other Universities, and explicitly focused on deep space applications [20]. As in these environments TCP cannot be used because of excessive RTTs, ION distribution also contains an implementation of Licklider Transport Protocol (LTP), which was designed to offer reliable service in environments characterized by very long delays, and can be suitably used as convergence layer in DTN architecture [11], [12]. Although some features, like DTN node naming, have been specifically designed for space applications, ION software can be used in other environments as well. Moreover, it offers a good interoperability with DTN2 nodes. ION is written in C and currently runs on various Linux platforms, OS/X, FreeBSD, Solaris, VxWorks, and RTEMS.

The ION source code is available as open source from the Open Channel Foundation [21]. The latest release is 2.3 and includes implementations of Contact Graph Routing and several convergence-layer adapters, including TCPCL (interoperable with DTN2), UDPCL (likewise interoperable with DTN2) and LTPCL. ION configuration and use appears somewhat more complex than DTN2; however, it offers some features of particular interest here, like intermittent links, which have not yet been implemented in DTN2. It should be noted, however, that scheduled links in ION require the use of LTP at convergence layer. Moreover, ION offers limited support of bundle fragmentation.

4.3 DTNperf_2

DTNperf is a client-server evaluation tool designed to assess goodput and to provide logs in DTN bundle layer architectures [22]. It is named after the famous Iperf application, widely used to test TCP and UDP performance in ordinary (i.e. non DTN) networks, and it is included in the official DTN2 package released by DTNRG. As DTNperf versions 2.x are significantly improved with respect to previous 1.x versions, they are called DTNperf_2, to stress this difference. The latest DTNperf_2 versions are available for downloading from the "bleeding edge" DTN2 version using Mercurial [23]. They are under an open-source license (Apache License 2.0).

DTNperf is intended to complement other debugging and testing tools included in DTN2, like DTNping (the DTN equivalent of "ping"), DTNsend and DTNrecv (to create, send and receive one bundle), or basic applications, like DTNcat (to send standard input data to another DTN node) or DTNcp (to copy a file between DTN nodes). By contrast, however, and like Iperf, DTNperf is focused on performance evaluation in terms of goodput. Moreover, it allows the user to easily collect the informative DTN "status reports" sent by DTN nodes, (i.e. sent, forwarded, received, custody accepted, delivered, deleted, etc), which are essential in the study of bundle transmission on complex DTN networks.

DTNperf_2 is written in C language, to maintain full compatibility with the DTN2 bundle layer reference implementation APIs. A version also compatible also with ION is envisaged but at present has not been developed. A distinctive feature of DTNperf_2 is examined in detail below because of its relevance in our tests.

4.4 DTNperf_2 Transmission Window

The first release of DTNperf, like other DTN tools, did not allow the source to send more than one bundle at a time, i.e. it was necessary to wait for the reception of an "acknowledgment" of the bundle sent before starting the transmission of a new bundle. This resulted in an obvious goodput ceiling of one bundle per RTT, and a less obvious additional delay for each intermediate DTN nodes due to the store and forward transmission mechanism. To overcome these limitations, which had a significant impact on goodput [8], DTNperf_2 introduced a transmission window that allows multiple bundle transmissions. The length of the Tx window, W, represents the maximum number of bundles that can be concurrently in-flight (i.e. sent but not acknowledged yet). By default, bundle acknowledgments are represented by the "delivery status report" [6], [7], ("status delivered", in short) sent by the receiver node. The DTNperf_2 transmission window is similar to TCP transmission window [8], with the difference that in-flight bundles can be non-consecutive to cope with the non-ordered delivery of the bundle protocol.

5 Numerical Results

In this section, experimental results obtained according to the scenarios presented in 3.2 and 3.3 are discussed. The experiments were carried out by means of a DTN testbed consisting of five GNU/Linux OS machines, with either DTN2 or ION installed. The rationale for the concurrent use in the same testbed of two Bundle protocol implementations, although on different machines, is to take advantage of both advanced DTNperf_2 features, including multiple bundle transmission (W>1), bundle logs, and bundle reordering, and ION link management capabilities, such as scheduled links and transmission speed regulation. General assumptions and scenario characteristics are summarized in Table 1.

Table 1. General assumptions and scenario characteristics

Characteristic	Value
LEO-ground station link type	Intermittent (10 min every 100 min); first contact 5 min after transfer start (Earth observation case). Intermittent (10 min every 50 min); first contact 5 min after transfer start (data mule case).
Ground stations-other terrestrial nodes link type	Wired link, always available, 100 Mbit/s, negligible delays.
LEO-ground station RTT	130 ms
LEO-ground station Bandwidth	1 Mbit/s (symmetric)
LEO-ground station PER	Not present
Number of ground stations	1 (Earth observation case) 2 (data mule case)
Number of total contacts between LEO and ground stations	2 (Earth observation case) 3 (data mule case)
Max contact volume	75 MB
File to transfer	80 MB (Earth observation case) 20 MB (data mule case)
Bundle size	200 kB
Bundle number	400 (Earth observation case) 100 (data mule case)
DTNperf_2 transmission window, W	200 (Earth observation case) 100 (data mule case)
Custody option	ON (all nodes)

5.1 Earth Observation Scenario

Here we assume that the LEO satellite takes images of Earth and, as soon as passes over the ground station, sends them toward the control center (Fig. 2-a). The corresponding testbed topology is shown in Fig. 3. It is worth noting that the LEO satellite has both a DTN2 and an ION node on board. The first acts as DTNperf_2 source (client), while the second is necessary to establish an LTP scheduled link with the ground station. Note that the use of LTP on scheduled links is mandatory in ION.

According to Table 1, maximum contact volume on the LEO-ground station link, obtained using full-speed transfer for the entire contact time, is 75 Mbyte. Depending on file length, file transfer can be completed in one or more passes. In the first case the transfer is quite simple and can be completed as soon as the LEO satellite comes into line of sight with its ground station. The second case is more interesting, and therefore is the sole considered here. Assuming an 80 MB file transfer, two satellite passes are necessary. At the control center, the DTNperf_2 server application, running on a DTN2 machine, has to reassemble the transmitted file by collecting and reordering all arriving bundles.

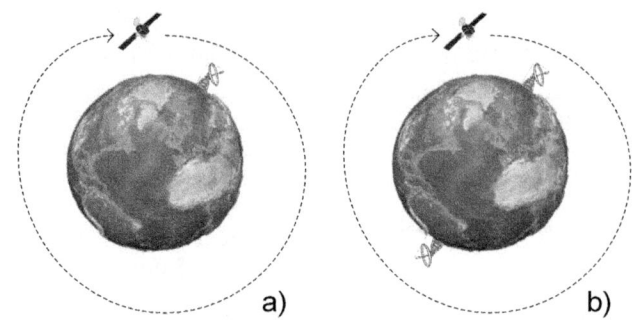

Fig. 2. Experimental cases: a) Earth observation, b) data mule

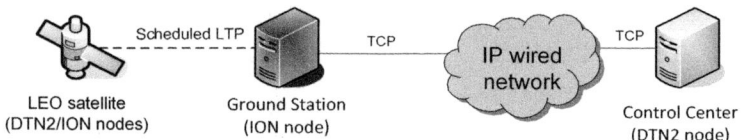

Fig. 3. Earth observation topology

Some details of bundle transfer are given in Fig. 4, taken from DTNperf_2 logs. At time zero the first bundles are transferred by the DTNperf_2 client to the Bundle protocol of the source DTN2 node and from here to the ION node inside the LEO satellite, where are taken into custody waiting for satellite link availability (first part of the "SENT/Custody on LEO" series). In order not to exceed the ION node storage limits (about 60 MB), we used a $W=200$ DTNperf_2 transmission window, which limits to 40 MB (half of the file) the amount of data to be stored on the ION node. As soon as the LEO-ground station link becomes available (at 300 s from time zero), bundles start to be progressively transferred to the ground station (at 1 Mbit/s) and from there to the control center (at 100 Mbit/s). Bundle deliveries are immediately confirmed to the DTNperf_2 client on the source by "delivered" status reports. At the sender side, the arrival of each status report ("DELIVERED ACK" series in the chart) triggers a corresponding sliding of the DTNperf_2 transmission window, thus allowing the remaining 200 bundles to be progressively sent and then taken into custody by the ION node inside the LEO satellite (second part of the "SENT/Custody on LEO" series). When the LEO link closes (at 900 s) all the bundles (400) have been sent by the source and taken into custody by the LEO ION node, but only a part (344, i.e. 68.8 MB) have actually been transferred to the control center as yet. Consequently, it is necessary to wait for the second contact (at 6300 s) to transmit the bundles still in custody (56, i.e. 11.2 MB) and complete the file transfer. LEO link availability is highlighted through horizontal segments in the figure. As a final remark, note that the transmission of 68.8 MB on the first pass, given a theoretical contact volume of 75 MB, is an excellent result, as link utilization efficiency is greater than 0.9.

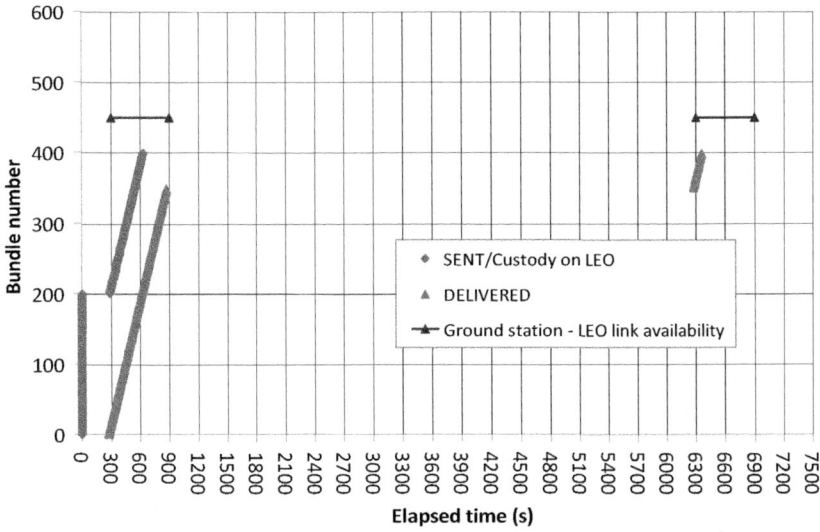

Fig. 4. Earth observation: bundle transmission logs

5.2 Data Mule Scenario

Here (Fig. 2-b) a 20 MB data transfer from two terrestrial nodes connected via one LEO satellite is considered. The LEO satellite forwards data from the first ground station to the second, alternately in line of sight with the satellite. The corresponding testbed topology is shown in Fig. 5.

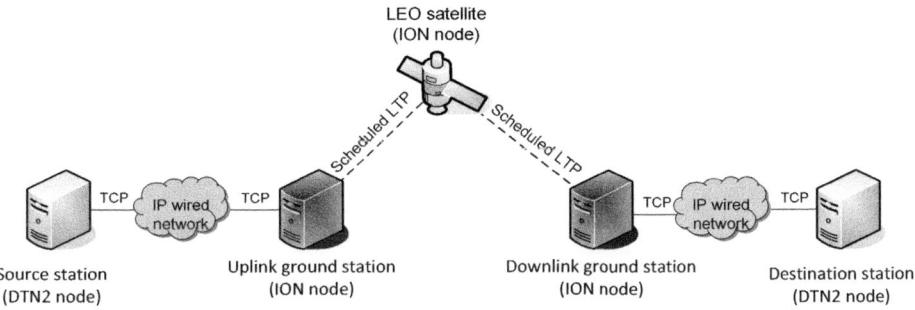

Fig. 5. Data mule topology

The file dimension has been assumed here lower than the maximum contact volume (75 MB as before, Table 1), which allows the file to be transferred in a single pass. Therefore, when LEO passes over the first ground station it is able to get the entire file; then, as soon as it is in line of sight with the second ground station, the file is transferred entirely toward the destination station. A second and last contact with the first ground station has the sole purpose of transmitting bundle acknowledgments (i.e., "delivered" status reports) to the source station running the DTNperf_2 client. As in the previous case, in the destination station the DTNperf_2 server application reassembles the transmitted file by collecting and reordering the arriving bundles.

Bundle transfer is illustrated in Fig. 6. At time zero all bundles are immediately transferred by the DTNperf_2 client to the source station bundle layer ("SENT" series) and from here to the first ground station, where they wait in custody for the next satellite link contact. "Custody" status reports generated by both the source and the first ground station are not shown, as they would overlap the "SENT" series. When the LEO satellite passes over the first ground station (300 s after time zero), bundles are progressively transferred on board (at 1 Mbit/s) and taken into custody ("Custody on LEO" series). The transfer time is about 160 s. When the satellite comes into line of sight with the second ground station (at 3300 s), bundles are downloaded (at 1 Mbit/s) and transferred at high speed (100 Mbit/s) to destination, which, in turns, sends back "delivered" status reports. The "DELIVERED" series in the chart represents here the time at which bundles are actually delivered (this information is contained in the "timestamp" field of "delivered" status reports). On their way back, the "delivered" reports have to stay on board the satellite until the next pass on the first ground station (at 6300 s), when they can finally be transferred to the source ("DELIVERED ACK" series). Their transfer time is almost instantaneous (vertical slope in the chart) because status reports consist of only few tents of bytes.

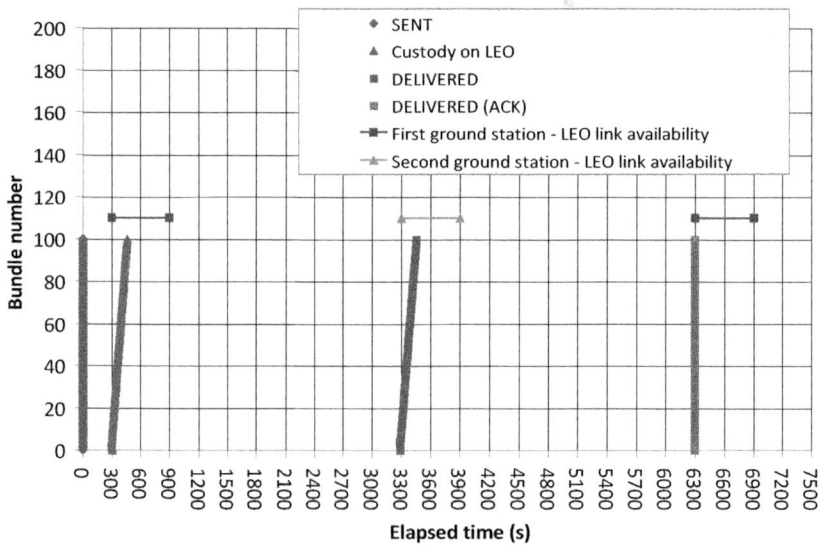

Fig. 6. Data mule: bundle transmission logs

6 Conclusions

In this paper the advantages of DTN when applied to LEO satellite communications have been assessed, considering both single satellites and constellations. In the former case, some preliminary results, obtained on a real testbed based on GNU/Linux machines are discussed. The tests were performed using both DTN2 and ION Bundle protocol implementations, and the DTNperf_2 evaluation tool. In both the applications considered, namely Earth observation and data mule communications, the results show the feasibility and the advantages of DTN in LEO satellite communications. In fact, both cases are characterized by intermittent connectivity on scheduled intervals, a challenge that would prevent the use of ordinary file transfer protocols and TCP/IP stack, but which is effectively tackled by DTN, as shown in the paper.

References

1. Burleigh, S., Hooke, A., Torgerson, L., Fall, K., Cerf, V., Durst, B., Scott, K., Weiss, H.: Delay-tolerant networking: an approach to interplanetary Internet. IEEE Communications Magazine 41(6), 128–136 (2003)
2. Farrell, S., Cahill, V., Geraghty, D., Humphreys, I., McDonald, P.: When TCP Breaks: Delay- and Disruption- Tolerant Networking. IEEE Internet Computing 10(4), 72–78 (2006)
3. McMahon, A., Farrell, S.: Delay- and Disruption-Tolerant Networking. IEEE Internet Computing 13(6), 82–87 (2009)
4. Fall, K., Farrell, S.: DTN: an architectural retrospective. IEEE Journal on Selected Areas in Commun. 26(5), 828–836 (2008)
5. DTNRG web site, http://www.dtnrg.org/wiki (last visited January 11, 2010)
6. Cerf, V., Hooke, A., Torgerson, L., Durst, R., Scott, K., Fall, K., Weiss, H.: Delay-Tolerant Networking Architecture. IETF RFC 4838 (April 2007)
7. Scott, K., Burleigh, S.: Bundle Protocol Specification. IETF RFC 5050 (November 2007)
8. Caini, C., Cornice, P., Firrincieli, R., Lacamera, D.: A DTN Approach to Satellite Communications. IEEE Journal on Selected Areas in Communications, special issue on Delay and Disruption Tolerant Wireless Communication 26(5), 820–827 (2008)
9. Ivancic, W., Eddy, W.M., Stewart, D., Wood, L., Northam, J., Jackson, C.: Experience with Delay-Tolerant Networking from Orbit. Int. J. of Satell. Commun. And Networking 28(5-6), 335–351 (2010)
10. Allman, M., Paxon, V., Stevens, W.: TCP Congestion Control. IETF RFC 5681 (September 2009)
11. Burleigh, S., Ramadas, M., Farrell, S.: Licklider Transmission Protocol —Motivation. IETF RFC 5325 (September 2008)
12. Ramadas, M., Burleigh, S., Farrell, S.: Licklider Transmission Protocol —Specification. IETF RFC 5326 (September 2008)
13. Wood, L., McKim, J., Eddy, W., Ivancic, W., Jackson, C.: Using Saratoga with a Bundle Agent as a Convergence Layer for Delay-Tolerant Networking. IETF Internet draft, work in progress, http://tools.ietf.org/id/draft-wood-dtnrg-saratoga (last visited January 11, 2010)

14. Border, J., Kojo, M., Griner, J., Montenegro, G., Shelby, Z.: Performance Enhancing Proxies Intended to Mitigate Link-Related Degradations. IETF RFC 3135 (June 2001)
15. ETSI TR 102 676: Satellite Earth Stations and Systems (SES); Broadband Satellite Multimedia (BSM): Performance Enhancing Proxies (PEPs)
16. Caini, C., Firrincieli, R., Cruickshank, H., Marchese, M.: Satellite Communications: from PEPs to DTN. In: Proc. of ASMS 2010, Pula, Italy, pp. 62–67 (September 2010)
17. Iridium website, http://www.iridium.com (last visited January 11, 2010)
18. Globalstar website, http://www.globalstar.com (last visited January 11, 2010)
19. DTN2 Reference Implementation, http://www.dtnrg.org/wiki/Code (last visited January 11, 2010)
20. Burleigh, S.: Interplanetary Overlay Network (ION) an Implementation of the DTN Bundle Protocol. In: The Proc. of 4th IEEE Consumer Communications and Networking Conference, pp. 222–226 (2007)
21. ION code, http://www.openchannelfoundation.org/projects/ION, (last visited January 11, 2010)
22. Caini, C., Cornice, P., Firrincieli, R., Livini, M.: DTNperf_2: a Performance Evaluation tool for Delay/Disruption Tolerant Networking. In: Proc. of ICUMT 2009 (E-DTN session), St.-Petersburg, Russia, pp. 1–6 (October 2009)
23. DTNperf_2 source code: DTN2 Mercurial repository, http://dtn.hg.sourceforge.net/hgweb/dtn/DTN2 (last visited January 11, 2010)

Asymmetric Spray and Multi-forwarding for Delay Tolerant Networks

Yue Cao, Haitham Cruickshank, and Zhili Sun

Centre for Communication System Research,
University of Surrey, Guildford, GU2 7XH, UK
{Y.Cao,H.Cruickshank,Z.Sun}@Surrey.ac.uk

Abstract. The framework of Delay Tolerant Networks (DTNs) has received an extensive attention from academic community because of its application ranging from Wireless Sensor Networks (WSNs) to interplanetary networks. It has a promising future in military affairs, scientific research and exploration. Due to the characteristic of long delay, intermittent connectivity and limited network resource, the traditional routing algorithms do not perform well in DTNs. In this paper, our proposed algorithm is based on an asymmetric spray mechanism combining with the concept of message classes. For each message class, a corresponding forwarding queue is designed and these queues are scheduled according to their priorities. Together with other designed assistant functions, our proposed algorithm outperforms other state of the art algorithms in terms of delivery ratio, overhead ratio, average latency as well as energy consumption.

Keywords: Delay Tolerant Networks, Routing Algorithms.

1 Introduction

The TCP/IP protocol has played an important role in the development of Internet. Specifically, it works under the assumptions such as contemporaneously end-to-end connectivity, relatively short round trip time and low error rate. However, this is impossible for some challenge networks including wildlife tracking, Vehicle Ad hoc Networks (VANETs), interplanetary networks, military networks, pocket switched networks, underwater networks and rural Internet. Generally, these are intermittently connected because of the sparse network density or high mobility.

Delay Tolerant Networks (DTNs) [1] are designed to cope with these challenges. It makes use of scheduled, predicted and opportunistic connectivity, forms a store and forward overlay network to provide custody based message oriented transfer. Routing is the main challenge in DTNs since its characteristic prevents the traditional routing techniques from working effectively. Up to now, many existing routing algorithms in DTNs have been proposed to enable message delivery in such challenge environment. Delivery ratio as the main performance objective is always taken into account. However, the performance of these algorithms creates more contention in terms of the network resource and more energy consumption even if they can achieve a high delivery ratio.

G. Giambene and C. Sacchi (Eds.): PSATS 2011, LNICST 71, pp. 199–212, 2011.

In general, the routing protocols must make a tradeoff between maximizing the message delivery ratio and minimizing resource consumption. On one hand, the ideal approach is to use the single copy for successful delivery. However on the other hand, the effective way to maximize the message delivery is to enlarge the number of message copies in the networks. Therefore, the feasible approach to reduce the overhead but maintain the high delivery ratio is to intelligently replicate the message.

The main contribution of this paper is the design of an algorithm to achieve high delivery ratio with low overhead as well as the relatively less latency and energy consumption. Our proposed algorithm mainly implements an asymmetric spray approach to promote the message dissemination to the intermediate node which more likely encounters the destined node with the guarantee that the message can be delivered before its expiration time. Based on the characteristic of messages, we classify them into three classes. For each class, a corresponding forwarding queue is proposed. In particular these queues are dynamically scheduled according to the defined priorities.

In the following section, we briefly review the taxonomy of unicast routing algorithms in DTNs, then in section 3 we present our algorithm. The simulation results are presented in section 4 followed by the conclusion section.

2   Related Work

Excluding the assistance of additional infrastructure, the taxonomy of unicast routing algorithms in DTNs are mainly classified into three families, which are single copy utility forwarding, multi copy naive replication and hybrid families.

2.1   Single Copy Utility Forwarding Family

The algorithms in this family use only one copy, which means the message carrier does not keep the copy of the forwarded message after the successful transfer. The earlier stage algorithms in [2] focus on the delay of each link state and requires a global information to route the message based on the shortest path. Social networks as a new research area proposed in recent years utilize the encounter relationship of pairwise nodes [3]. Other parameters such as energy, movement speed, network density and location can also be used for routing decision. For example the Context Aware Routing (CAR) [4] utilizes the residual energy, variation of network topology for the DTNs based routing algorithm. In particular, if the contemporaneously end-to-end connectivity is currently available, then the routing function shifts to the traditional routing protocol such as Destination-Sequenced Distance Vector (DSDV) to forward the message, otherwise it adopts the context information to select the candidate node for the DTNs based routing algorithm. Nevertheless, these routing algorithms do not work effectively in the sparse scenario where the message lifetime is quite limited since the only one copy can not guarantee the effective delivery.

2.2 Multi Copy Naive Replication Family

The simplest algorithm is Direct Delivery [5], which only replicates the message if the current carrier encounters the destination. It is considered as a degraded naive replication based algorithm. The Epidemic as the earliest multi-copy based algorithm is proposed in [6]. In detail, each node does not implement the routing decision but just replicates the message to encountered node unless it already carries this message. Provided that the buffer resource and bandwidth is large enough, Epidemic theoretically guarantees the lowest latency for maximum delivery ratio. Nevertheless, the contention due to the limited resource in reality is the main limitation of the scalability. The Spray-and-Wait [7] combines the diffusion speed of Epidemic with the simplicity and thriftiness of Direct Delivery. For each message, an initial number of copy tickets is defined to limit the number of replication, which enables them to be sprayed at each encounter opportunity with the guarantee at least one of them can be delivered. Intermediate node carrying the message of which the copy ticket is one performs the Direct Delivery.

2.3 Hybrid Family

The algorithms in this family utilize the advantage of single copy utility forwarding and naive replication based algorithms, which aims to improve the overhead ratio as well as acceptable delivery ratio. The Prophet [8] integrates the property of replication and prediction based forwarding. The current carrier replicates the message to the candidate node with higher encounter probability. In addition, it also uses the transitivity to enhance the congestion avoidance. The core concept of the MaxProp [9] protocol is a ranked list of the carried messages based on a cost for each destination. The cost is an estimate of virtual end to end route failure possibility, initially the possibility for each pair of nodes is uniformly distributed and updated according to the incremental averaging. Two thresholds are defined to calculate the drop and transmission priority for message. In addition, MaxProp uses acknowledgment to inform the intermediate node to clear out the existing copies of the delivered messages. The Spray-and-Focus [10] aims to optimize the Spray-and-Wait in the wait phase. Instead, it forwards the message with single copy to the candidate node with transitive recent encounter time rather than just wait. Nevertheless, its performance is strongly affected by the specific mobility characteristic.

3 Our Proposed Algorithm

The overall function flowchart of the proposed algorithm is illustrated in Fig.1 and its specific functions are introduced in the following subsections.

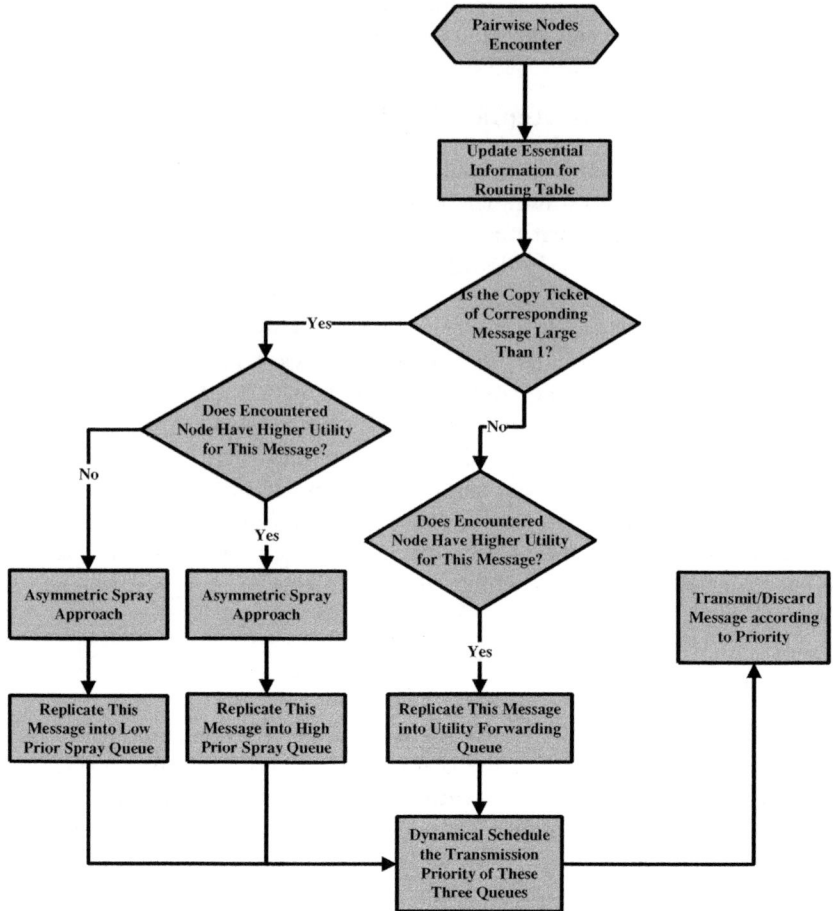

Fig. 1. Function Flowchart

3.1   Definition of the Utility Metric

Traditionally, the main problem of designing an efficient routing algorithm in DTNs is how to obtain the network topology information. Due to the limited property of device, it is difficult to obtain this information by the broadcasting mechanism used in MANETs. Some techniques in related work assume that the partial history information can predict the future encounter opportunity. However, it does not comprehensively take into account the mobility pattern. Assuming the future mobility pattern is known in advance is unreasonable in DTNs, thus our algorithm is designed based on these assumptions but makes use of the history information.

Considering the mobility factor, we address three conventional metrics between pairwise encountered nodes N_i and N_j, which are history encounter count $C_{i,j}$, history encounter duration $D_{i,j}$ and history inter meeting time $I_{i,j}$ where $i, j \in S$ and S is the total number of nodes in the network. We assume DTN node does not strictly move with a cyclic mobility pattern. To this end, we propose an cumulative formula to smooth the effect of large variation within the number of encounters, where the utility $U_{i,j}$ is defined as:

$$U_{i,j} = \frac{\sum_{k=2}^{C_{i,j}} (\frac{D_{i,j}^{k-1}}{I_{i,j}^{k}})}{(C_{i,j} - 1)} \qquad (1)$$

In detail, $D_{i,j}^{k-1}$ is the encounter duration before the kth encounter and it is valid after the first encounter. With the time elapsing this property is useful since the node experiences a large number of encounters is more likely to successfully route the message to the final destination than those who have infrequent encounters.

Normally, for each encounter opportunity the pairwise nodes would update their local routing information which contains a set of $U_{i,j}$ for the nodes they encountered before. Nevertheless, to estimate the delivery potential based on the local knowledge is unreasonable since it ignores the factor of its history encountered nodes.

Therefore, we propose an approach to help each node to improve this limitation. For instance, when pairwise nodes encounter, firstly both of them would calculate and update their $U_{i,j}$ for each other. Afterwards, they would also add their local routing information to each other for the purpose of extending the knowledge. To this end, they can obtain the knowledge from their neighbors' history encounter information. Based on the above analysis, an improved utility $U'_{i,j}$ is proposed:

$$U'_{c,d} = \frac{\sum_{k=1}^{n} U_{c(k),d} + U_{c,d}}{count + 1} \qquad (2)$$

where n is the number of history encountered nodes of current carrier c, $c(k)$ is the history encountered node of c, d is the corresponding destination node. In detail, $count$ is initialized with zero and increased by one if $c(k)$ contains the $U_{c(k),d}$ for d. Therefore, the local node would obtain an abstract average knowledge for d not only based on itself but also based on the history information from its neighbors by (2). As an example in TABLE 1, the $U'_{6,2}$ based on the view of N_6 is calculated as:

$$U'_{6,2} = \frac{0.6 + 0.3 + 0}{2 + 1} = 0.3 \qquad (3)$$

3.2 Asymmetric Spray Approach

Binary Spray-and-Wait as a classic algorithm has been proved and used in many scenarios because of its acceptable high delivery ratio and relatively low overhead ratio. However, it does not take into account the delivery potential of the candidate node. Each node just naively sprays half number of copy tickets to any encountered node. Based on our improved utility defined in the previous

Table 1. Example of Routing Table of N_6

current carrier N_6	$U_{6,1}$	0.4
	$U_{6,5}$	0.2
	$U_{6,4}$	0.0
history encountered N_1 of N_6	$U_{1,6}$	0.4
	$U_{1,2}$	0.6
	$U_{1,4}$	0.3
history encountered N_5 of N_6	$U_{5,2}$	0.3
	$U_{5,8}$	0.1
	$U_{5,6}$	0.2

subsection, we assume that each node has a certain potential to route the message towards its destination. Therefore, on one hand, to equally spray the copy ticket might not be reasonable since to spray the half number of copy tickets to the encountered node with less $U'_{i,j}$ would waste some encounter opportunity. Relatively, on the other hand, to unequally spray the copy ticket without any consideration is also infeasible. To this end, we propose a novel copy ticket spray approach based on binary Spray-and-Wait. For each message M of which the destination is N_d and with T copy tickets carried by N_i, if node N_i has a lower $U'_{i,d}$ for this message destination than $U'_{j,d}$ of the encountered node N_j, N_i sprays more copy ticket of message M to N_j and keeps less copy ticket by itself. The specific process is illustrated in Algorithm 1. Provided that $U'_{i,d}$ is larger than or equal to $U'_{j,d}$, inherently, it is appropriate for the current carrier to keep the original copy ticket until it encounters a better candidate node. Nevertheless, this behavior might result in higher latency since the specific future prediction of mobility is independent of our assumption. As such, the poor candidate node might encounter another better candidate node in the future. Therefore, the current carrier sprays less number of the copy tickets to the encountered node with lower $U'_{j,d}$.

3.3   Multi-forwarding Approach

High Priority Spray Queue ($HPSQ$): Upon the asymmetric spray approach, for each encounter of pairwise nodes, the current carrier N_i will check whether the encountered N_j has a higher improved utility for the destination of the carried message M. Besides, it also checks whether the copy ticket of M is larger than one. If any message M accords with the above two conditions, then N_i replicates M to N_j and pushes it into $HPSQ$. Basically, with the asymmetric spray mechanism, more copy ticket of this message would be sprayed to the candidate node which potentially moves towards the destination for efficient delivery.

Algorithm 1. Asymmetric Spray Approach

Input:
    current carrier: N_i
    encountered node: N_j
    carried messages in N_i: M with its destination N_d and copy ticket T
    improved utility for N_d: $U'_{i,d}$, $U'_{j,d}$
Output:
1. **for** each M **do**
2.   **if** $U'_{i,d} < U'_{j,d}$ **then**
3.   **if** $T > 2$ **then**
4. N_i sprays M with $math.ceil(\frac{T}{2})+math.round(\frac{U'_{j,d}}{U'_{j,d}+U'_{i,d}})$ to N_j
5. N_i keeps M with $math.ceil(\frac{T}{2.0})-math.round(\frac{U'_{j,d}}{U'_{j,d}+U'_{i,d}})$
6.   **else**
7. N_i sprays M with $(\frac{T}{2})$ to N_j
8. N_i keeps M with $(\frac{T}{2})$
9.   **end if**
10.   **else**
11.   **if** $T > 2$ **then**
12. N_i sprays M with $math.ceil(\frac{T}{2.0})-math.round(\frac{U'_{i,d}}{U'_{j,d}+U'_{i,d}})$ to N_j
13. N_i keeps M with $math.ceil(\frac{T}{2})+math.round(\frac{U'_{i,d}}{U'_{j,d}+U'_{i,d}})$
14.   **else**
15. N_i sprays M with $(\frac{T}{2})$ to N_j
16. N_i keeps M with $(\frac{T}{2})$
17.   **end if**
18.   **end if**
19. **end for**

Low Priority Spray Queue ($LPSQ$): If the encountered node has a smaller improved utility for the destination than the current carrier, N_i would try its best to spray the copy ticket of all the carried messages to the encountered node even it spays the less copy ticket. These messages with more than one copy ticket are pushed into $LPSQ$.

Utility Forwarding Queue (UFQ): Regarding the message of which the copy ticket is equal to one. It cannot be sprayed but performed as the utility based replication mechanism. For each M destined to N_d carried by N_i, this message is replicated to the encountered node only if $U'_{j,d} > U'_{i,d}$. Accordingly, this message is pushed into the UFQ.

3.4 Scheduling the Priority of Queues

Inherently, the messages with multi-copy ticket in $HPSQ$ and $LPSQ$ should be scheduled prior to the messages with single copy ticket in UFQ. The main reason is that for messages in $HPSQ$ and $LPSQ$, they are sprayed with the dedicated copy ticket. If their life time expire, the worst case is that the messages with

---

Algorithm 2. Multi-Forwarding Approach

Input:
    current carrier: N_i
    encountered node: N_j
    carried messages in N_i: M with its destination N_d and copy ticket T
    improved utility for N_d: $U'_{i,d}$, $U'_{j,d}$

Output:
1. **for** each encounter between N_i and N_j **do**
2.   **for** each M in N_i **do**
3.   **if** N_j already has M **then**
4. M is skipped for replication
5.   **else if** $U'_{i,d} < U'_{j,d}$ and $T > 1$ **then**
6. N_i replicates M according to asymmetric spray and puts into $HPSQ$
7.   **else if** $U'_{i,d} < U'_{j,d}$ and $T = 1$ **then**
8. N_i replicates M and puts into UFQ
9.   **else if** $T > 1$ **then**
10. N_i replicates M according to asymmetric spray and puts into $LPSQ$
11.   **end if**
12.   **end for**
13. **end for**

Algorithm 3. Multi-Forwarding Approach

Input:
    priority for $HPSQ$: SP_{HPSQ}
    priority for $LPSQ$: SP_{LPSQ}

Output:
1. **for** each message transfer **do**
2.   **if** $SP_{HPSQ} \geq SP_{LPSQ}$ **then**
3.   schedule $HPSQ$ until $HPSQ$ is empty
4.   then schedule $LPSQ$ until $LPSQ$ is empty
5.   then schedule UFQ until UFQ is empty
6.   **else**
7.   schedule $LPSQ$ until $LPSQ$ is empty
8.   then schedule $HPSQ$ until $HPSQ$ is empty
9.   then schedule UFQ until UFQ is empty
10.   **end if**
11. **end for**

---

maximum copy ticket are cleared out from the buffer space. In this case it might degrade the delivery ratio.

Nevertheless, in order to schedule the priority between $HPSQ$ and $LPSQ$, we define a metric called scheduling priority SP for these two queues.

$$SP = \frac{\sum_{k=1}^{m}[MU(c,d) * MCT]}{m} \tag{4}$$

where m is the number of messages in the queue. After each message transfer, the current carrier will check the current SP of these two queues. To this end, as it is proposed in the DTNs RFC, we classify the messages into three classes which are bulk, normal and expedited. In the meanwhile they are processed according to the specific forwarding policies and scheduled according to their priorities respectively.

3.5 Transmit/Discard Message According to Priority

The main motivation to define the message priority MP is to transmit the most appropriate message at each encounter opportunity. Totally different from the algorithms in traditional networks focus on delay, herein we propose to address the delivery potential of each message based on the view of the corresponding node.

The priority of message in this $HPSQ$ is defined as:

$$MP_{HPSQ} = MU(e, d) * MCT \tag{5}$$

It is based on the view of the encountered node e for the destined node d. In detail, MCT is the message copy ticket and MU is the message utility that is defined as $U'_{e,d}$, therefore these messages are scheduled according to the improved utility based on the view of the encountered node. For example, if encountered node has a higher $U'_{e,d}$ for the destination of M_1 than M_2, then M_1 is allocated with higher priority than M_2 if both of them are sent to this encountered node.

The main difference between the priority of messages in $HPSQ$ and $LPSQ$ is that the messages in $LPSQ$ are scheduled based on $U'_{c,d}$ of the current carrier c and their copy tickets. It is defined as:

$$MP_{LPSQ} = MU(c, d) * MCT \tag{6}$$

For messages with one copy ticket that are processed by UFQ, their priorities are defined as:

$$MP_{UFQ} = \frac{MU(e, d)}{Message\ Lifetime} \tag{7}$$

If the message with higher $MU(e, d)$ but with very limited lifetime, it is regarded to be the emergent message. To this end, as the priority proposed, the message which has high potential to be delivered based on the view of the encountered node e and low lifetime is always guaranteed to exist in the networks, which plays a positive effect on maximizing the delivery ratio.

Normally, the storage is limited in the restricted scenario and accordingly each node can not carry all the messages. Hence a reasonable buffer management function is essential. The carried messages are classified into multi copy ticket based and single copy ticket based, then they are pushed into different bins respectively and discarded according to DP defined as:

$$DP = MU(c, d) * MCT \tag{8}$$

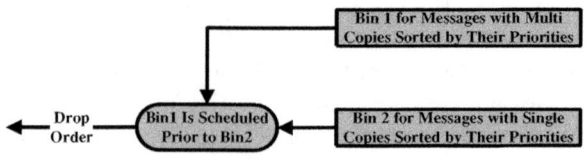

Fig. 2. Drop Priority

Normally the messages in the bin for multi copy ticket based are firstly discarded from the lowest priority. If there are no more messages in this bin, then the messages in the bin for single copy ticket based are discarded from the lowest priority. As it is illustrated in Fig.2, consideration behind this approach is that the message with lowest copy ticket and lowest delivery potential based on the view of current carrier would be more useless since most of its copy has been sprayed to a better candidate node. The messages with one copy ticket are assumed can be delivered with higher potential since they are replicated based on the improved utility. Therefore they are discarded with lowest priority once there are no more messages with multi copy ticket.

To further reduce the redundant transmissions, the destination will generate an acknowledgement of which the size can be ignored compared with the size of message when it successfully receives this message, and this acknowledgement will be flooded to the entire network. Intermediate nodes receive this acknowledgement will check their buffer and discard the message which has been successfully received.

4   Simulation Results

The simulation results are evaluated by Opportunistic Network Environment (ONE) [11]. The scenario area is 15.3 km^2 with 126 mobile nodes configured with different variable speeds. In particular each node has an interest to visit some places rather than randomly select the next point based on the route. We evaluate the Spray-and-Focus (SaF), binary Spray-and-Wait (SaW), Epidemic, Prophet and MaxProp for comparison. Energy function is also integrated into all these algorithms. For the purpose of fairness, the initial number of copies for SaF, SaW is set to 13, which is a recommended value between 10% and 15% nodes in the scenario. We address delivery ratio, overhead ratio, average latency and total residual energy for performance evaluation. Specifically, the delivery ratio and overhead ratio are defined as (9) and (10), total residual energy is measured by the sum of the residual energy of each node.

$$Delivery\ Ratio = \frac{Delivered\ Messages}{Generated\ Messages} \tag{9}$$

$$Overhead\ Ratio = \frac{Relayed\ Messages - Delivered\ Messages}{Number\ of\ Delivered\ Messages} \tag{10}$$

Table 2. Simulation Configurations

Simulation Time	12 Hours
Bandwidth	2Mb/s
Transmission Range	10m
Buffer Size	10MB
Number of Nodes	126
Message Size	200kB-2MB
Message Generation Interval	30s
Message Lifetime	240 Minutes
Initial Energy per Node	850mA/h
Transmission Energy per Node	51.47mA/h
Scanning Energy per Node	38.61mA/h
Scenario Mobility	Helsinki City Model

4.1 Effect of Buffer Size

In Fig.3(a), both Epidemic and Prophet achieve the lowest delivery ratio because they do not take into account the utilization of network resource. SaW and SaF limit the number of replication for message, thus the contention regarding the bandwidth and buffer space is alleviated. Compared with MaxProp, which is regarded as a preeminent one for comparison, our algorithm achieves higher delivery ratio particular when the buffer size increases.

With respect to the overhead ratio in Fig.3(b), the overhead ratio of our proposed algorithm is close to SaW if the buffer space is large enough. SaF requires more transmission during the focus phase whereas SaW just implements Direct Delivery in its second phase, this results in a higher overhead ratio of SaF compared with SaW. Even if MaxProp is well designed with buffer management function, our proposed algorithm still outperforms MaxProp.

In Fig.3(c), our proposed algorithm also achieves the lowest average latency compared with other algorithms. Particularly, as we discuss in previous section, our asymmetric spray mechanism plays the important role on this good performance. Another contribution comes from the scheduling approaches, which aims to transmit the most appropriate message at each encounter opportunity.

Energy issue as a new issue has been taken into account in DTNs. According to the result in Fig.3(d), Inherently, our algorithm performs a utility replication approach to route the message if its copy ticket is equal to one, which occupies more network resource and might abort some messages due to the mobility factor. Therefore, it requires the retransmission for the messages which have been aborted, then it consumes more energy than SaW and relatively similar energy as SaF.

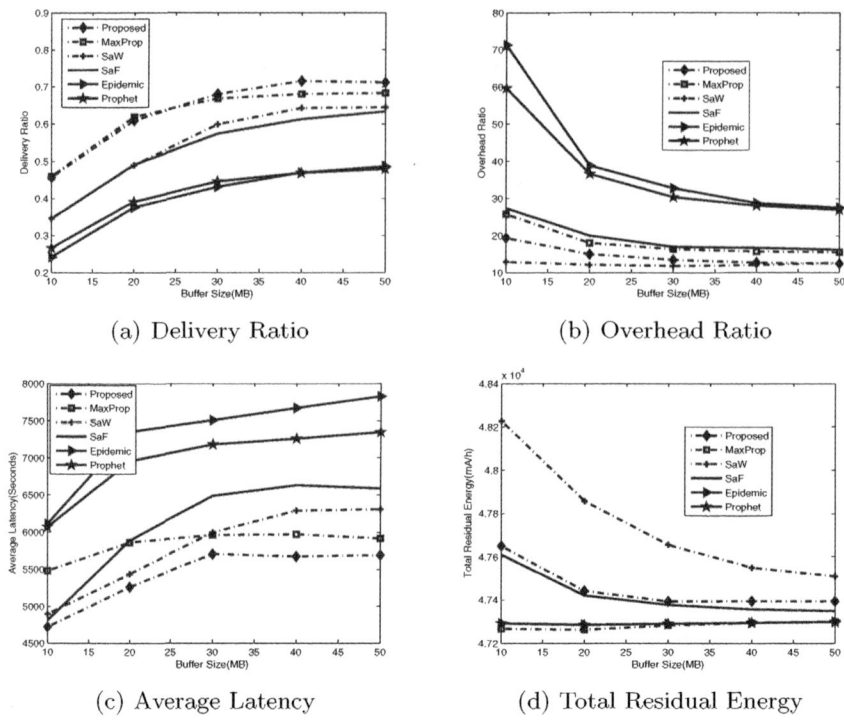

(a) Delivery Ratio   (b) Overhead Ratio

(c) Average Latency   (d) Total Residual Energy

Fig. 3. Effect of Buffer Size

4.2   Effect of Message Lifetime

In this section, we fix the buffer size as 50MB but vary the value of message lifetime.

When the message TTL increases in Fig.4(a), our algorithm still outperforms other algorithms. MaxProp with a dedicated buffer management also performs well compared with SaF and SaW since they are not designed with any buffer management function. Due to the limited resource, Epidemic and Prophet perform worse.

The inherent characteristic of our algorithm determines its overhead would be higher than SaW in Fig.4(b). However the difference is quite close if the message TTL is increased, this is because the asymmetric spray approach works significantly since it can spray the message to the candidate node before the expiration time.

With respect to the average latency in Fig.4(c), our algorithm achieves the lowest latency, which is similar to the result affected by the buffer size.

Because of the large message lifetime, the messages in the buffer space might be discarded in case of the replication algorithms. Thus the current carrier would require more transmission for the messages which have been cleared from the buffer, which results in more energy consumption. According to the result in

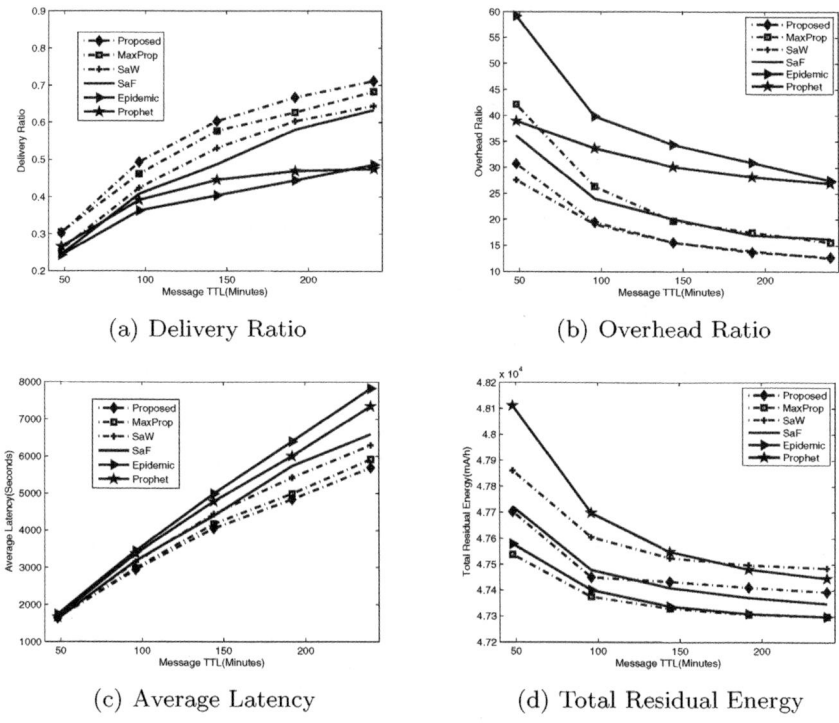

(a) Delivery Ratio   (b) Overhead Ratio

(c) Average Latency   (d) Total Residual Energy

Fig. 4. Effect of Message Lifetime

the Fig.4(d), based on the the overall performance, SaW and SaF consume less energy since they adopt less number of replication. Relatively, Prophet does not achieve an acceptable delivery ratio even if it consumes the least energy. Our algorithm balances the energy consumption and the deliver ratio, thus it is energy efficient.

5   Conclusion

The ability to efficiently route message and appropriately select the candidate node through intermittently connected networks is critically important in DTNs. Most of the algorithms in hybrid family achieve high delivery ratio but still with relatively high overhead ratio. Besides, the limited network resource degrades the performance due to the contention of the buffer space and bandwidth usage. With a novel multi-forwarding model based on the dynamic message classification and an asymmetric spray scheme, our proposed algorithm outperforms other state of the art algorithms in terms of message delivery ratio, overhead ratio and average latency with lower energy consumption as well.

References

1. Fall, K., Farrell, S.: DTN: an Architectural Retrospective. IEEE Journal on Selected Areas in Commun. 26(5), 828–836 (2008)
2. Jain, S., Fall, K., Patra, R.: Routing in a Delay Tolerant Network. In: SIGCOMM 2004, Portland (September 2004)
3. Daly, E., Haahr, M.: Social Network Analysis for Routing in Disconnected Delay-Tolerant MANETs. In: MOBIHOC, Montréal, Québec, Canada (2007)
4. Musolesi, M., Masclo, C.: CAR: Context Aware Adaptive Routing for Delay Tolerant Mobile Networks. IEEE Transaction on Mobile computing 8(2), 246–260 (2009)
5. Grossglauser, M.: Tse DNC.: Mobility increases the capacity of ad hoc wireless networks. IEEE/ACM Transactions on Networking 10(4), 477–486 (2002)
6. Becker, D.: Epidemic routing for partially connected ad hoc networks. Technique Report, CS-2000-06, Department of Computer Science, Duke University, Durham, NC (2000)
7. Spyropoulos, T., Psounis, K., Raghavendra, C.: Effcient routing in intermittently connected mobile networks: The multiple-copy case. IEEE/ACM Transactions on Networking 16(1), 77–90 (2008)
8. Lindgren, A., Doria, A., Schelén, O.: Probabilistic Routing in Intermittently Connected Networks. In: SIGMOBILE Mobile Computing Communications Review (July 2003)
9. Burgess, J., Gallagher, B., Jensen, D.: MaxProp: Routig for Vehicle-Based Disruption Tolerant Networking. In: IEEE INFOCOM 2006, Barcelona (April 2006)
10. Spyropoulos T., Psounis, K., Raghavendra, C.: Spray and Focus: Effcient mobility Assisted routing for Heterogeneous and Correlated Mobility. In: PerCOM, Newyork (March 2007)
11. http://www.netlab.tkk.fi/tutkimus/dtn/theone/

A Proactive DOS Filter Mechanism for Delay Tolerant Networks

Godwin Ansa, Haitham Cruickshank, and Zhili Sun

Centre for Communications Systems Research, University of Surrey, GU2 7XH, England
{g.ansa,h.cruickshank,z.sun}@surrey.ac.uk

Abstract. Denial of Service (DOS) attacks are a major threat faced by all types of networks. The effect of DOS in a delay tolerant network (DTN) is even more aggravated due to the scarcity of resources. Perpetrators of DOS attacks in DTN-like environments look beyond the objective of rendering a target node useless. The aim of an attacker is to cause a network-wide degradation of resources, service and performance. This can easily be achieved by exhausting node or link resources and partitioning the network. In this paper we seek to provide a proactive approach in making the DTN authentication process robust against DOS. Our aim is to make security protocols which provide mandatory DTN security services resilient to DOS attacks. The overall objective is to make it hard to launch a DOS attack and ensure the availability of DTN services. A DTN-cookie mechanism has been proposed to quickly identify and filter out illegitimate traffic.

Keywords: Denial of service, attacker, delay tolerant network, resource exhaustion, DTN-cookie.

1 Introduction

Delay tolerant networking is fast becoming an area of great research interest. Where there is no direct link between a source and a destination, a node in one region can pass a message to another node in a remote region using store-and-forward message switching technique. Store-and-forward message switching technique or asynchronous message passing [1] as illustrated in Fig. 1 requires that the integrity of a message is verified by an intermediate node before it is forwarded.

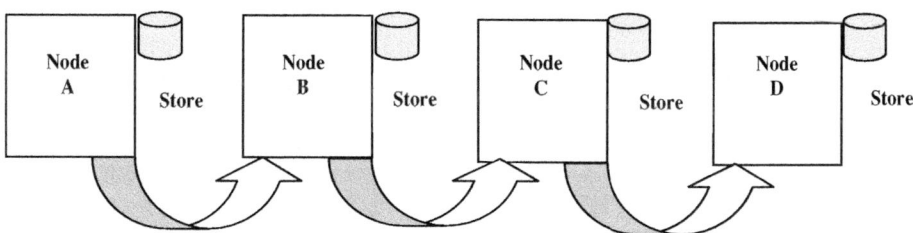

Fig. 1. DTN Store-and-Forward Message Switching

G. Giambene and C. Sacchi (Eds.): PSATS 2011, LNICST 71, pp. 213–226, 2011.
© Institute for Computer Sciences, Social Informatics and Telecommunications Engineering 2011

In Wireless Sensor Networks (WSNs), messages are forwarded to a destination through intermediate nodes which act as routers. Sensor nodes are resource-constrained in memory, CPU cycles, battery power, and bandwidth [2]. It is therefore expedient that DTN-enabled sensor nodes forward or take into custody only bundles that are authentic and still within their useful lifetime.

Denial of service attacks poses a major threat to availability because it prevents an entity or network from fulfilling its functions by disabling or degrading the services it provides [2], [3]. A classification of DOS attacks by the Computer Emergency Readiness Team (CERT) groups DOS into three main categories [4]: Destruction or alteration of configuration information, physical destruction or alteration of network components and consumption of scarce, limited, or non renewable resources. Our main focus is to protect the scarce resources of the DTN which include communication contact time, battery power, bandwidth, CPU cycles, disk space and memory from exhaustion.

We present a DTN scenario with two Wireless Sensor Networks and a third network where the sensor data are processed. Each network has a security gateway which acts as an interconnection point to facilitate inter-regional communications. The three networks are bridged using a satellite as relay node. Within each region opportunistic message forwarding or short-range radio communication such as Bluetooth can be used. Data mules can be deployed to collect sensitive sensor data in a more scheduled manner.

Each sensor network is depicted as shown in Fig. 2, is divided into X number of domains by the security gateway. This is done during the initial set-up phase and subsequently during periodic refreshments of the regional secret key and nonce values. Each domain comprises a Group Head (GH) node, one or more security-aware nodes and numerous sensor nodes.

The remainder of the paper is structured as follows. Section 2 gives a brief overview of related work on DOS defence mechanisms in different networks. The network threats, design objectives, networking and security requirements are presented in section 3. Section 4 provides design details of the DOS-filter mechanism for both intra-regional and inter-regional scenarios. It also enumerates the design assumptions, and describes in-depth the use of loose time synchronization of the security gateways. A detailed evaluation of the proposed mechanism is carried out in section 5. Section 6 concludes the paper and gives a summary of our contribution towards DOS resilience in DTN.

2   Related Work

A clever attacker can exploit the strong security of a system or network to launch a protocol-based DOS attack through flooding and resource exhaustion. A security protocol is prone to this type of DOS attack if the server commits memory or computational resources during the client authentication process. In terrestrial networks, a number of solutions have been proposed to tackle this problem. An initial work by Dwork and Noar [5] in tackling the junk mail problem proposed the use of cryptographic puzzles where a sender is required to compute a puzzle for every message sent. The cost of this technique is negligible for normal users when compared to mass mailers. The client puzzle idea was extended to connection depletion attacks such as the TCP SYN flood attack by Juels and Brainard [6].

Another technique which combats protocol-based DOS is the Internet Security Association and Key Management Protocol (ISAKMP) defined by IPSec and derived from the PHOTURIS protocol. It is an anti-clogging technique where a client is required to return a server generated cookie. See [4] and [7] for more information on the ISAKMP specification. Meadows [8] proposed a formal framework for network DOS. The idea is to gradually strengthen the authentication process as the protocol executes by introducing a weak authentication phase prior to signature verification. Leiwo et al. [9] suggest that allocation of server resources can only take place after client authentication, and that a client's workload must be greater than that of the server.

The aforementioned DOS mitigating solutions proposed for terrestrial systems are only suitable for low-delay well-connected networks but unsuitable for delay tolerant networks. Sensor nodes are resource-limited in battery power and computational capabilities and will not be able to solve the cryptographic puzzles; the ISAKMP cookie requires a number of message exchanges during client authentication which is infeasible in DTN. The round-trip delay, broadcast nature of the satellite channel and the wireless communication medium makes the scenario described in section 1 prone to eavesdropping, interception of cookie values and masquerade.

The authors of [10] define a header extension field with no related trailer field and three ciphersuites for the specification. In their definition, a cookie value can be a long random number whose length is determined by the implementation. They assert that longer cookies are stronger and harder to guess but consume more bandwidth.

3   Threat Analysis, Design Objectives, Networking and Security Requirements

This section provides a detailed threat analysis for the DTN scenario described in section 1. It outlines the design objectives and the networking and security requirements

3.1 Threat Analysis

The DTN scenario presented in section 1 is prone to eavesdropping due to the wireless communication medium and the broadcast nature of the satellite channel. The depicted scenario is also susceptible to bundle content modification, masquerade, and denial of service attack. The Bundle Security Protocol (BSP) specification [11] states that bundles have to be validated at a security-aware node for authenticity and integrity. The BSP defines four security blocks for this purpose, for more details see [11] and [12]. To protect the scarce resources of the DTN it is mandatory that bundles are BAB-protected and validated. The aim is to ensure that network resources are used solely for forwarding authentic bundles with valid lifetimes.

A clever attacker can exploit this requirement to flood the network with small-sized BAB-protected bundles or flood a target node directly. Since the BABs on the bundles are fake, a security-aware node will waste its resources (CPU and battery)

trying to verify the bundles. Victim nodes can become congestion points and legitimate bundles with no access to an alternative next hop node might be dropped if they expire on transit.

3.2 Security Objectives

Our primary objective is to make DTN security protocols resilient to DOS attacks launched through resource exhaustion by ensuring that the authentication process is robust and light-weight. The DOS filter mechanism should detect and discard malicious traffic as early as possible, detect and discard bundles whose headers have been modified, ensure that unauthorized entities do not gain control of the DTN infrastructure.

3.3 Networking and Security Requirements

It is imperative that a DOS filter mechanism for DTN should be able to withstand significant node mobility, run efficiently on resource-limited nodes like sensors and be resilient to delays which can be in the order of minutes, hours or days. The mechanism should support varying data rates and withstand changes in contact times. It should also be able to operate efficiently in the absence of an end-to-end path between source and destination.

In terms of security requirements, we restrict security processing to computationally capable nodes. To ensure freshness, we use nonce and timestamps to thwart the replay of old and expired bundles. Every bundle is checked for integrity to prevent bundle content modification during transit. Bundles are authenticated to ensure that they originate from legitimate sources and are still within their useful lifespan.

4 Design of a DOS Filter Mechanism for DTN Environments

A DOS filter mechanism should have a detection, classification and response element to be highly effective [3]. As design requirements, the DTN-cookie generation process must be simple and fast; the DTN-cookie value must be random and hard to forge, discourage Transport Layer Security (TLS) style of negotiation or handshake, the verification of the DTN-cookie should provide a weak authentication phase which is light-weight. Strong authentication can only take place if the weak authentication phase is successful; bundles that fail the weak authentication should be silently dropped.

Each sensor network is divided into X number of domains by the security gateway. This is done during the initial set-up phase and subsequently during periodic refreshments of the regional secret key and nonce values. Each domain comprises a Group Head (GH) node, one or more security-aware nodes and numerous sensor nodes. A GH node makes decisions on behalf of other nodes within its domain. It determines the Network Threat Level (NTL). The GH node and security-aware nodes act as data aggregation points for sensed data to ease management and coordination. These are more powerful in terms of computational and storage capabilities and act as security-sources and security-destinations for less-capable or IDless sensor nodes.

Bundle size within the DTN is fixed at 64KB for ease of processing. A node can only interact with the security gateway and other nodes within the same region. Inter-regional communication is gateway-to-gateway via the satellite whose pass is scheduled and can be predicted. The satellite acts only as a relay node for inter-regional communications. We define a new Administrative (AD) bundle called Alert for the dissemination of security information within each DTN region. An AD bundle has an expedited Class of Service (CoS) with priority of 1, while a data (D) bundle has a priority of 2 or 3 equivalent to Normal or Bulk CoS respectively.

The primary objective is to make DTN security protocols resilient to DOS attacks launched through resource exhaustion by ensuring that the authentication process is securely robust and light-weight.

4.1 Design Assumptions

- a node has bounded resources that could possibly be exhausted by a clever attacker
- the computational resource of the attacker is very large
- the attacker can compute efficiently pseudo-random functions and MACs in record time but does not possess the network secrets
- the attacker is assumed to have the ability to replay, modify, transmit, and receive bundles.
- the attacker has the ability to execute the protocol
- the attacker does not take over a legitimate node and in the process steal keying material and sensed data
- in inter-regional communications, the attacker is assumed to be a rogue router with enormous processing and sending capabilities and can predict the pass/schedule of the satellite
- trust is established during initial registration of a security- aware node with the security gateway
- the protocol is between two communicating entities
- a bi-directional communication asymmetry between a pair of nodes is assumed

A secure Key Management mechanism is required for the distribution of nonce seeds and cryptographic secret keys.

4.2 Intra-regional DOS Mitigation

Fig. 2 depicts the intra-regional DOS scenario in a DTN wireless sensor network which shows sensor nodes, Group Heads (GHs), security-aware (SA) nodes, an attacker and a security gateway. All in-bound and out-bound traffic must pass through the security gateway.

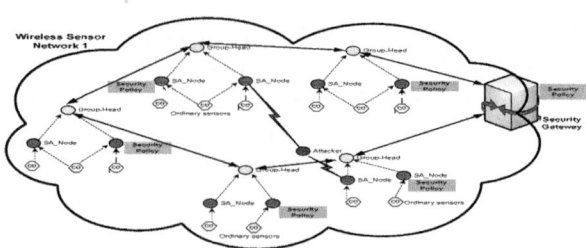

Fig. 2. Intra-Regional DOS Scenario

A generic DTN bundle is made up of the primary block and the payload block. Additional blocks such as the BAB, PIB and PCB can be added to provide security to the traffic. This is depicted in Fig. 3 and more details on security blocks can be found in [11] and [12]. To provide DOS-resilience in DTN, we propose a new security extension block called a DTN-COOKIE which adds a weak authentication phase and has less computational overheads at the security-aware nodes.

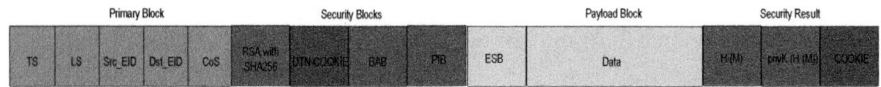

Fig. 3. A DTN Bundle with Security Extension Blocks

- TS: the timestamp which is a concatenation of a bundle creation time and a monotonically increasing sequence number which is unique for every new bundle from a Source Endpoint Identifier (EID)
- LS: the bundle lifespan or expiration time. The LS value of a bundle can be in minutes, hours or days
- Src_EID: the Source EID of a bundle, and we assume that each EID is a singleton
- Dst_EID: the Destination EID i.e. the entity for which the bundle is destined
- RSAwithSHA256: represents the ciphersuite and gives an idea to what security blocks are in use
- M: the bundle payload
- H (M): h is the hash value derived by passing the payload M through the function H. H is a cryptographic hash function such as MD5, SHA1 or SHA256. We will be using SHA256 as the underlying hash function to the signature and MAC algorithms
- pubKXi: the public key of node Xi
- privKXi: the private key of node Xi

The BSP specification [11] provides minimal protection against DOS attacks. DTN nodes simply drop bundles that fail the authentication and access control checks. We have identified resource exhaustion as a simple means of launching DOS attacks and causing availability problems in DTN. For the intra-regional DOS scenario, we propose three variants of the DTN-cookie which can be dynamically chosen based on the perceived Network Threat Level (NTL).

$$DTN\text{-cookie} = h = H\ ((Timestamp \mid Src\_EID_x) \mid p\text{-RNG}\ (IV)) - v1$$

The Initialization Vector (IV) is known only to registered nodes of the region. The IV value is used to seed the pseudo-Random Number Generator (p-RNG); the result is a random long integer value. A concatenation of the timestamp and bundle source EID provides a unique bundle identifier. This unique bundle identifier is concatenated with the random long integer. It is then hashed using a one-way hash function H (SHA-256 algorithm) to produce a fixed-length hash value h. It is the hash value h that is appended to a bundle as DTN-cookie. The IV is changed periodically by the regional security gateway to ensure freshness.

$$DTN\text{-cookie} = h = H\ ((Timestamp \mid Src\_EID_x)\ Xor\ p\text{-RNG}\ (IV)) - v2$$

The second variant of DTN-cookie (v2) is the result when we perform a bit-wise Exclusive-OR operation on Timestamp|Src_EID$_x$ and p-RNG (IV) which results in the flipping of the bits. | is the concatenation operator, p-RNG (IV) is the same as in v1. This variant of DTN-cookie has more randomness and provides a stronger DOS solution.

$$DTN\text{-cookie} = HMAC\ ((Timestamp|Src\_EID_x)\ Xor\ p\text{-PNG}\ (IV), K_{RS}) - v3$$

The third variant of DTN-cookie (v3) is derived in the same way as v2 with SHA-256 as the underlying hash function. The only difference is that the result of the operation is hashed with a regional secret key K_{RS} to produce a fixed-length MAC which we append to every bundle. The mode of generation of the secret key, the use of p-RNG and bit-wise Exclusive-OR operation inputs more randomness to the DTN-cookie. The secrecy of the IV and the key makes the DTN-cookie hard to forge. These values are changed periodically by the security gateway to prevent compromise and ensure freshness.

When a bundle arrives at a security-aware node, the Bundle Protocol Agent (BPA) examines the bundle to determine if it is from a legitimate source and not expired, if it is a Data, or Alert bundle. Next the BPA tries to determine the perceived Network Threat Level (NTL) associated with the bundle. It does this by looking at the DTN-COOKIE Block. The DTN-COOKIE Block contains the NTL indicator (where Low = 1, Mild = 2, Severe = 3). Based on the NTL indicator, the BPA is able to choose which ciphersuite to use to verify the DTN-cookie. The DTN-COOKIE Block has a trailer block with the security result of the DTN-cookie computation as payload. The BPA can also use the Class of Service parameter to deduce the type of bundle it is dealing with by looking into the CoS field in the bundle primary block.

The perceived Network Threat Level (NTL) is determined by the GH node of the affected domain in a localized fashion. Every security-aware node keeps a Node Misbehaviour List (NML). If a SA node records three failed authentication entries against a node within the timeframes shown in Table 1, a Misbehaviour Alert notification is sent to the GH node in the affected domain. The Misbehaviour Alert bundle carries a DTN-cookie that matches the SA's perceived NTL. Entries on the NML beyond 120 minutes are flushed to create space and save memory.

Table 1. Network Threat Level and Associated DTN-Cookie Variants

Network Threat Level (NTL)	NTL Classification	Timeframe (minutes)	DTN-cookie Type
NTL 1	Low	51 - 120	DTN-cookie (v1)
NTL 2	Mild	31- 50	DTN-cookie (v2)
NTL 3	Severe	0 - 30	DTN-cookie (v3)

Table 1 shows NTL values and the associated DTN-cookie variants. This helps security-aware nodes to determine which DTN-cookie variant to use. Bundles arriving with lower NTL values will be processed as long as the node's EID is legitimate. The NTL threshold value reverts back to LOW if there are no Alert updates from a GH node within 24 hours. This is a measure designed to save power and make the DOS mechanism dynamic and adaptive to changing Network Threat Levels.

Egress filtering is enforced at security gateways to help ensure that no malicious or attack traffic leaves the region. The purpose is to prevent an attacker within the network from spoofing any source_EID in a bid to launch a DOS attack. To achieve this, the source_EID must belong to a valid node within the region. The egress filtering policy at security gateways requires all out-bound bundles to have a BAB. Also rate limiting techniques can be used to police the network interface at the security gateway to prevent an attacker from flooding it with bogus bundles. A step-by-step process of providing DOS-resilience within a DTN region is shown in Fig. 4.

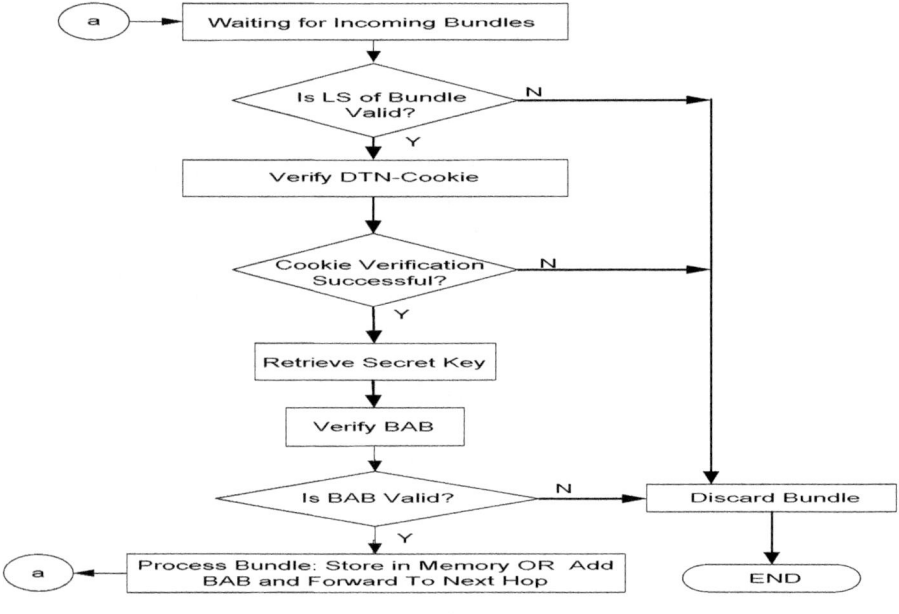

Fig. 4. Intra-Regional Dos Mitigation Flow Diagram

4.3 Inter-regional DOS Mitigation

Protecting inter-regional communications against DOS attacks is very vital to the survivability of a DTN and guarantees the availability of its services. A detailed description of the scenario is given in section 1. The satellite segment of the proposed solution is there to support inter-regional communications between two or more remote or isolated regions. In this scenario, the satellite is only a relay node to provide connectivity [4]. Large round trip times (RTT), large bandwidth delay product, burst errors on coded satellite links and variable RTT have impact that affect transport layer and application performance [4], [15].

The high altitude of the satellite provides a large terrestrial coverage which allows security gateways to send and receive messages to and from the satellite. Satellites are not affected by limitations such as line of sight range of ground-based nodes [15]. For more details on the advantages of satellite communications see [15]. The proposed solution for DOS mitigation in this scenario is similar to that for the intra-regional scenario. We assume that the security gateways are workstations with enormous storage, CPU processing and power capabilities. We use v3 as the proposed DTN-cookie to provide weak authentication in this scenario.

Ingress filtering is enforced at the security gateways to ensure that before a bundle is processed, it must originate from a legitimate and known source. Ingress filtering is a "good neighbour" policy based on mutual cooperation among gateways to thwart source address spoofing DOS attacks. An attacker may decide to spoof the source_EID of a legitimate gateway in order to mount a DOS attack. Such attacks will be thwarted during the verification of the DTN-cookie since the attacker does not know the secret of the network. The same steps shown in Fig. 4 apply to inter-regional communications with subtle differences. Before a bundle is processed any further, the gateway will have to verify the DTN-cookie. The composition of the DTN-cookie used for inter-regional communications is defined below where K_S is the inter-regional secret key.

$$\text{DTN-cookie} = \text{HMAC} \left((\text{Timestamp}|\text{Src\_EID}_x) \text{ Xor p-RNG (IV)}, K_S \right)$$

The DTN-cookie is derived in the same way as specified for the intra-regional scenario. The only difference is that in the inter-regional scenario, we use variable nonce values in different timeslots to seed the p-RNG. This solution requires the security gateways to be loosely time-synchronized. Communication time is divided into timeslots of 2 hours interval with each timeslot having its associated nonce value such that $(0\text{-}2) : S_1$, $(2\text{-}4) : S_2$, ...:..., $(20\text{-}22) : S_{11}$, $(22\text{-}24) : S_{12}$ where $(0\text{-}2)$, $(2\text{-}4)$, ..., $(20\text{-}22)$ and $(22\text{-}24)$ are the timeslots and S_1, S_2..., S_{11}, S_{12} represent the associated nonce values. Section 4.3.1 provides a detailed description of the process. The ingress filtering policy requires all in-bound bundles to have a BAB, PIB and PCB block. The BAB should be used for integrity and sender-side authentication.

We define the BAB to be a digital signature using asymmetric ciphersuite. Neighbour authentication in inter-regional communications does not make sense since the gateways belong to different regions [12]. Inter-regional communications should be gateway-to-gateway given the limited power budget of wireless mobile nodes. This is to enhance the survivability of the network and prevent mobile rogue routers from keeping the satellite busy with fake or mal-formed bundles during each pass of the satellite. Payload encryption is required in gateway-to-gateway communications for both maintenance traffic and data bundles. Also rate limiting techniques can be used to police the network interfaces at the security gateways to guard against flooding attacks.

4.3.1 Time-Synchronization of Security Gateways

Time-synchronization is an important aspect to be considered when designing a mechanism to provide DOS-resilience in DTN. In section 5.2 of [4], the authors elaborate on the use of timestamps and the need for time synchronization. We assume initial pre-shared symmetric keys between SGW_{HQ}, SGW_{WSN1} and SGW_{WSN2}. The security gateways have a common view of time (say UTC) irrespective of their time zones. Also, the p-RNG functions at the security gateways have a uniform initial seed value (S_o). Communications among the gateways is initiated by the SGW_{HQ} by sending two different S_1 nonce values to SGW_{WSN1} and SGW_{WSN2}. The nonce is the bundle payload and is encrypted using the public key of SGW_{WSN1} and SGW_{WSN2}. The SGW_{HQ} signs the bundles using its private key, calculates the DTN-cookie, appends it to the bundles and sends to the gateways.

At the WSN-SGWs, the timestamp and sender EID are retrieved from the bundle and based on the pre-shared symmetric keys (K_S) between the SGW_{HQ}, SGW_{WSN1} and SGW_{WSN2}. The DTN-cookie is computed and compared to that on the received bundle. The bundle is silently dropped if the DTN-cookie verification is unsuccessful. On the other hand, if the verification is successful we proceed to test the integrity of the BAB (digital signature). Each SGW_{WSN} uses the public key of the SGW_{HQ} to verify the signature. If the signature verification fails the bundle is dropped because its content is considered modified on transit. If the verification of signature is successful, we proceed to decrypt the payload. Each SGW_{WSN} uses its private key to decrypt the payload which is the new reference nonce for communications. Attackers within the satellite's coverage are able to see every communication since the satellite uses a broadcast channel. To prevent eavesdropping of the nonce value, we encrypt the payload. We define a bound for the generation of nonce values as follows: $0 < S_i < 999999$ where i is a positive integer. If the S_i value generated is greater than the defined upper-bound, the entire seed generation process is started all over again.

A security gateway with data to send first chooses a random number within the pre-defined bound which it uses as seed to the p-RNG function. The result of this operation is a nonce which it sends to a destination gateway. This is done following the steps described earlier above. The gateways remain synchronized using previous nonce values as seed to the p-RNG function. Where initial nonce equals S_o, $S_1 = $ p-RNG (S_o), $S_2 = $ p-RNG (S_1), $S_3 = $ p-RNG (S_2) and so on thereby forming a hash-chain of nonce values.

$$DTN\text{-}cookie = HMAC\ ((Timestamp|EID_{HQ\text{-}SGW})\ Xor\ p\text{-}RNG\ (S_i),\ K_S)$$

One important property of one-way hash chains is that intermediate values can be recomputed using subsequent values in the chain. Bundles that arrive after their timeslot can still be processed if they are not expired. Also SGW_{WSN1} and SGW_{WSN2} can communicate with each other via the satellite and can remain synchronized by following the steps as described.

5 Analysis and Evaluation of Design

We have critically analyzed the solutions proposed for the terrestrial Internet and other networks and found them unsuitable for DTN. Our design minimizes the number of roundtrips required during entity authentication by discouraging TLS-like handshake. The use of NTL indicators and their associated DTN-cookies in the intra-regional scenario makes the proposed solution dynamic, energy-efficient, and provides DOS-resilience in a localized fashion. Our solution for the inter-regional scenario uses variable nonce values based on a hash-chain of previous nonce values. The security gateways are assumed to be powerful workstations that are loosely time-synchronized with enormous processing, computational, and storage capabilities.

The proposed design is light-weight since the process of generating the DTN-cookie is simple and fast and requires less CPU processing cycles and power. Due to the limited storage and computation capacities of sensor nodes, we use symmetric cryptography for our proposals. The DTN-cookie variants use simple cryptographic primitives such as hash functions and MACs because they require less computational requirements. Symmetric cryptography and hash functions are four orders of magnitude faster than public-key cryptography and digital signatures. The DTN-cookie variants are random and hard to forge because a cryptographically secure pseudo-random number generator in conjunction with fixed or variable seed values are used to generate the nonce. The fixed/variable seed values and the secret keys (K_S, K_{RS}) are prerequisites for computing a valid nonce and DTN-cookie which are unknown to the attacker.

A unique feature of the DTN-cookie is the concatenation of the timestamp and source_EID to produce a unique bundle identifier useful for thwarting replay attacks and preventing old or expired bundles from circulating the network. Any attempt to change the timestamp field will invalidate the bundle during the DTN-cookie verification.

The design is similar to fail-stop protocols described in the work of Gong and Syverson [13], but different because it introduces a weak authentication phase prior to strong authentication. The design also follows a proposed framework by Meadows [8] where a server gradually gains assurance of the client's intentions at every step during protocol execution. By providing a weak authentication phase, the design is able to quickly identify and discard bogus bundles from an attacker.

The v1 and v2 DTN-cookie variants use SHA-256 as hash function. SHA-256 is a 256-bit hash function which uses 32-bit words and provides 128 bits of security against collision attacks [14]. The hash operation produces a fixed-length DTN-cookie which saves memory, CPU processing and provides integrity. As a requirement, H can be applied to a block of data of any size, and it is relatively easy to compute H(x) for any x. For any given value h it is computationally infeasible to find $y \neq x$ such that H(y) = H(x) (weak collision resistance). Finally it is computationally infeasible to find any pair (x, y) such that H(x) = H(y) (strong collision resistance) [14]. The v1 and v2 DTN-cookie variants have all these properties in-built.

The third DTN-cookie variant uses HMAC, a mechanism for message authentication and uses SHA-256 and a secret key. The cryptographic strength of HMAC is dependent on the properties of the underlying hash function and the bit length of the key. On average an attack will require $2^{(k-1)}$ attempts on a k-bit key. The amount of effort needed for a brute-force on a MAC algorithm can be expressed as min $(2^k, 2^n)$. The key and MAC lengths should satisfy the relationship min (k, n) $\geq$ N, where N is in the range of 128 bits [14]. The irreversibility property of the one-way hash function and the secrecy of the symmetric keys (K_S, K_{RS}), makes the proposed DTN-cookie hard to forge.

Attacks which aim at substituting or tampering with the bundle payload are thwarted during the signature verification phase which is triggered if the weak authentication succeeds. The v3 variant of DTN-cookie proposed for the inter-regional scenario is light-weight, random, hard to forge and is a much stronger mechanism. The computation of a one-way hash chain of nonce values is lightweight. Hash-chain based authentication requires time-synchronization at granularities which might require special hardware [16]. Our design proposes the security gateways to be loosely-time synchronized and we assume that storage at the security gateways is large.

Table 2. Comparison of DTN-Cookie Variants

DTN-cookie Type	Complexity	Robustness	Processing	Energy Efficiency	Resilience (security)
DTN-cookie (v1)	Low	Yes	Low	Yes	$2^{n/2} = 128$
DTN-cookie (v2)	Medium	Yes	Medium	Yes	$2^{n/2} = 128$
DTN-cookie (v3)	High	Yes	High	Yes	$2^k, 2^{n/2} = 128$

Table 2 compares the three DTN-cookie variants in terms of complexity, robustness of the mechanisms, processing requirements, power-saving and resilience to attacks. The v1 variant is the least complex while v3 is the most complex of the three variants. The three mechanisms have been designed to be adaptive to the prevailing NTL and any bundle whose DTN-cookie fails to authenticate is dropped. This makes the proposed solution very robust.

In terms of processing needs v3 is more computationally demanding than v1 and v2. HMAC has a higher energy cost than hash functions and can be as high as 96%. The three variants (v1, v2 and v3) are still more energy efficient in terms of verification efficiency when compared to digital signatures and public-key cryptography. Since DTN is an overlay network, we transfer all security processing to wireless mobile nodes (sinks) and adopt the concept of security clusters, domains and a hierarchical based model to conserve battery power.

Apart from the cryptographic properties of the proposed solution, a number of operational security measures have also been proposed. Setting the bundle size to a reasonable uniform length of 64KB is adequate for the application scenario, the bundles are easy to verify, the network is protected from memory exhaustion and waiting times are drastically reduced. Egress and ingress filtering rules at the security gateways help ensure that only bundles with legitimate EIDs and valid lifetimes are processed while illegitimate bundles are dropped.

6 Conclusions

Network DOS is a threat which can degrade network performance and the availability of DTN services. Implementing strong security does not imply that a network is attack-proof; instead it exposes it to resource exhaustion which degrades performance at resource-constrained nodes. It is therefore not advisable to use strong cryptographic algorithms for these nodes with limited resources. A more efficient approach is to begin with weak authentication which is more efficient and light-weight and gradually progress to stronger authentication mechanisms.

In this paper, we have identified resource exhaustion as a simple means by which an attacker can launch DOS in DTN. We have proposed the use of DTN-cookies for both scenarios and based on our evaluation, the proposed solution is considered lightweight, efficient, hard to forge, and incurs less overheads in terms of computation, communication, power, and bandwidth. The proposed solution is highly random since inputs such as the timestamp, nonce, and symmetric secret keys are generated through very random processes. In summary, the proposed mechanism can proactively filter out attack bundles and make the DTN resilient to DOS attacks.

As future work, we will focus on the compromised nodes problem and how to identify, isolate and mitigate their effects in the network. We will implement these designs through simulations and emulation using the ONE Simulator and DTN2 Reference Implementation (RI) respectively.

References

1. Bindra, H., Sangal, A.: Considerations and Open Issues in Delay Tolerant Network's (DTNs) Security. Wireless Sensor Network Scientific Research Journal, 635–648 (2010)
2. Raymond, D.R., Midkiff, S.F.: Denial-of-Service in Wireless Sensor Networks: Attacks and Defences. IEEE Pervasive Computing 7(1), 74–81 (2008)
3. Loukas, G., Öke, G.: Protection Against Denial of Service Attacks - A Survey. The Computer Journal 53, 1020–1037 (2010)
4. Ansa, G., Johnson, E., Cruickshank, H., Sun, Z.: Mitigating Denial of Service Attacks in Delay-and Disruption-Tolerant Networks. In: Sithamparanathan, K., Marchese, M., Ruggieri, M., Bisio, I. (eds.) Psats 2010. LNICST, vol. 43, pp. 221–234. Springer, Heidelberg (2010)
5. Dwork, C., Naor, M.: Pricing via Processing or Combating Junk Mails. Springer, Heidelberg (1998)
6. Juels, A., Brainard, J.: Client Puzzles: A Cryptographic Countermeasure Against Connection Depletion Attacks. In: Proc. Network and Distributed Systems Security Symposium, pp. 151–165 (1999)
7. Maughan, G., Schertler, M., Schneider, M., Turner, J.: Internet Security Association and Key Management Protocol (ISAKMP), RFC 2408 (1998)
8. Meadows, C.: A Formal Framework and Evaluation Method for Network Denial of Service. In: Proc. IEEE Computer Security Foundations Workshop (1999)
9. Leiwo, J., Aura, T., Nikander, P.: Towards Network Denial of Service Resistant Protocols. In: Proc. IFIP TC11 Conference Proceedings, vol. 175, pp. 301–310 (2000)
10. Farrell, S., Ramadas, M., Burleigh, S.: RFC5327: Licklider Transmission Protocol – Security Extensions Network Working Group (2008)

11. Symington, S., Farrell, S., Weiss, H., Lovell, P.: Bundle Security Protocol Specification, Draft-irtf-dtnrg-bundle-security-17 (2010)
12. Ivancic, W.D.: Security Analysis of DTN Architecture and Bundle Protocol Specification for Space-Based Networks. In: IEEE Aerospace Conference, Big Sky Montana (2010)
13. Gong, L., Syverson, P.: Fail-stop Protocols: An Approach to Designing Secure Protocols. In: Proc. of IFIP DCCA-5, Illinois (1995)
14. Krawczyk, H., Bellare, M., Canetti, R.: HMAC: Keyed-Hashing for Message Authentication. In: Crypto 1996, pp. 1–15 (1996)
15. Sterbenz, J. P.G., et al.: Survivable Mobile Wireless Networks: Issues, Challenges, and Research Directions. In: Proceedings of the 1st ACM Workshop on Wireless Security WISE 2002 (2002)
16. Yang, H., Luo, H., Ye, F., Zhang, L.: Security in Mobile Ad hoc Networks: Challenges and Solutions. IEEE Wireless Communications 11(1), 38–47 (2004)

Confirmed Delivery Multicast Protocol in Delay-Tolerant and Disruptive Networks

Nedo Celandroni, Erina Ferro, and Alberto Gotta

National Research Council of Italy (CNR), ISTI Institute
CNR Research Area, Via Moruzzi 1, 56124 Pisa, Italy
name.surname@isti.cnr.it

Abstract. This paper evaluates the performance of the NORM multicast transport protocol, when used in mobile satellite channels that behave as in delay-tolerant and disruptive networks (DTN). Comparisons are made between multicast transmissions with and without interleaver when NORM is used in "confirmed delivery" mode.

Keywords: DTN, mobile satellite channel, NORM multicast protocol, confirmed delivery, FEC, interleaver, performance evaluation.

1   Introduction

Communications in delay-tolerant networks (DTN) are characterized by intermittent connectivity (congested links or disrupted links), long and variable delays, unbalanced traffic, and high error rates [1]. A DTN is a network of regional networks, which bases its communications on the asynchronous message forwarding paradigm (called *bundle*) implemented by the Bundle protocol (BP) [2], which lies between the transport and upper layers, thus augmenting to eight the number of the OSI layers. DTN concepts, such as asynchronously connected nodes, store-and-forwarding principles, and routing for opportunistic connectivity-based networks, can be applied in scenarios where mobile satellite links are present.

An example of this type of scenarios is shown in Figure 1: a data server (the data source) has (directly or indirectly) access to a wide-area broadcast medium, such as a DVB geostationary satellite link, that covers the area where the final receivers (moving people) are located. It is unrealistic to assume that each final receiver has the capability to receive data directly from the broadcast medium. For reasons of cost-efficiency and power conservation, receivers on broadcast links would rather be established as routers (the vans or the helicopters, in Fig. 1) that store and forward data either towards other intermediate routing nodes, thus acting as bent pipe repeaters, or towards the final receivers (people) via a more efficient terrestrial wireless technology. This is an example of an application that operates in a group-based manner, thus requiring efficient network support for group communications (multicast transmissions).

G. Giambene and C. Sacchi (Eds.): PSATS 2011, LNICST 71, pp. 227–235, 2011.

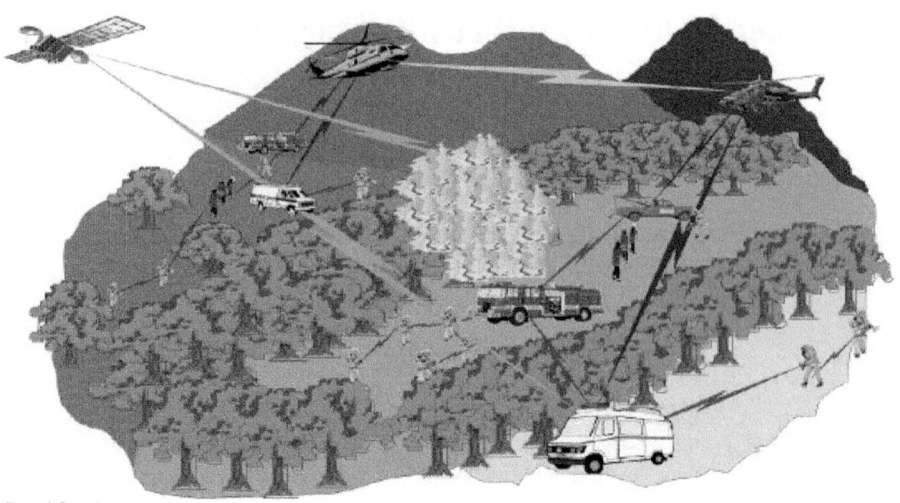

Fig. 1. An example of scenario where mobile satellite links behave as DTNs

While multicasting in the Internet and mobile ad hoc networks has been extensively studied, in DTNs it is a considerably different and challenging problem, due to the DTN unique characteristic of frequent partitioning. Therefore, traditional multicast methods proposed for the Internet (e.g., MOSPF [3] and DVMRP [4] or mobile ad hoc networks (e.g., AMRoute [5] and ODMRP [6] are not suitable for DTNs. First, it is difficult to maintain the connectivity of a source-rooted multicast tree (or mesh) during the lifetime of a multicast session. Second, data transmissions suffer from large end-to-end delays along the communication tree because of the repeated disruptions caused by periodically broken branches. Third, the traditional approaches may fail to deliver a message when the probability of link unavailability becomes high (e.g. ~80%). DTN bundle protocol accesses the underlying transport networks by using Convergence Layers (CL), which map existing protocol suites to a common set of functions.

Currently, the bundle communication protocol supports the FLUTE [7] multicast transport protocol thanks to the Uni-DTN [8] convergence layer, while it does not support the NORM multicast transport protocol [9]. The main difference between FLUTE and NORM is that NORM can operate in both the "confirmed delivery" and "silent" mode, the last one being the equivalent to the unidirectional modality supported by NORM. In [10], we defined a convergence layer, named DT-NORM, which allows the bundle protocol to support the NORM transport protocol. NORM protocol, as defined in the RFC 5740, supports the FEC (forward error correction) coding but it does not implement the interleaving, which is another powerful tool (when used together with FEC) to reduce the probability of error packets in channels like the ones we consider.

In this paper, we present the results of the performance of NORM when used with and without the interleaving technique. When used with interleaving, it is shown a dramatic reduction in the residual (after decoding and de-interleaving) packet loss

(RPL), over a mobile satellite channel in rural and suburban environments, which behaves as in a DTN. We stress that reducing the residual packet loss implies reducing the number of retransmission requests to the sender; as a consequence, the end-to-end delivery delay of a file is reduced.

2 The Channel Model Assumed

The mobile satellite channel has been modelled as a two-state Gilbert channel [11], by means of a discrete time Markov chain (DTMC) model, defined by a transition matrix:

$$\mathbf{T} = \begin{pmatrix} 1-b & b \\ g & 1-g \end{pmatrix}.$$

The stationary probability vector $\mathbf{P}^{\infty}$ is a normalized eigenvector relative to the dominant eigenvalue $\lambda=1$, which characterizes the matrix $\mathbf{T}$; we have:

$$\mathbf{P}^{\infty} = \left[\frac{g}{b+g}, \frac{b}{b+g} \right].$$

In [11], the mobile satellite channel has been modelled for four environments: urban, suburban, rural, and highway. We selected mobile suburban and highway environments only since rural mobile channel has fading characteristics close to the suburban one, while urban fading is too adverse and other wireless technologies (like WIFI, WIMAX or HSDPA) are reasonably preferable in fully connected cities. The numeric values of the elements of the transition matrix $\mathbf{T}$ are reported in Table 1 for the two different types of environment considered.

Table 1. Channel parameters of the DTMC model in suburban and highway environments

Environment	Transition matrix		Stationary probabilities
Suburban	0.995671	0.004329	0.799308
	0.017241	0.982759	0.200692
Highway	0.998069	0.001931	0.899306
	0.017241	0.982759	0.100694

3 The Advantage of Using the Interleaver

The sender segments NORM data into symbols and transforms them in FEC coding blocks before transmission. In NORM, a FEC encoding symbol directly corresponds to the payload of a "segment". When systematic FEC codes are used, data symbols are sent in the first portion of a FEC encoding block and are followed by parity symbols generated by the encoder. These parity symbols are generally sent in response to repair requests, but some of them may be sent proactively in each encoding block in order to reduce the volume of feedback messages. When non-systematic FEC codes are used, all symbols sent consist of FEC encoding parities. In this case, the receiver needs to receive a sufficient number of symbols to reconstruct the original user data for the given block (FEC decoding).

Interleaving is an additional feature necessary to reduce the correlation between loss events, as in disruptive channels characterized by memory and high loss rates. Figure 2 shows how the interleaver works: the sub-matrix k x n contains the allocation for each segment of data. Each row i {$i=1...m$} is a Reed Solomon Erasure (RSE) codeword constituted by k information segments and r parity segments; each column j {$j=1...n$} is fragmented into m units that are the payload of as many NORM segments.

Figure 3 shows how a finite interleaver can almost perform as an ideal interleaver, in terms of RPL in the two chosen (rural and suburban) environments. The two parameters characterizing the mobile satellite channel model for each environment are the average packet loss (PL) and the channel correlation factor $\rho=1-b-g$ [12]. These two parameters are sufficient to define the DTMC process, which models the channel behavior, as presented in [11], [12], [13], [14].

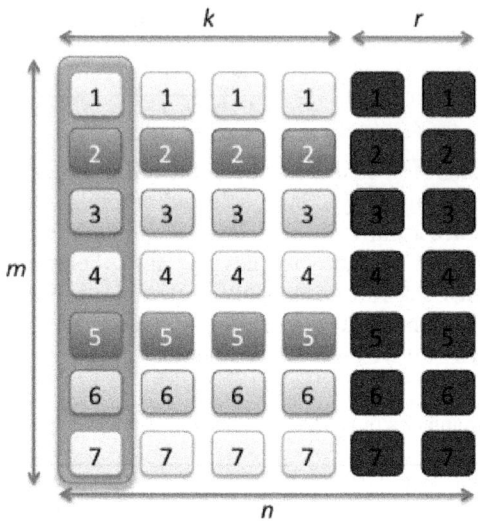

Fig. 2. RSE FEC with interleaver: m=7, n=6, k=4, r=2

Results reported in Fig. 3 have been obtained with the analytical model developed in [14] and applied to a DVB-S2 satellite network with mobile users. The satellite mobile environment has the same characteristics of a DTN network (unpredictable delays, intermittent connectivity, unbalanced traffic, and high error rates).

We assumed a packet size of 1024B, which includes the NORM header of 48B. 40B for UDP/IP has to be considered for the IP multicast addressing. The transmission rate is 256Kbps, which results in a sampling time of 0.03325s (one packet per sample) for DTMC process. For what concerns NORM features, we have assumed, likely to NORM, a proactive redundancy, by choosing an RSE (42, 32) and an RSE (42, 37) code in suburban and highway environments, respectively. In Fig. 3, *uncRPL* stands for "uncorrelated residual packet loss", i.e. the residual packet loss after RSE decoding in a channel where errors are totally uncorrelated, while RPL stands for "residual packet loss" after decoding, in case of a correlation factor ρ between channel losses.

The choice of the optimal proactive redundancy as a function of the channel error process would allow providing, a priori, the resource demand (in terms of bandwidth and processing power) and the scalability limit of a multicast distribution satellite system for disruptive channels.

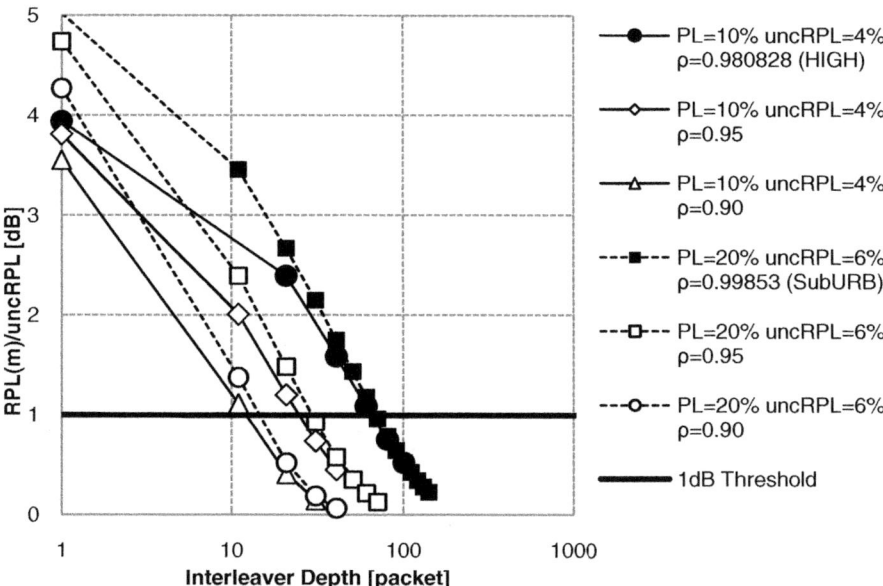

Fig. 3. Performance of a finite interleaver depth, in terms of residual packet loss relative to an ideal interleaver, for different values of the channel correlation factor ρ and for different values of the average channel packet loss (PL)

In details, the scalability, in terms of the maximum number of users within a multicast group, is severely limited by the RPL, which impacts on each receiver and reduces the performance of the whole group [15]. In fact, reliable multicast protocols adapt both the delivery rate and the error recovery parameters according to the receiver that experiences the worst channel conditions in terms of packet loss. That is, the residual packet loss is harsh to the scalability of a multicast distribution system, since the "equivalent packet loss", seen by the sender, is given by the overlap of loss events experienced by each receiver.

For these reasons, counteracting disruptive error processes is a very challenging issue in multicast mobile satellite networks.

4 Performance of NORM with and without Interleaving

The results of this section have been obtained by using the same environment types and the same considerations done for getting results reported in Fig. 3; the confidence interval is 5% at 95% level. NS2 2.34 has been adopted as a simulation environment and it has been patched (v. 1.4b3) by adding the NORM protocol developed by the US Naval Research Laboratory, which is downloadable from [16]. We assumed the sender has to transmit 10 files of 2.5MB each. An intermediate node acts as a multi-spot satellite, which realizes a multicast distribution by replicating the incoming data flow toward the spots that cover the subscribing mobile receivers. The NORM sender adopts no congestion control mechanism, and an information bundle of 32 and 37 packets for suburban and highway environments, respectively. A pre-computed redundancy of 10 and 5 packets, respectively, is proactively appended to the information bundle transmitted. The resulting codeword is a block of 42 packets in both cases. The coding rates, corresponding to n-k/n ratios, are just greater than the respective values of packet loss in the two environments. Such a condition would be sufficient to conveniently reduce the RPL, if the errors were uncorrelated; in case of correlated errors, instead, the parity length has to face the burst error length. In order to compensate for such a deficiency, our results show how a near-ideal interleaver works; in fact, such an interleaver allows to benefit from few parities and

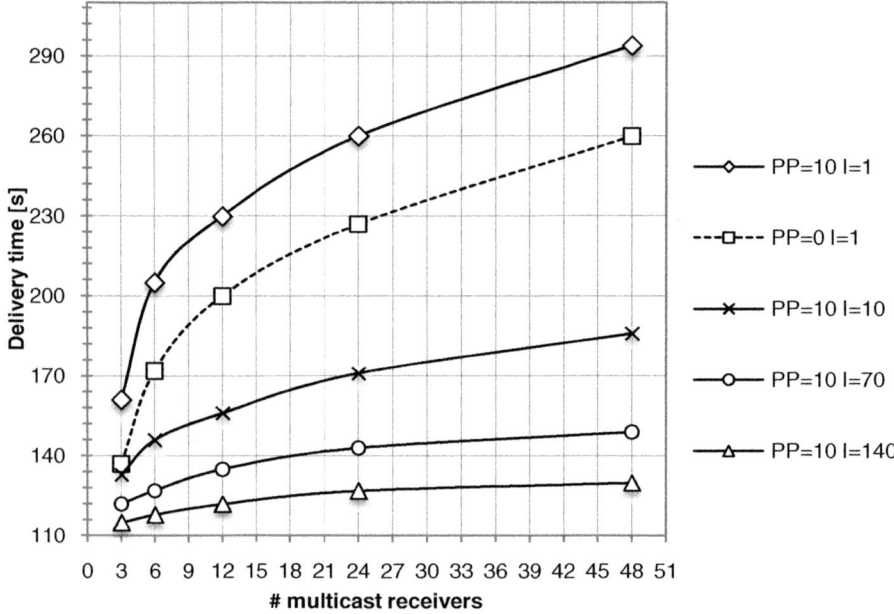

Fig. 4. NORM performance with and without proactive parity (PP) RSE (42, 32), in terms of delivery time, for different values of both the interleaver depth (I) and the number of multicast receivers in suburban environment

short coding blocks, which limit the number of recovery cycles of the NORM error recovery algorithm. In case of no NACK reception, i.e. a silent delivery, the proactive parity probably produces a waste of bandwidth (may be too much redundancy used), even if it minimizes the delivery time. Instead, just few recovery cycles (ideally one only) can represent a good tradeoff between delivery time and bandwidth overhead.

Simulation results reported in Figures 4 and 5 show the combined performance of interleaving, proactive parity, and reactive error recovery adopted in NORM, in suburban and highway environments, respectively. The term of comparison is the average delivery time, i.e. the interval between the reception of the first and the last packet of a file by all receivers.

Dotted lines refer to the NORM usage with no proactivity parity and no interleaving (I) add-on. In case of a few packets of proactive parity PP=10, PP=5 and average error burst length of 58 packets in both the environments, the delivery time is even underperforming, since error bursts are longer than the parity length, on average. However, the combination of PP with interleaving is outperforming the case without PP and in case of quasi-ideal interleaving (i.e. I=140 packets in suburban and 70 packets in highway) NORM robustly behaves even when the number of multicast receivers scales.

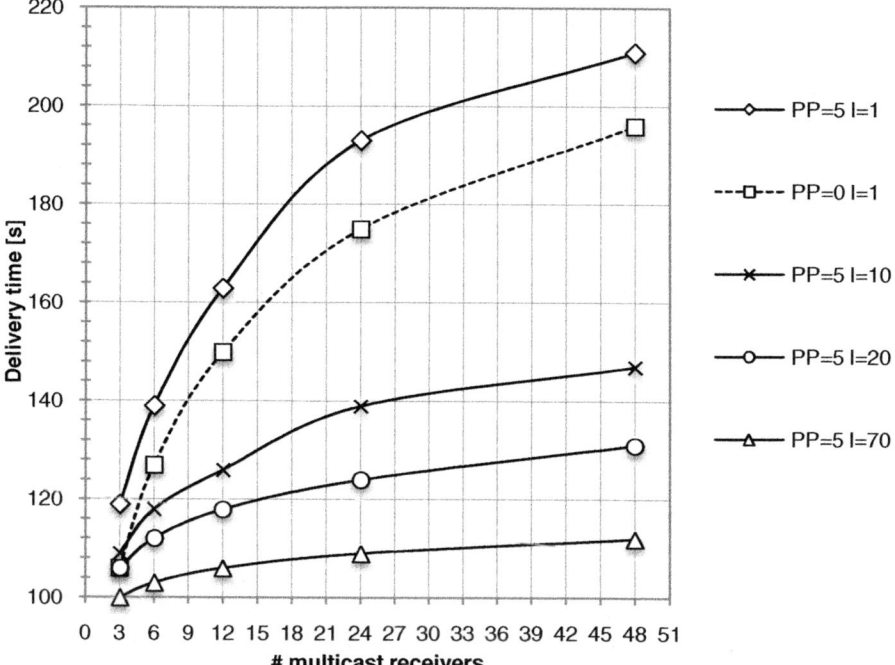

Fig. 5. NORM performance with and without proactive parity (PP) RSE (42, 37), in terms of delivery time, for different values of both the interleaver depth (I) and the number of multicast receivers in highway environment

In all the interleaved cases, an extra buffering time has to be accounted for, since the interleaver must be fulfilled before starting the decoding process. This time interval, in practice, occurs between the subscription of the file and the reception of the first packet. Interleaving delay is given by $I(n-1)pt$, where pt is the packet time equal to $0.03325s$ in our simulations. The interleaver delay may rise up to 191s, referring to the worst case in suburban PP=10 and I=140 of Fig. 4. Thus, the download time, i.e. the time between the notification of the file transmission (notified by the sender through a command message and acknowledged by the receivers) and the end of the file delivery, is given by the delivery time and the interleaving delay.

5   Conclusions and Future Works

The combined adoption of FEC and interleaving fosters the NORM's performance in terms of scalability and delivery time, in land mobile satellite environments. Interleaving allows limiting the packet error rate on the channel by means of a small amount of proactive parity. However, the interleaver depth is closely related to the channel correlation factor, which is responsible for bursts of errors. Scaling on the transmission rate results in longer error bursts, which require a deeper interleaver. In delay-tolerant networks, interruptions due to unpredictable obstacles, fading events, or, more generally, channel interruptions can last for unpredictable times; to counteract blockages that overcome a certain threshold, the interleaver may be so long to be no more convenient, as it reduces the transmission time but it may sensibly increase the user's perceived downloading time, because the adoption of the interleaver implies a pre-buffering time. This consideration encourages deepening our studies in order to optimize the tradeoff among bandwidth, end-to-end delivery time, and required computing power.

Acknowledgements. Work funded by ESA (European Space Agency) in the framework of the SatNEx-III research activities

References

1. (RFC 4838) Delay-Tolerant Networking Architecture (April 2007)
2. (RFC 5050) Bundle Protocol Specification (November 2007)
3. Moy, J.: Multicast extensions to OSPF. IETF RFC 1584 (1994)
4. Waitzman, D., Partridge, C., Deering, S.: Distance vector multicast routing protocol (DVMRP). IETF RFC 1075 (1988)
5. Xie, J., Talpade, R.R., Mcauley, A., Liu, M.Y.: AMRoute: ad hoc multicast routing protocol. Mobile Networks and Applications 7(6), 429–439 (2002)
6. Bae, S.H., Lee, S.-J., Su, W., Gerla, M.: The design, implementation, and performance evaluation of the on-demand multicast routing protocol in multihop wireless networks. IEEE Network, 70–77 (January 2000)
7. (RFC 3926) FLUTE - File Delivery over Unidirectional Transport (October 2004)
8. (draft00) Uni-DTN: A DTN Convergence Layer Protocol for Unidirectional Transport draft-kutscher-dtnrg-uni-clayer-00.txt

9. (RFC 5740) NACK-Oriented Reliable Multicast Transport Protocol (NORM) (September 2009)
10. Ferro, E., Celandroni, N., Gotta, A., De Cola, T., Roseti, C., Caini, C., Cruishunk, H., Giambene, G., Bisio, I., Cello, M.: Applications of DTN. SatNEx. Deliverable TN 3 2-final (2010)
11. Scalise, S., Ernst, H., Harles, G.: Measurement and modeling of the land mobile satellite channel at Ku-Band. IEEE Trans. Vehic. Technol. 57(2), 693–703 (2008)
12. Yee, R., Weldon, E.J.: Evaluation of the performance of error-correcting codes on a Gilbert channel. IEEE Transactions on Communications 43(8), 2316–2323 (1995)
13. Celandroni, N., Davoli, F., Ferro, E., Gotta, A.: Video Streaming Transfer in a Smart Satellite Mobile Environment. International Journal of Digital Multimedia Broadcasting IJDMB 2009 Article ID 369216, 12 (2009) doi:10.1155/2009/369216
14. Celandroni, N., Gotta, A.: Performance Analysis of Systematic Upper Layer FEC Codes and Interleaving in Land Mobile Satellite Channels. IEEE Transactions on Vechicular Technology 60(4), 1887–1894 (2011), doi:10.1109/TVT.2011.2122253
15. Nonnenmacher, J., Biersack, E.W., Towsley, D.: Parity-based loss recovery for reliable multicast transmission. IEEE/ACM Transactions on Networking 6(4), 349–361 (1998)
16. http://downloads.pf.itd.nrl.navy.mil/norm/

Blind Demodulation Framework for Satellite Signals

Markus Flohberger, Wilfried Gappmair, and Otto Koudelka

Institute of Communication Networks and Satellite Communications
Graz University of Technology, Austria
Markus.Flohberger@TUGraz.at

Abstract. With the growing amount of satellite traffic an efficient usage of the existing resources, i.e. geostationary arc and available frequency bands, is mandatory. Therefore appropriate monitoring techniques are of paramount importance for reasonable operation of satellite networks. The increase in available computational power enables signal processing tasks that were not even thinkable a decade ago. So it is proposed to apply blind demodulation techniques, suited for implementation on software-defined radio (SDR) platforms, for carrying out the required monitoring tasks. The incorporated algorithms are presented, including performance comparison and remarks on efficient implementation. Finally, the demodulation capabilities of the introduced framework are assessed in terms of the error vector magnitude (EVM).

Keywords: synchronization, blind demodulation, software-defined radio, satellite communications.

1 Introduction

An automated blind demodulation framework can alleviate significantly the monitoring tasks, required for maintaining the quality of service (QoS) in satellite networks. Apart from the advantage of online traffic detection and evaluation of carrier characteristics, the framework can be used to generate a remodulated version of the received signals. This feature may be used for identifying interfering signals or aiding measurements relying on correlation techniques, since subtraction of the remodulated carriers can lead to an improvement of the post-correlation signal-to-noise ratio (SNR). An in-depth discussion of the blind demodulation framework and appropriate applications will be available in [1].

For reasonable transmission of data in communication systems, it is inevitable to determine the most important parameters of the receiving signal. The evaluation of these parameters is termed parameter estimation and synchronization [2]. In general, nominal parameters are available at the receiver side allowing the demodulator to focus on estimation of residual carrier frequency offset, carrier phase error, optimum timing instant and, finally, the transmitted data symbols. However, this is not the case for non-cooperative environments, e.g. signal interception or monitoring systems. A totally blind procedure necessitates carrier

G. Giambene and C. Sacchi (Eds.): PSATS 2011, LNICST 71, pp. 236–249, 2011.

detection and evaluation of additional parameters [3], e.g. symbol rate, SNR, modulation scheme and baseband pulse shape.

The type of modulation is limited to linear modulation schemes used in satellite communications, i.e. phase shift keying (PSK) and amplitude phase shift keying (APSK) [4]. Although not frequently applied in satellite communications, due to the sensitivity to nonlinear distortion, rectangular quadrature amplitude modulation (QAM) is considered as well. Moreover, it is assumed that the baseband pulse is of root-raised cosine (RRCOS) type which is established as the de facto standard in satellite communications. Finally, for successful carrier detection, restrictions on the possible range of symbol rates have to be applied. Since modularity is a key aspect of the proposed framework, new features such as modulation schemes, baseband pulses or synchronization algorithms, can be added in an easy way.

For geostationary earth orbit (GEO) satellites, attenuation and additive white Gaussian noise (AWGN) can be considered as the main impairments of the transmitted signals [5], leading to the signal model introduced in Section 2. The spectral carrier detection procedure and the subsequent blind demodulation stages are briefly discussed in Section 3. Since the accuracy of all estimated parameters can not be presented in a suitable way, an appropriate measure for evaluating the overall demodulation performance has to be found. It turned out that the EVM of the remodulated signals might be applied for this reason, as verified by simulation in Section 4. Final remarks and future research topics conclude this paper in Section 5.

2   Equivalent Baseband Model

If the nonlinear impact of the high power amplifier (HPA) is neglected, the received multi-carrier signal $r(t)$ can be stated as follows:

$$r(t) = \sum_q u_q(t) + \sqrt{E_n}\, w(t) \tag{1}$$

with the carrier $u_q(t)$ expressed as

$$u_q(t) = \sqrt{E_{s,q}}\, e^{\jmath(2\pi f_{c,q}t + \theta_q)} \sum_i c_{i,q}\, h_q(t - iT_q - \tau_q) \tag{2}$$

The subscript q denotes parameters belonging to the q-th carrier. The center frequency of the specific carrier is denoted by $f_{c,q}$. Let $w(t)$ be the complex zero-mean AWGN with independent real and imaginary part. Together with the noise power E_n and the signal power $E_{s,q}$, the SNR of the q-th carrier is defined by $\gamma_{s,q} = E_{s,q}/E_n$. Furthermore, $c_{i,q}$ are the transmitted data symbols, which are assumed to be independent and identically distributed (i.i.d.) and normalized to unit energy such that $E[|c_{i,q}|^2] = 1$. So far, seven modulation schemes are supported: BPSK, QPSK, 8-PSK, 16-QAM, 64-QAM, 16-APSK, and 32-APSK. The symbol period T_q is the reciprocal value of the symbol rate, i.e. $T_q = 1/f_{d,q}$.

The phase offset is expressed by $\theta_q \in [-\pi, \pi)$ and the symbol timing offset by $|\tau_q| \le T_q/2$. The baseband pulse is described by $h_q(t)$ and is a function of the roll-off factor α_q, with $0 \le \alpha_q \le 1$.

In Fig. 1 the power spectrum R_n of an example scenario relying on the above signal model is shown, obtained by sampling the multi-carrier signal $r(t)$ appropriately.

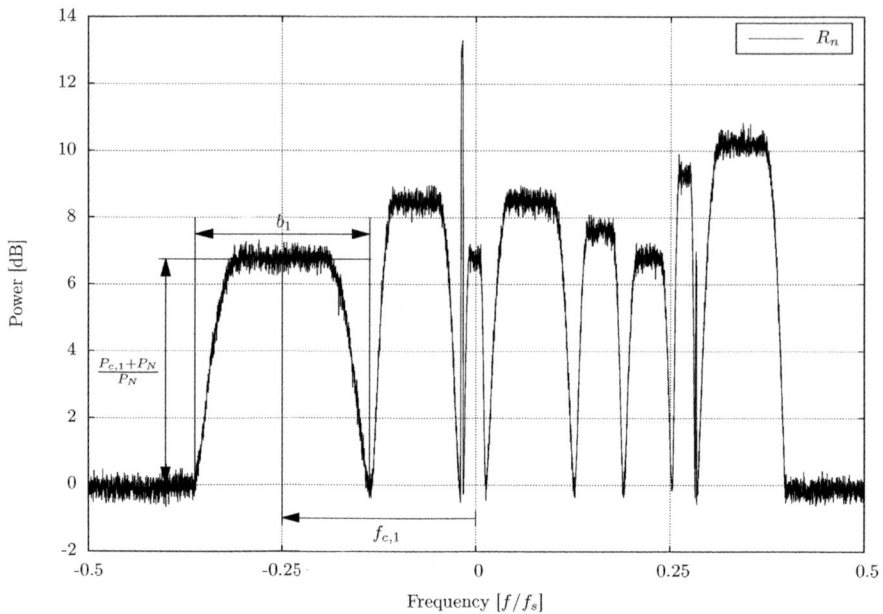

Fig. 1. Power spectrum R_n of the example scenario

The received signal is separated into single carriers that are converted down to baseband and applied to the matched filter (MF). Thus the resulting sampled signal at the filter output can be formulated as

$$x_{k,q} \triangleq r(t) \otimes h_q^*(-t)\big|_{t=kT_s} \approx e^{j(2\pi k \Delta f_q T_q/N_{s,q} + \theta_{k,q})} s_{k,q} + \sqrt{E_n} n_{k,q} \qquad (3)$$

with the signal component

$$s_{k,q} = \sqrt{E_{s,q}} \sum_i c_{i,q}\, g_q((k/N_{s,q} - i - \epsilon_q)T_q) \qquad (4)$$

In this context, Δf_q expresses the residual frequency offset after down conversion; T_s denotes the sampling period, i.e. $T_s = 1/f_s$, so that the oversampling ratio of the q-th carrier is given by $N_{s,q} = T_q/T_s = f_s/f_{d,q}$; $n_{k,q}$ is the (non-white) Gaussian noise sample shaped by the receiving filter $h_q^*(-t)$. The overall baseband pulse shape $g_q(t) = h_q(t) \otimes h_q^*(-t)$ results in a raised cosine (RCOS) to guarantee absence of inter-symbol interference (ISI), where $\otimes$ denotes convolution. It is common practice to normalize the timing offset τ_q by the symbol period T_q expressed by ϵ_q.

3   Blind Demodulation Framework

An overview of the proposed blind demodulation framework is illustrated in Fig. 2. In the following, a brief description of the particular stages is provided.

3.1   Carrier Detection

The detection of carriers is performed in the frequency domain using a well-averaged periodogram [6]. Additional smoothing mitigates fluctuations in the spectrum, but still does not filter out narrow-band carriers exhibiting low signal power.

The detection procedure starts with the estimation of the noise power level P_N by inspection of the histogram derived from the power spectrum R_n [7]. In the following, R_n is scanned for values which are a pre-defined threshold above the estimated noise floor P_N. To mitigate fluctuations close to the threshold a type of hysteresis has to be applied. By this means, candidates for possible carriers inside the observed bandwidth can be detected. For further improving the detection reliability, restrictions with respect to symbol rate, carrier spacing and SNR are applied.

Afterwards, pre-estimates for carrier center frequency $f_{c,q}$, occupied bandwidth b_q, symbol rate $f_{d,q}$, SNR $\gamma_{s,q}$ and roll-off factor α_q are derived from the spectrum. The determined characteristics are used for extracting the particular carrier $r_{k,q}$ from the multi-carrier signal r_k by band-pass filtering, down conversion and decimation. Moreover, these pre-estimates are used as prior knowledge in the subsequent blind demodulation stages for enhancing the performance significantly. As from now, the carrier index q is dropped for the sake of readability.

3.2   Symbol Rate Estimation

Existing estimation schemes from the open literature may not work properly, if baseband pulse shaping is applied [8]. Moreover, hardly any performance results are available for non-integer oversampling ratios. Therefore, an enhanced version of the method described in [9] is applied and the behavior for non-integer oversampling ratios is investigated as well.

By applying a nonlinearity to r_k and appropriate post-processing, a spectral component is generated at the symbol rate f_d. The significance of the spectral line is degraded by decreasing SNR γ_s and roll-off α, small observation intervals and non-integer oversampling ratios N_s. Two measures can be taken to counteract this behavior: (i) narrow-band pre-filtering of the signal to remove unwanted noise; (ii) application of a coarse symbol rate estimate from the carrier detection stage for narrowing the spectral search range.

Finally, the enhanced algorithm is compared to the methods proposed in [9] (standard), [8] (filter bank) and [10] (cyclic correlation). An estimation is termed successful, if the estimation error is smaller than the spectral resolution. The observation interval is assumed to be $L = 16384$ samples, oversampled with $N_s = 4$ this results in 4096 symbols. The residual frequency offset is set to $\Delta fT = 0.001$. The necessary periodogram is calculated using $M = 13$ overlapping sample

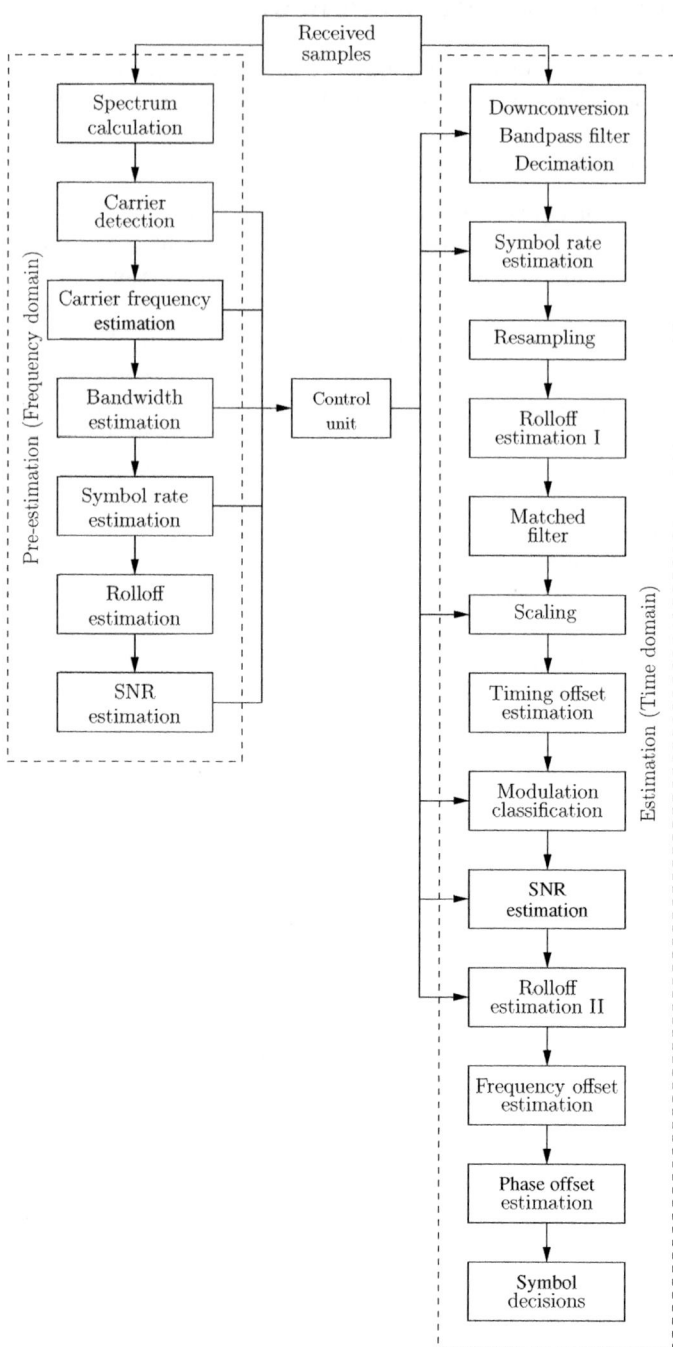

Fig. 2. Overview of the blind demodulation framework

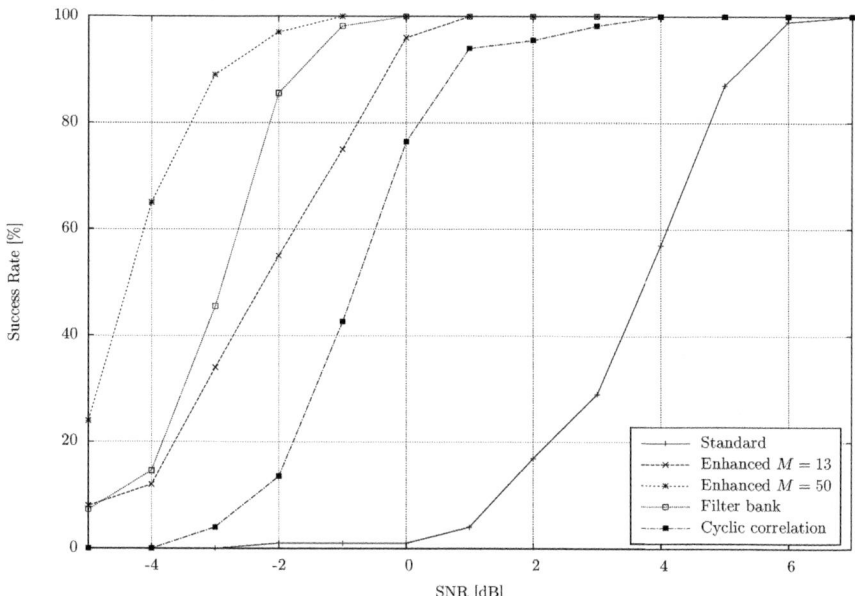

Fig. 3. Evolution of the success rate for symbol rate estimation

blocks of length $N = 4096$. The evolution of the success rate is illustrated in Figure 3. It can be seen that the success rate of the enhanced algorithm is superior to the standard approach, only outperformed by the filter bank method. For the sake of completeness, it is shown that increasing the observation interval to $M = 50$ overlapping blocks leads to a considerable improvement of the success rate.

To handle the computational complexity in the subsequent demodulation stages, the oversampling ratio N_s is reduced to the minimum required value. For arbitrary rate changes, a Lagrange type interpolator is used [11,12]. Afterwards, the signal is applied to the MF. For maximizing the SNR at the filter output, transmitting and receiving filter must exhibit the same roll-off. However, it turned out that a slight mismatch results only in a minor SNR degradation. So, either the roll-off is determined by using the scheme described in [7] or set to a reasonable value, taking a possible degradation into account.

3.3   Rescaling

Rescaling of r_k is mandatory for several processing stages, e.g. design of filter co-efficients in the timing tracker, partitioning used for SNR estimation, frequency and phase recovery or, finally, for the decision of the transmitted data symbol drawn from a multi-level modulation scheme. Using the SNR pre-estimate and the second-order moment M_2, an appropriate scaling factor a_s can be derived.

As investigated in [13], for oversampled signals the second-order moment at the MF output is given as

$$M_2 = E[|x_k^2|] = (1 - \alpha/4)E_s + E_n \tag{5}$$

So the signal power at the MF output can be expressed as $E_s' = (1 - \alpha/4)E_s$. Using the relationship for the SNR, i.e. $\gamma_s = E_s/E_n$, in Equation 5, the signal power ahead of the MF becomes

$$E_s = \frac{\gamma_s M_2}{\gamma_s(1 - \alpha/4) + 1} \tag{6}$$

Now for proper scaling the signal has to be simply divided by $a_s = \sqrt{E_s}$.

3.4   Timing Offset Estimation

Due to the importance of the symbol timing recovery stage, non-data-aided (NDA) schemes are compared in terms of performance, complexity and applicability to blind operation. As representatives of efficient feedback (FB) schemes, the well established Gardner detector (GA) [14], its extended version (xGA) [15] and the detector by Moeneclaey and Batsele (MB), reviewed from a different perspective in [16], are considered. The Oerder and Meyr (OM) device [17] and two estimators (E_1 and E_2), based on the MB detector [18], are selected as feedforward (FF) alternatives.

Basically, OM and E_1 can be operated in a totally blind manner, i.e. no knowledge of modulation type and roll-off α is required, whereas the GA algorithm requires knowledge of α for the optimum design of the loop filter. Additionally, for xGA, MB and E_2, the used signal constellation is necessary. The application of different parameters may lead to suboptimal behavior.

For assessing the computational complexity, the minimum required oversampling ratio N_s plays an important role. When assuming equidistant sampling, E_1 requires at least $N_s = 4$ samples per symbol, OM $N_s = 3$, E_2, GA and xGA $N_s = 2$, whereas MB can be operated at baud rate. It should be pointed out that acceptable performance can only be achieved by applying two iterations for E_1 and even three iterations for E_2.

In spite of their different nature, subsequent feedforward and feedback schemes are compared in terms of jitter performance. Therefore the timing error is set to $\epsilon = 0.0$, the FF estimator length to $L = 100$ samples and the equivalent noise bandwidth of the FB methods to $B_L T = 0.005$. As required for bandwidth efficient communication systems, the roll-off is set to $\alpha = 0.2$. Due to the limited space, results for GA are omitted at all and MB FB is selected as representative for MB-based methods.

The evolution of the normalized mean square error (MSE) is illustrated in Fig. 4. For comparison purposes, the modified Cramer-Rao lower bound (MCRLB) is plotted as the theoretical limit of the jitter variance [2]. For signal constellations with constant modulus it becomes obvious that OM and GA exhibit poor performance for small excess bandwidths. In this case the remaining

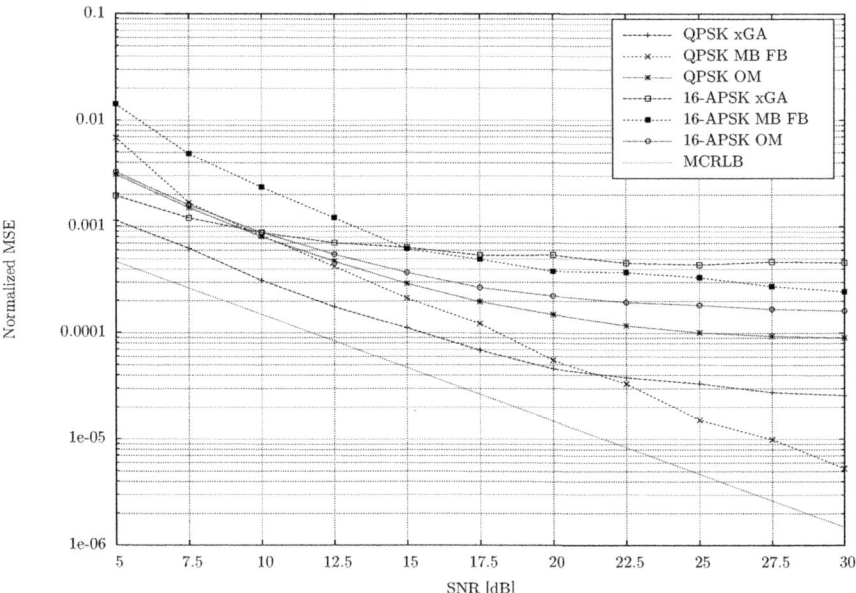

Fig. 4. Normalized MSE of FF and FB algorithms for symbol timing recovery

algorithms feature a significantly lower MSE, especially the MB-based schemes
lack the self-noise floor in the high SNR range at all. The situation changes dramatically when multi-level constellations are used. A significant increase of the
MSE can be observed for xGA and MB-based algorithms, whereas the degradation for GA and OM is considerably smaller. These results suggest using the OM
for symbol timing recovery; however, if required, the most appropriate method
can be selected after classification of the modulation scheme.

After successful recovery of the optimum timing instant by Lagrange interpolation, a feature-based modulation classifier supporting the considered modulation schemes is applied [19].

3.5 SNR Estimation

Although a robust measure for the SNR is available from the pre-estimation
stage, there are three reasons for repeated estimation: (i) the accuracy of the
refined SNR estimate may be superior; (ii) a failure of the symbol timing recovery
stage might be detected by a significant drop in the estimated SNR; (iii) powerful
SNR estimation can be used to estimate the roll-off in an alternative manner,
since the SNR at the MF output is a maximum for a properly designed receiving
filter.

The moment-based M_2M_4 estimator [20] exhibits good performance for signal constellations with constant envelope; however, it deteriorates completely for
multi-level schemes, e.g. QAM and APSK, in the medium-to-high SNR range.
An algorithm using partitioned subsets of the original signal constellation is

presented in [21]. This approach exhibits good performance in the medium-to-high SNR range, but it degrades for small SNR values due to a wrong assignment of the symbols to the subsets. Finally, in [22] an estimator using also the eighth-order moment M_8 is presented to achieve satisfactory performance for multi-level constellations. Moreover, the algorithm allows tuning for specific SNR values. Thus, a combined estimator is furnished, consisting of the standard M_2M_4 estimator for the low SNR range, the partitioned method in the medium-to-high SNR range and the M_8 estimator for the overlapping area when performance is insufficient.

3.6   Carrier Frequency/Phase Estimation

Frequency and phase estimation is carried out by the Rife and Boorstyn (RB) [23] and the Viterbi and Viterbi (VV) [24] algorithm. Both schemes apply a nonlinearity to the received symbols x_n to generate a harmonic with frequency and phase being multiples of the true offsets. Therefore, the searched offsets can be extracted by appropriate measures.

The above procedures work properly for PSK. However, problems arise for multi-level constellations, since no promising way is available for stripping off the modulation from the symbols x_n. To overcome the problem, strategies relying on partitioning [25] and optimal nonlinear transformations [26,27] are applied. The used partitioning approach turns out to be closely related to the linear approximations of the complex transformations. Since only subsets of the entire symbol alphabet are used for computation, the observation interval must be increased appropriately to achieve the same estimator length as in the unpartitioned case.

To allow reliable phase recovery in presence of residual frequency offsets, a tracker should follow the acquisition stage. Therefore, a second-order tracking loop with an appropriate filter design in terms of damping ζ and equivalent noise bandwidth B_LT must be implemented.

For subsequent symbol decisions, an issue arises which is unique for APSK. The symbols located on the inner circle correspond to a QPSK constellation with the symbols rotated by $\pi/4$ relative to the axis. In contrast, the outermost circle for 16-APSK exhibits an ambiguity of $\beta = \pi/6$ and for 32-APSK of $\beta = \pi/8$. So the inner QPSK constellation may remain tilted after phase recovery, which will be problematic during symbol decision. For this reason a second iteration of phase estimation is performed using the symbols on the inner circle. Direct addition of the obtained QPSK offset $\Delta\hat{\theta}$ to the initial estimate $\hat{\theta}$ using the outermost circle would lead to inferior performance due to the increased sensitivity to noise effects. However, the knowledge that $\Delta\hat{\theta}$ has to be a multiple of the ambiguity β of the outermost circle can be used to obtain acceptable accuracy.

In the following, the performance of the partitioned RB algorithm relying on a VV nonlinearity is assessed by simulation. The estimator length is set to $L = 1024$ symbols, with a zero-padding factor of $k_{zp} = 4$ leading to an FFT length of 4096 points. The roll-off is set to $\alpha = 0.35$ and the VV parameter to $\mu = 1$, since it produces the most promising results. Finally, the frequency offset is assumed to be $\Delta fT = 0.01$. The evolution of the mean square error (MSE) is

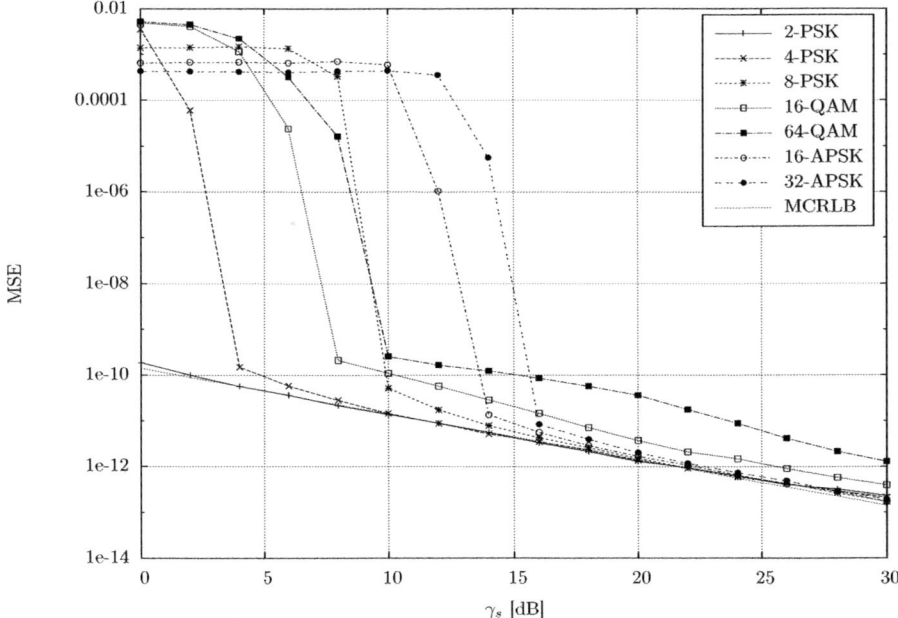

Fig. 5. MSE of estimated frequency error

depicted in Fig. 5. A threshold effect caused by identifying the wrong spectral line can be observed very clearly in the low SNR range. Moreover, the poor performance of both APSK schemes should be emphasized, but no promising alternative algorithm was found up to now.

4   Simulation Results

The obtained decisions can be used to form a remodulated signal $\hat{u}_k$ as

$$\hat{u}_k = \sqrt{\hat{E}_s} e^{\jmath(2\pi k \hat{f}_c T_s + \hat{\theta})} \sum_i \hat{c}_i \, \hat{h}((k/\hat{N}_s - i - \hat{\epsilon})\hat{T}) \tag{7}$$

It can be seen that, when assuming ideal synchronization, the above equation corresponds to a discrete version of the received single-carrier signal stated in Equation 2.

The EVM Λ of the remodulated signal $\hat{u}_k$ can be used to assess the performance of the blind demodulation framework. The latter is usually specified in dB and defined as

$$\Lambda = \frac{\sum_{k=0}^{N-1} |u_k - \hat{u}_k|^2}{\sum_{k=0}^{N-1} |u_k|^2} \tag{8}$$

For simulation results, the received carriers are assumed to be oversampled by $N_s = 4$, impaired by time/phase offset and the center frequency chosen arbitrarily for each iteration. The carriers are detected and demodulated blindly to

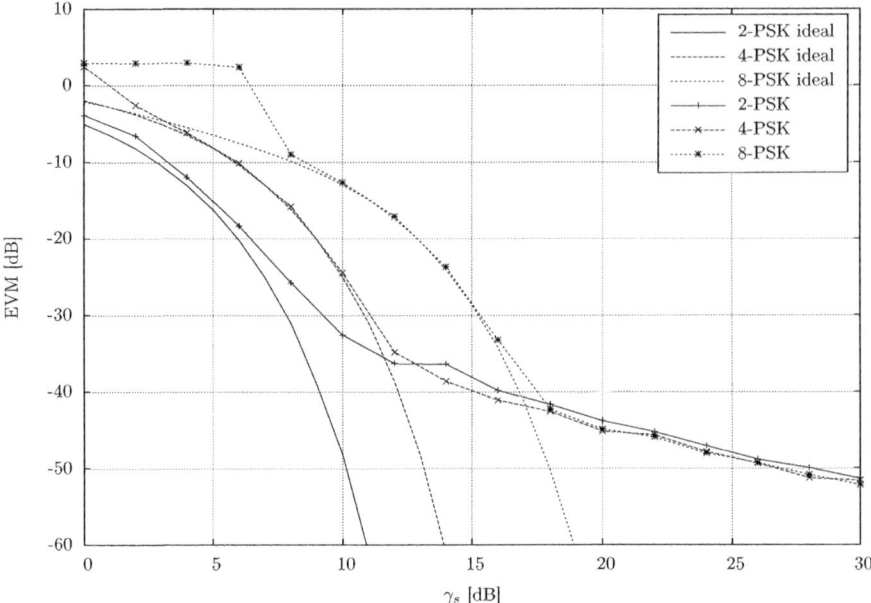

Fig. 6. EVM evolution for PSK

obtain decisions which are remodulated for EVM evaluation. For proper carrier detection, $M = 75$ blocks of length $N = 2^{16}$ overlapping by 75 % are processed for generation of a well-averaged periodogram. After successful parameter estimation, a block of $N = 2^{14}$ samples is applied for remodulation. The EVM evolution for PSK is illustrated in Fig. 6. For comparison reasons the EVM relying on ideal synchronization is included as well. It can be seen that in the medium SNR range the remodulation performance is dominated by decision errors and thus very close to the ideal EVM. In contrast, for large SNRs the EVM exhibits an error floor caused by residual estimation errors. Moreover, in the low SNR range the EVM increases significantly. This threshold effect is produced by malfunction of the blind demodulation framework. The main impacts are due to the frequency estimator degradation for higher order and multi-level modulation schemes as well as the SNR restriction applied for modulation classification.

The EVM evolution for multi-level constellations is depicted in Fig. 7. Basically, very similar effects as for PSK can be observed. The ideal curves are shifted in direction of higher SNRs due to the increased probability of false decisions. The noise floor in the high SNR region is approximately increased by one order of magnitude. The solid frequency estimation performance of QAM enables the threshold effect to occur at smaller SNR values as for 8-PSK. Interestingly, 16-QAM and 64-QAM exhibit the same EVM characteristic in the very low SNR range. The reasons for this behavior are the SNR restriction feature of the used modulation classifier and the similarity of both signal constellations in terms of quadrature symmetry. Finally, the pronounced threshold effect for APSK

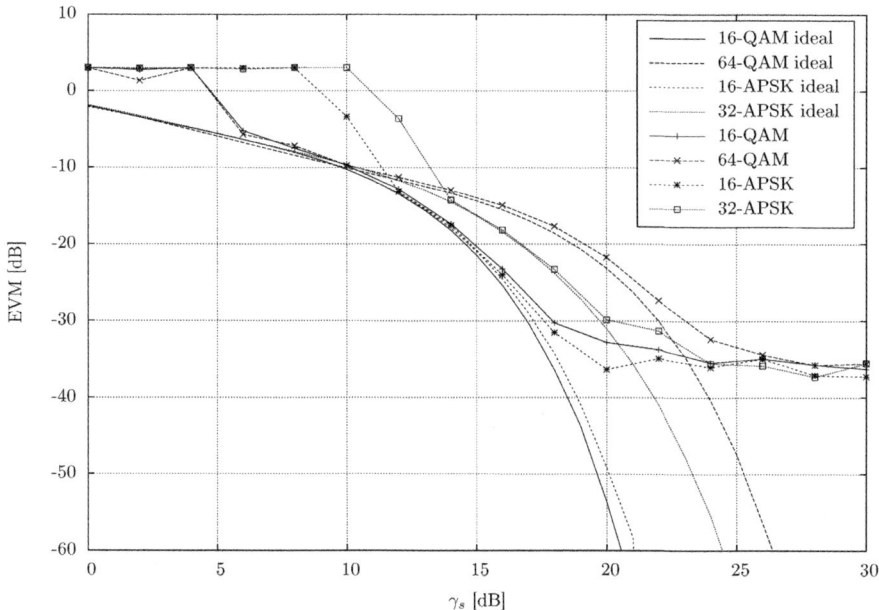

Fig. 7. EVM evolution for QAM and APSK

constellations becomes apparent. The latter is caused by the poor frequency estimator performance mentioned previously.

5 Conclusion and Outlook

A versatile blind demodulation framework especially suited for satellite signals is presented. The latter can be implemented on SDR platforms for carrying out monitoring tasks imposed to satellite operators. After setting the stage by definition of the used signal model, carrier detection and required blind demodulation stages are discussed. An attractive feature is the remodulation of the obtained symbol decisions. This capability can be used for improving the present SNR situations which would be most welcome for measurements relying on correlation techniques, e.g. evaluation of the cross-ambiguity function (CAF).

Since the number of possible input signal scenarios is enormous, the need for future work suggests itself. In this context, detection and handling of time-division multiple access (TDMA) signals would be most attractive. Since synchronization of the latter might be problematic for NDA algorithms, usual communication standards could allow data-aided (DA) estimation as well. Furthermore, the implementation of forward error correction (FEC) schemes could improve the remodulation success rate in the low SNR range significantly. Finally, alternative algorithms for frequency offset estimation of APSK signals have to be found, since the currently implemented RB algorithm exhibits inferior performance.

References

1. Flohberger, M.: Advanced Satellite Monitoring Using Blind Demodulation Techniques. Ph.D. thesis, Graz University of Technology, Austria, to be published (2011)
2. Mengali, U., D'Andrea, A.N.: Synchronization Techniques for Digital Receivers. Plenum Press, New York (1997)
3. Treichler, J., Larimore, M., Harp, J.: Practical Blind Demodulators for High-Order QAM Signals. Proc. IEEE 86, 1907–1926 (1998)
4. ETSI EN 302 307: Second Generation Framing Structure, Channel Coding and Modulation Systems for Broadcasting, Interactive Services, News Gathering and other Broadband Satellite Applications (2005)
5. Evans, B.: Satellite Communication Systems, 3rd edn. Institution of Engineering and Technology, London (1999)
6. Welch, P.D.: The Use of Fast Fourier Transform for the Estimation of Power Spectra: A Method Based on Time Averaging Over Short, Modified Periodograms. IEEE Trans. Audio Electroacoust. 15, 70–73 (1967)
7. Xu, H., Zhou, Y., Huang, Z.: Blind Roll-Off Factor and Symbol Rate Estimation Using IFFT and Least Squares Estimator. In: Int. Conf. on Wireless Communications, Networking and Mobile Computing (WICOM), Shanghai, China, pp. 1052–1055 (2007)
8. Yu, Z., Shi, Y.Q., Su, W.: Symbol-Rate Estimation Based on Filter Bank. In: Proc. IEEE Int. Symposium on Circuits and Systems (ISCAS), pp. 1437–1440, Kobe, Japan (2005)
9. Kueckenwaitz, M., Quint, F., Reichert, J.: A Robust Baud Rate Estimator for Noncooperative Demodulation. In: Proc. 21st Century Military Communications (MILCOM) Conf., Los Angeles, USA, vol. 2, pp. 971–975 (2000)
10. Ciblat, P., Loubaton, P., Serpedin, E., Giannakis, G.: Asymptotic Analysis of Blind Cyclic Correlation-Based Symbol-Rate Estimators. IEEE Trans. Inform. Theory 48, 1922–1934 (2002)
11. Gardner, F.M.: Interpolation in Digital Modems - Part I: Fundamentals. IEEE Trans. Commun. 41, 501–507 (1993)
12. Erup, L., Gardner, F.M., Harris, R.A.: Interpolation in Digital Modems - Part II: Implementation and Performance. IEEE Trans. Commun. 41, 998–1008 (1993)
13. Gappmair, W., Flohberger, M., Koudelka, O.: Moment-Based Estimation of the Signal-to-Noise Ratio for Oversampled Narrowband Signals. In: Proc. 16th IST Mobile and Wireless Communications Summit, Budapest, Hungary, pp. C4.1–1 – C4.1–4 (2007)
14. Gardner, F.M.: A BPSK/QPSK Timing-Error Detector for Sampled Receivers. IEEE Trans. Commun. 34, 423–429 (1986)
15. Gappmair, W., Cioni, S., Corazza, G.E., Koudelka, O.: Extended Gardner Detector for Improved Symbol-Timing Recovery of M-PSK Signals. IEEE Trans. Commun. 54, 1923–1927 (2006)
16. Flohberger, M., Gappmair, W., Koudelka, O.: Open-Loop Analysis of an Error Detector for Blind Symbol Timing Recovery Using Baud-Rate Samples. In: Proc. IEEE 4th Int. Workshop on Satellite and Space Communications (IWSSC), Toulouse, France, pp. 176–180 (2008)
17. Oerder, M., Meyr, H.: Digital Filter and Square Timing Recovery. IEEE Trans. Commun. 36, 604–612 (1988)
18. Flohberger, M., Gappmair, W., Cioni, S.: Two Iterative Algorithms for Blind Symbol Timing Estimation of M-PSK Signals. In: Proc. IEEE 5th Int. Workshop on Satellite and Space Communications (IWSSC), Siena, Italy, pp. 8–12 (2009)

19. Flohberger, M., Gappmair, W., Koudelka, O.: Modulation Classifier for Signals Used in Satellite Communications. In: Proc. IEEE 5th Advanced Satellite Mobile Systems (ASMS) Conf., Cagliari, Italy, pp. 198–202 (2010)
20. Pauluzzi, D.R., Beaulieu, N.C.: A Comparison of SNR Estimation Techniques for the AWGN Channel. IEEE Trans. Commun. 48, 1681–1691 (2000)
21. Gappmair, W., Koudelka, O.: Moment-based SNR Estimation of Signals with Non-Constant Envelope. In: Proc. 3rd Advanced Satellite Mobile Systems (ASMS) Conf., Hersching, Germany, pp. 301–304 (2006)
22. Álvarez Díaz, M., López-Valcarce, R., Mosquera, C.: SNR Estimation for Multi-level Constellations Using Higher-Order Moments. IEEE Trans. Signal Process. 58, 1515–1526 (2010)
23. Rife, D.C., Boorstyn, R.R.: Single-Tone Parameter Estimation from Discrete-Time Observations. IEEE Trans. Inform. Theory 20, 591–598 (1974)
24. Viterbi, A.J., Viterbi, A.M.: Nonlinear Estimation of PSK-Modulated Carrier Phase with Application to Burst Digital Transmission. IEEE Trans. Inform. Theory 29, 543–551 (1983)
25. Morelli, M., D'Andrea, A.N., Mengali, U.: Feedforward Estimation Techniques for Carrier Recovery in 16-QAM Modulation. In: Broadband Wireless Communications. Springer, Heidelberg (1998)
26. Duryea, T., Sari, I., Serpedin, E.: Blind Carrier Recovery for Circular QAM Using Nonlinear Least-squares Estimation. Digital Signal Processing 16, 358–368 (2006)
27. Wang, Y., Serpedin, E., Ciblat, P.: Optimal Blind Nonlinear Least-Squares Carrier Phase and Frequency Offset Estimation for General QAM Modulations. IEEE Trans. Wireless Commun. 2, 1040–1054 (2003)

Channel Estimation on the Forward Link of Multi-beam Satellite Systems

Michael Bergmann[1], Wilfried Gappmair[1], Carlos Mosquera[2], and Otto Koudelka[1]

[1] Institute of Communication Networks and Satellite Communications
Graz University of Technology, Austria
{michael.bergmann,gappmair,koudelka}@tugraz.at
[2] Department of Signal Theory and Communications
University of Vigo, Spain
mosquera@gts.tsc.uvigo.es

Abstract. Multi-beam concepts are an essential component of next generation broadband satellite systems. Due to aggressive design goals, full frequency reuse is suggested in this context so that appropriate interference mitigation techniques have to be applied on both forward and return links. In this respect, channel estimation is of paramount importance. Throughout this paper, we are focusing on channel estimation of the forward link with emphasis on orthogonal and non-orthogonal training sequences used for this purpose. By analytical and simulation results, it is confirmed that the former are best suited in terms of the obtained jitter performance. On the other hand, non-orthogonal codes are not restricted by their length, but it is shown that a linearly independent set of unique words is significantly affected by an amplification of the noise component – a result not available from the technical literature on this subject so far. Furthermore, it is demonstrated that a simple correlation procedure, which might be employed for any kind of non-orthogonal training sequences, produces a non-negligible jitter floor in the higher SNR regime, which is primarily given by the cross-correlation properties of the code.

Keywords: Channel estimation, multi-beam satellite systems, forward link, orthogonal and non-orthogonal training sequences.

1 Introduction

Compared to terrestrial solutions, the major benefit of satellite communications is that all users can be served at the same cost factor within the coverage zone. In this context, it is well known that the potential of a single-beam system is limited in terms of availability and throughput – in contrast to an approach based on multiple or even full *frequency reuse* [1]. Therefore, with regard to next generation broadband satellite systems, it is suggested to consider a multi-beam technique, i.e., the coverage zone is served via a number of spot beams, suitably shaped by the antenna feeds forming part of the payload. Usually, the latter are controlled by a properly designed beamforming matrix [2]. In case the beam pattern must be modified, e.g., driven by the requirements of some mobile service, solely the matrix needs to be reorganized accordingly;

G. Giambene and C. Sacchi (Eds.): PSATS 2011, LNICST 71, pp. 250–259, 2011.
© Institute for Computer Sciences, Social Informatics and Telecommunications Engineering 2011

for reasons of flexibility and cost, it is intended in the near future to do that on ground [3].

Nevertheless, the multi-beam concept is closely related to a major problem: adjacent spots are heavily affected by *interference*, in particular if they employ the same frequency band. A fairly conservative answer for this sort of impairment is to employ a frequency reuse factor $f_R > 1$, i.e., adjacent user cells are served by different bands. There is no doubt that this decreases the spectral efficiency. As a consequence, upcoming satellite systems will be operated at full frequency reuse, i.e., $f_R = 1$, which means that all beams share the same spectrum, whereas the interference problem is to be mitigated by appropriately selected baseband algorithms. Most promising in this respect are linear or nonlinear precoding schemes on the forward link [4] and powerful interference cancellation techniques on the return link [5], [6].

Throughout this paper, we are focusing on the *forward link*, i.e., transmission of data from the centralized satellite gateway (GW) to the user terminals (UTs). Due to the fact that the GW controls the forward link (star topology), we assume a TDM-based (symbol-synchronous) concept. But no matter which kind of precoder we are going to implement at the end, for proper operation the corresponding channel estimates have to be known by the GW station. In this context, it makes sense to work with a data-aided approach [7], i.e., training sequences or unique words will be specified between GW and UTs. In the sequel, different solutions for channel estimation on the forward link will be investigated in detail by exploring the jitter variance as the most important figure of merit in this context.

The remainder of the paper is organized as follows. In Section 2, we introduce the signal model used for analytical and simulation work. Several channel estimators are discussed in Section 3, mainly by concentrating on the discrepancy between orthogonal and non-orthogonal training sequences. Numerical results are presented in Section 4, and conclusions are drawn in Section 5.

2 Signal Model

In the following, the forward link is considered to consist of an ideal satellite uplink (no fading or frequency-selective impacts) and an impaired downlink with full frequency reuse. It is assumed that the latter consists of $K = 100$ beams, pointing to 100 user cells, henceforth meant to serve 100 UTs simultaneously, i.e., a single UT per beam/cell at a given time. This is exemplified in Fig. 1 showing the 3 dB contour lines[1].

The 100 beams are formed by an antenna array with $N = 155$ feeds; each beam integrates 20 feeds tuned appropriately in amplitude and phase so as to achieve the required beamforming. Fig. 2 depicts a contour plot of the 155 feeds indicated by a separate number each; the contour lines represent the 3 dB attenuation limit of the respective feed. It is to be noticed that in our context channel estimation will be related to *feeds*, not beams, since it is supposed that interference mitigation techniques based on feeds will be more powerful [8], simply because $N \gg K$, i.e., more degrees of freedom in the design of precoding and cancellation algorithms.

---

[1] This antenna model has been provided by the European Space Agency (ESA) in the framework of SatNEx-III for a study on next generation broadband satellite systems; such systems are primarily characterized by a multi-beam concept, data transmission in the Ka or even Q/V band, and the application of adaptive coding and modulation (ACM) strategies [11].

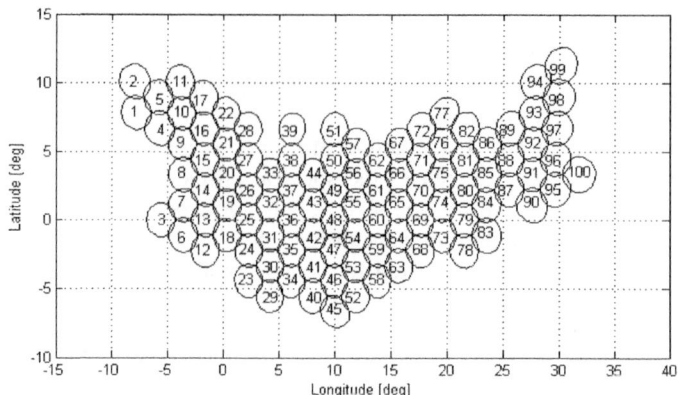

Fig. 1. Beam pattern (3 dB) generated by the satellite antenna

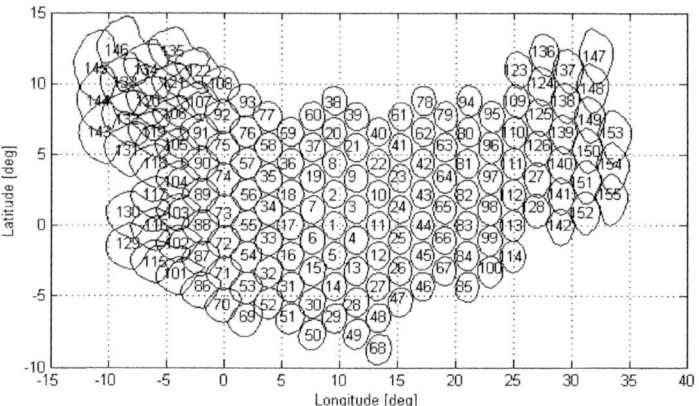

Fig. 2. Contour plot (3 dB) of the 155 satellite feeds

As already mentioned in the introductory section, a unique word (UW) is used to estimate the channel matrix $\mathbf{H} = \mathbf{WGD}$:

- **D** is the $N \times N$ matrix including transmitter GW effects, uplink, and on-board repeater chains; normally, **D** can be assumed as diagonal with the m-th entry representing the gain between feed m and GW station.
- **G** is the $K \times N$ feeder matrix; the entry (m, n) represents the gain between user link m and antenna feed n.
- **W** is the $K \times K$ diagonal fading matrix on the user downlink; the m-th entry represents the gain for user m.

In our signal model, the m-th line of **H**, i.e., $\mathbf{h}_m = (h_{m,1}, h_{m,2}, \ldots, h_{m,N})$, includes $N = 155$ complex-valued numbers characterizing the satellite downlink to the m-th UT, simply denoted by UT_m and as such sketched in Fig. 3. Considering this diagram, it is obvious that the estimation of each element in $\mathbf{h}_m$ suffers from a different signal-to-noise ratio (SNR). For example, the signal power impinging from feed 119 at the UT_{48} (see Figs. 1 and 2) strongly differs from the signal power from feed 3, whereas noise power is the same for both; this has to be taken into account, when accuracy in terms of jitter variance is an issue throughout the estimation process.

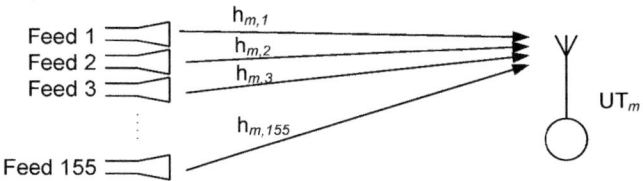

Fig. 3. Satellite downlink for the m-th user terminal

By inspection of Fig. 3, it is clear that the received signal at UT_m, i.e., $\mathbf{y}_m$ with $m = 1, 2, \ldots, K$, is given by

$$\mathbf{y}_m = \mathbf{h}_m \mathbf{C} + \mathbf{w}_m \tag{1}$$

where matrix $\mathbf{C}$ contains the complete set of $N = 155$ UWs as row vectors of length L and $\mathbf{w}_m$ denotes the sequence of zero-mean white Gaussian noise samples with variance $\sigma_w^2 = N_0/E_s$, where E_s/N_0 symbolizes the SNR per symbol.

3   Channel Estimation on the Forward Link

As already mentioned in the introductory section, the channel matrix $\mathbf{H}$ must be known to the GW station for interference mitigation on the forward link, e.g., realized by an appropriately selected precoding algorithm [9], [10]. This might be achieved by a separate calibration network; a more elegant and less expensive solution for this purpose, however, is the usage of the communication network as such. To this end, the estimates of $\mathbf{h}_m$, $m = 1, 2, \ldots, K$, will be collected by UT_m and reported to the GW via the return link.

The estimate of $\mathbf{h}_m$ is straightforwardly obtained by post-multiplying (1) with the Moore-Penrose pseudo-inverse of $\mathbf{C}$ expressed as $\mathbf{C}^+ = \mathbf{C}^H (\mathbf{C}\mathbf{C}^H)^{-1}$, where $\mathbf{C}^H$ denotes the Hermitean of $\mathbf{C}$. As a consequence, we arrive at

$$\hat{\mathbf{h}}_m = \mathbf{y}_m \mathbf{C}^+ = \mathbf{h}_m + \mathbf{e}_m \tag{2}$$

with the error vector evaluated as

$$\mathbf{e}_m = \mathbf{w}_m \mathbf{C}^+ . \tag{3}$$

3.1   Orthogonal UW Sequences

Orthogonal sequences are best suited for channel estimation since they do *not* produce interference noise. The reason is that the pseudo-inverse of $\mathbf{C}$ is in this case furnished by the Hermitean transpose $\mathbf{C}^H$, i.e., $\mathbf{CC}^H = \mathbf{I}$ so that (3) becomes $\mathbf{e}_m = \mathbf{w}_m \mathbf{C}^H$. Having a closer look to $e_{m,i}$, which denotes the i-th entry of the error vector $\mathbf{e}_m$, we have that

$$e_{m,i} = \sum_{k=1}^{L} w_{m,k} \, c_{k,i}^{H} \tag{4}$$

where $c_{k,i}^{H}$ is the (k, i)-th element in $\mathbf{C}^H$; normally, the codes are binary, i.e., the corresponding entries in $\mathbf{C}$ are given by $c_{k,i} = \pm 1$, so that $c_{k,i}^{H} = \pm 1/L$. In this respect, simply by taking into account that $E[|w_{m,k}|^2] = \sigma_w^2$, the total jitter variance for orthogonal sequences is straightforwardly derived as

$$\sigma_H^2 = \sigma_w^2 \sum_{i=1}^{L} |c_{k,i}^{H}|^2 = \frac{1}{L \, E_s/N_0}. \tag{5}$$

It is to be noticed that the frequently used Walsh-Hadamard codes are all of length $L = 2^n$, $n = 1, 2, 3, \ldots$ Therefore, randomly generated non-orthogonal sequences of arbitrary length L will be considered in the sequel.

3.2 Non-orthogonal UW Sequences

Basically, the pseudo-inverse $\mathbf{C}^+$ can be computed, if the UWs constituting matrix $\mathbf{C}$ are *linearly independent*. A necessary condition in this regard is that the length of the training sequence is equal or greater than the number of feeds, i.e., $L \geq N$. Then, according to (4), the i-th entry of the error vector $\mathbf{e}_m$ becomes

$$e_{m,i} = \sum_{k=1}^{L} w_{m,k} \, c_{k,i}^{+} \tag{6}$$

where $c_{k,i}^{+}$ is the (k, i)-th element in $\mathbf{C}^+$. Recalling again that the noise samples $w_{m,k}$ are zero-mean white Gaussian, the total jitter variance for linearly independent sequences is given by

$$\sigma_R^2 = \sigma_w^2 \sum_{k=1}^{L} |c_{k,i}^{+}|^2. \tag{7}$$

On the other hand, if $L < N$, the pseudo-inverse does *not* exist. In this case, we simply resort to a correlation procedure according to

$$\hat{\mathbf{h}}_m = \mathbf{y}_m \mathbf{C}^H \tag{8}$$

where the $N \times L$ matrix $\mathbf{C}$ necessarily includes some *linearly dependent* training sequences in form of row vectors. Moreover, it is to be observed that for linearly independent sequences, no matter if they are orthogonal or non-orthogonal, the jitter variance decreases with increasing values of $E_s/N_0 = 1/\sigma_w^2$, which is easily verified by inspection of (5) and (7). Confirmed by simulation results in the next section, this is in striking contrast to the correlation method producing a *jitter floor* for non-orthogonal codes, which is given by

$$\sigma_F^2 = \left| \sum_{m \neq k} \sum_{i=1}^{L} h_{m,i} c_{m,i} c_{k,i}^{*} \right|^2. \tag{9}$$

4 Numerical Results

Since $N = 155$ channels (feeder links) must be distinguished, Walsh-Hadamard (WH) codes require a length of $L = 2^8 = 256$ symbols in order to properly estimate the channel matrix $\mathbf{H}$. Using this sort of UWs, Fig. 4 illustrates the simulation results for estimation accuracy in terms of the mean square error (MSE) for both magnitude (full dots) and angle of $h_{m,k}$ (open dots); with regard to the former, the values normalized by $|h_{m,k}|$ are shown and verified by the analytical relationship in (5): for higher E_s/N_0 values, the jitter variance for amplitude and phase of $h_{m,k}$ is given by $\sigma_H^2/2$ (solid line); note that the curve for phase estimates starts to flatten at $E_s/N_0 < -30$ dB, since they become more and more equally distributed between $\pm\pi$ so that the variance is given by $\pi^2/3$.

For a non-orthogonal code, based on randomly (RN) generated but linearly independent UWs with $L = 156$, Fig. 4 depicts the jitter performance of $h_{m,k}$ estimated at UT_{48} according to the scenario presented in Figs. 1 and 2. It can be seen that, in contrast to the ideal case (dashed line) indicating an assumed orthogonal code (OC) with $L = 156$, both angle and magnitude exhibit a significant degradation of the jitter variance given by $\sigma_R^2/2$ (dashed-dotted line, simulation results for magnitude/phase visualized by solid/open triangles). This is explained by the fact that, compared to WH sequences, the estimation procedure via RN schemes causes a non-negligible amount of additional noise introduced by $\mathbf{C}^+$ – to the best of the authors' knowledge, a result not discussed so far in the open literature. Finally, for $E_s/N_0 = 40$ dB, Fig. 5 shows the top view of the jitter performance as a function of the geographical location within the coverage zone (for normalization purposes, the strongest component of $\mathbf{h}_m$ is employed).

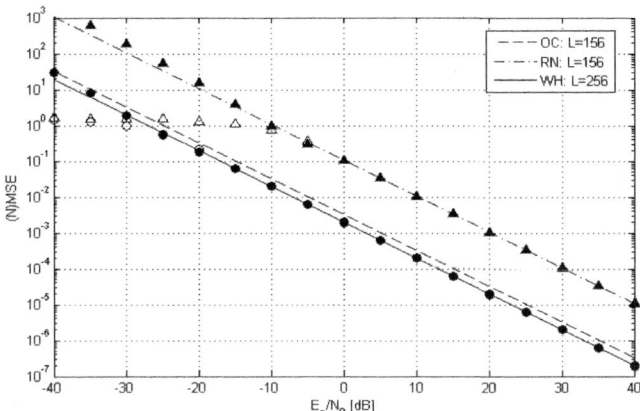

Fig. 4. Estimation accuracy for orthogonal and non-orthogonal codes achieved by the pseudo-inverse

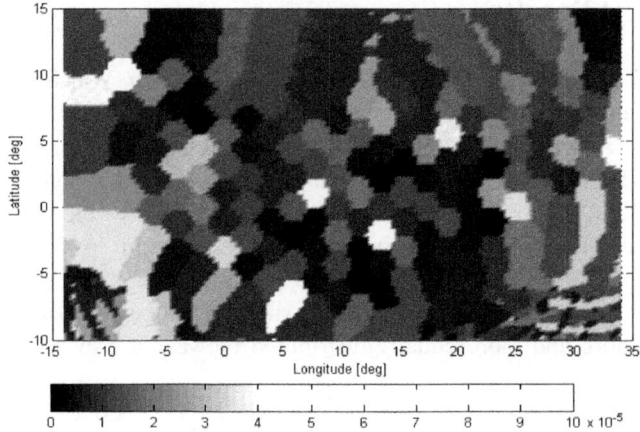

Fig. 5. Top view of estimation accuracy for non-orthogonal codes achieved by the pseudo-inverse ($E_s/N_0 = 40$ dB, $L = 156$)

For UT_{48} operated at $L = 78$, Fig. 6 shows the estimation accuracy using a set of RN sequences; but in contrast to the former case with $L = 156$, the UWs are now linearly dependent so that a pseudo-inverse does not exist. It can be seen that in the low SNR range, both the MSE of the angle (open dots/triangles) and the normalized MSE of the amplitude (full dots/triangles) are close to the theoretical limit $\sigma_H^2/2$ symbolizing an assumed orthogonal scheme (dashed/dashed-dotted line). Unfortunately, the estimation performance of the amplitude suffers from a significant jitter floor according to (9) in the higher SNR regime as can be observed by inspection of Fig. 6 (horizontal dashed/dashed-dotted line).

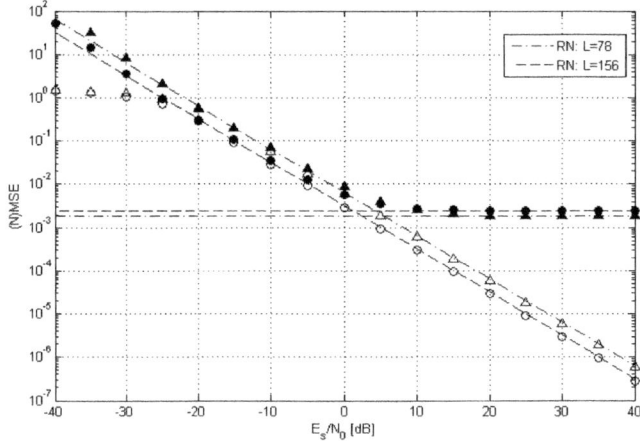

Fig. 6. Estimation accuracy for non-orthogonal codes achieved by correlation

Note also that the angle exhibits *no* jitter floor, a phenomenon not reported in previous publications on this topic so far. It is explained by the fact that all feeds transmit their assigned UW in a *symbol-synchronous* manner. Moreover, each of the sequences travels through the same physical path down to the UT, thus determined by the same attenuation and the same time shift. Taking this into account, it is clear that the correlation process does not affect the phase of the channel estimate so that, in contrast to the amplitude, no jitter floor is present.

It is important to understand that the jitter floor is *not* related to the use of linearly dependent sequences as such. Instead, it is caused by the correlation method, which had to be applied for linearly dependent UWs due to the lack of a pseudo-inverse. For the sake of completeness, Fig. 6 visualizes also the (normalized) MSE for an RN code composed of linearly independent vectors with length $L = 156$ (simulation results for magnitude/phase indicated by solid/open dots). It is also identified that the value of L plays a minor role so as to decrease the floor.

For $E_s/N_0 = 40$ dB and $L = 78$, Fig. 7 shows the top view of the jitter floor as a function of the geographical location. It obviously exhibits larger variations depending on the UTs' positions. By inspection of the plot, it can be seen that the major part of the coverage zone is affected by a value about 10^{-2}, only small areas have a floor lower than 10^{-3}. Although not shown due to limited space, this has been observed with other non-orthogonal codes as well, like Gold or Kasami schemes, as soon as they are applied to a correlation procedure for estimation purposes.

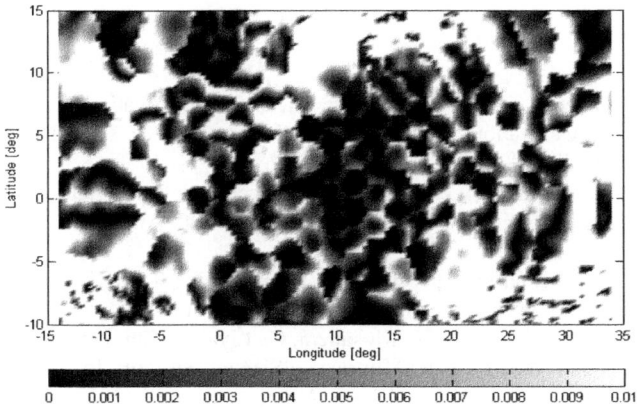

Fig. 7. Top view of estimation accuracy for non-orthogonal codes achieved by correlation ($E_s/N_0 = 40$ dB, $L = 78$)

5 Concluding Remarks

Assuming a multi-beam concept for next generation broadband satellite systems, several algorithms for channel estimation on the forward link have been investigated. By both analytical and simulation results, it could be confirmed that orthogonal training sequences – with Walsh-Hadamard codes as most prominent example in this respect – are best suited for this purpose. Due to the restricted application of the latter, non-orthogonal sequences were explored in this context. It turned out that a set of linearly independent vectors, which is basically only possible if their length is equal or greater than the number of links (antenna feeds), ends up in a significant degradation of the jitter performance due to an amplification of noise, introduced by the pseudo-inverse of the matrix containing the unique words. On the other hand, applying a correlation procedure to extract the channel estimates from a set of non-orthogonal codes (no matter if they are linearly independent or not), yields a jitter floor, mainly determined by the cross-correlation properties of the selected unique words.

Finally, it is to say that carrier frequency/phase and symbol timing have been assumed as perfectly established for channel estimation on the forward link. This is acceptable for UTs, which form already part of the network; in this case, the data following the unique word will be used for estimation and synchronization of carrier and timing, because they are not very much impaired by interferer noise due to precoding. Nevertheless, if a UT is new to the satellite system, then channel estimates, carrier and timing offsets must be jointly acquired. This resembles very much the situation met on the return link, which is subject of ongoing investigations.

Acknowledgments. The work was supported in part by SatNEx-III (Satellite Network of Experts) launched by the European Space Agency for advanced research in satellite communications (ESA Contract No. RFQ/3-12859/09/NL/CLP).

References

1. Gallinaro, G., et al.: Novel Intra-System Interference Mitigation Techniques & Technologies for Next Generations Broadband Satellite Systems. ESA Final Report, Contract No. 18070/04/NL/US (2008)
2. Angeletti, P., Gallinaro, G., Lisi, M., Vernucci, A.: On-Ground Digital Beamforming Techniques for Satellite Smart Antennas. In: Proc. 19th AIAA, Toulouse, France, pp. 1–8 (2001)
3. Angeletti, P., Alagha, N.: Space/Ground Beamforming Techniques for Emerging Hybrid Satellite Terrestrial Networks. In: Proc. 27th AIAA, Edinburgh, UK, pp. 1–6 (2009)
4. Alvarez-Diaz, M., Courville, N., Mosquera, C., Liva, G., Corazza, G.E.: Nonlinear Interference Mitigation for Broadband Multimedia Satellite Systems. In: Proc. 3rd Int. Workshop Satellite and Space Commun., Salzburg, Austria, pp. 61–65 (2007)
5. Millerioux, J.P., Boucheret, M.L., Bazile, C., Ducasse, A.: Iterative Interference Cancellation and Channel Estimation in Multibeam Satellite Systems. Int. J. Satell. Commun. Network 25, 263–283 (2007)
6. Di Cecca, F., Gallinaro, G.: Ground Beamforming and Interference Cancellation for TDMA Based Reverse-Link Access Schemes. In: Proc. Ka-Band Conf., Cagliari, Italy, pp. 1–8 (2009)
7. Mengali, U., D'Andrea, A.N.: Synchronization Techniques in Digital Receivers. Plenum Press, New York (1997)
8. Corrazza, G.E., et al.: Digisat Techniques Review: Hybrid Space-Ground Processing. ESA Technical Note, Contract No. RFQ/3-12859/09/NL/CLP (2010)

9. Caire, G., Debbah, M., Cottatellucci, L., De Gaudenzi, R., Rinaldo, R., Mueller, R., Galli-naro, G.: Perspectives of Adopting Interference Mitigation Techniques in the Context of Broadband Multimedia Satellite Systems. In: Proc. 23rd AIAA, Rome, Italy, pp. 1–5 (2005)
10. Cottatellucci, L., Debbah, M., Gallinaro, G., Mueller, R., Neri, M., Rinaldo, R.: Interfe-rence Mitigation Techniques for Broadband Satellite Systems. In: Proc. 24th AIAA, San Diego, CA, pp. 1–13 (2006)
11. Cioni, S., De Gaudenzi, R., Rinaldo, R.: Channel Estimation and Physical Layer Adapta-tion Techniques for Satellite Networks Exploiting Adaptive Coding and Modulation. Int. J. Satell. Commun. Network. 26, 157–188 (2008)

Interference Management versus Interference Cancellation: SATCOM Case

Lei Jiang and Maria-Angeles Vázquez-Castro

Dpt. Telecom. and Systems Engineering, Universitat Autónoma de Barcelona, Spain
{jiang.lei,angeles.vazquez}@uab.es
http://wirelesssatcom.uab.es/index.htm

Abstract. Interference management and interference cancellation are addressed in the satellite communication (SATCOM) system. We first introduce the two options for a multibeam payload: flexible (FLEX) where allocation is fully flexible in the frequency domain, and conventional on which beamforming (BF) is applied. Further, two optimization problems are proposed, P1: spectral mask vector design and power allocation in the FLEX SATCOM system by interference management, and P2: BF weighting vector design and power allocation in the BF SATCOM system by interference cancellation. Specifically, we provide a resource allocation algorithm for each system. The performance of the resource allocation algorithm is evaluated with asymmetrical traffic distribution models. The numerical results show that, by using interference cancellation, BF system with full frequency reuse provides the best performance. For the FLEX system, the bandwidth can be utilized more efficiently with larger frequency reuse factor and smaller bandwidth granularity.

Keywords: Multibeam satellite, resource allocation, interference management, and interference cancellation.

1 Introduction, Previous Work and Contribution

Current satellite communication (SATCOM) systems make use of frequency reuse by means of multiple beams grouped in clusters throughout the coverage. Power and bandwidth allocation should be optimized to adapt the asymmetric traffic distribution and channel conditions (e.g. at Ka band, where rain attenuation can be of dozens of dBs). In this paper, the co-channel interference of satellite system is formulated mathematically in both frequency and time-space domains. Two techniques will be analyzed: one is the interference management based on flexible (FLEX) system where the bandwidth allocation is fully flexible in the frequency domain, the other is the interference cancellation based on conventional system on which beamforming (BF) is applied. The payloads of flexible and conventional SATCOM systems refer to [1, 2].

Given the Channel State Information (CSI), the transmitter can manage the interference in frequency domain by splitting the total available bandwidth into numerous carriers and managing the carrier and power allocation for each

G. Giambene and C. Sacchi (Eds.): PSATS 2011, LNICST 71, pp. 260–273, 2011.

beam. The interference management has been discussed in [4, 5, 3, 6]. In [3], an axiomatic-based interference model for Signal to Interference plus Noise Ratio (SINR) balancing problem is proposed with individual target SINR per user, but it is focused on the terrestrial wireless communications. In [4], a power allocation policy is suggested to stabilize the system based on the amount of unfinished work in the queue and the channel state. In [5], the authors make an effort to design a tradeoff strategy between different objectives and system optimization. However, the co-channel interference is not taken into account. The authors in [6, 7] discuss power and carrier allocation problem in SATCOM system, but [6] only focuses on the return uplink and [7] discusses the beamforming vector design in terms of sum-rate.

The interference cancellation can be realized by precoding techniques, i.e. Dirty Paper Coding (DPC) and BF. It is well known from the literature [8, 9] on mobile terrestrial channel that for a given composite channel matrix, the sum-rate capacity is achieved by "Gaussian codes" and DPC. In fact, DPC is the optimal (capacity achieving) strategy in MIMO broadcast (BC) channels. However, DPC is difficult to implement in practical systems due to the high computational burden of successive encodings and decodings. BF is a suboptimal strategy that can serve multiple users simultaneously, but with reduced complexity relative to DPC. In BF, each user stream is precoded by a beamforming weight vector for reducing/eliminating the mutual interference among different streams. However, the precoding techniques increase the transmitted signal power (precoding loss). As discussed in [10], the precoding loss is negligible with larger constellations and is always bounded, i.e. for the QPSK constellation, the precoding loss is never greater than 1.5dB. Besides, it quickly falls towards 0dB for larger constellations.

Most of the interference management and interference cancellation techniques discussed in the above literatures are focused on the terrestrial communication systems. However, SATCOM systems have a different geometrical topology and the satellite payload poses a number of constraints not present in terrestrial systems [11]. Specifically, the contributions of this paper can be summarized as

- An unified system model formulation, which is valid both in frequency and space-time domains. As the system is interference-limited, this model allows us for the derivation of SINR for different interference countermeasure mechanisms in an unified way.
- Optimization of a SATCOM system by interference management with full flexibility in the frequency domain. In particular, we solve the problem P1: optimization of the spectral mask vector and power allocation under the individual SINR constraints (we will call this system "FLEX").
- Optimization of a SATCOM system by interference cancellation with beamforming in the conventional system [2], where the frequency bandwidth is reused by a subset of beams. In particular, we solve the problem P2: beamforming weight vector design and power allocation under the individual SINR constraints (we will call this system "BF").

The rest of the paper is organized as follows: in Section 2, the unified system model is formulated. In Section 3, the SINR expression in frequency domain is formulated, a spectral mask vector design and SINR balancing problem is proposed and solved. Section 4 formulates the SINR in time-space domain, and resource allocation optimization problem is proposed and solved in this section. Section 5 provides selected numerical results. Finally, the summarizing conclusions are presented in Section 6.

We adopt the following notations: bold uppercase letters denote matrices and bold lowercase letters denote vectors, $(\cdot)^T$ and $(\cdot)^H$ denote transpose and conjugate transpose, respectively, $(\cdot)^\dagger$ denotes pseudo-inverse, $\varepsilon(\cdot)$ stands for the expectation, $\lambda_{\max}(\cdot)$ denotes the maximum eigenvalue, $\upsilon_{\max}(\cdot)$ indicates the eigenvector related to the maximum eigenvalue, $\operatorname{diag}\{(\cdot)\}$ denotes a diagonal matrix with the elements $(\cdot)$ along its diagonal, and $\mathbf{I}_N$ denote the identity matrix of size $N \times N$.

2   System Model

In this section, an unified SATCOM system model for both multibeam FLEX system (in frequency domain) and BF system (in time-space domain) is formulated. In a multibeam satellite systems, the beamforming antenna generates K beams over the coverage area. Let us assume a MIMO BC model equipped with K transmit antennas and the ith user terminal with N_i receive antennas. For simplicity, we assume that all the users are homogeneous and experience independent fading. The signal received by a user i can be expressed as

$$\mathbf{y}_i = \mathbf{H}_i \mathbf{x} + \mathbf{n}_i, \quad i = 1, \ldots, K, \tag{1}$$

where $\mathbf{x} \in \mathbb{C}^{K \times 1}$ is the transmitted symbol from the satellite antennas, $\mathbf{H}_i \in \mathbb{C}^{N_i \times K}$ is the channel gain matrix to the ith user, $\mathbf{n}_i \in \mathbb{C}^{N_i \times 1}$ is zero-mean complex circular Gaussian noise with variance σ^2 at user i, and $\mathbf{y}_i$ is the received signal vector by user i. The transmitter has a power constraint $\varepsilon\{\mathbf{x}\mathbf{x}^H\} \leq P_{\text{tot}}$. We assume that each user is equipped with a single antenna, i.e. $N_i = 1$, for $\forall i$, this is a common assumption in the satellite scenario, e.g. in [12,1,2]. Depending on different system assumptions, e.g FLEX or BF, $\mathbf{x}$, $\mathbf{H}_i$, $\mathbf{n}_i$ and $\mathbf{y}_i$ have different dimension, size and meaning. We will discuss in detail the system model for both FLEX and BF systems in the following sections.

2.1   FLEX System in Frequency Domain

In this section we analysis the FLEX system model. The total available bandwidth, B_{tot}, is divided in N_c carriers providing carrier granularity of $B_c = B_{\text{tot}}/N_c$. Note that we assume Time-Division Multiplex (TDM). For a single carrier slot, the signal model can be expressed as equation (1). We are interested in the multiple carriers TDM mode, hence, for a specific beam i, the transmitted symbols over N_c carriers are defined as $\mathbf{s}_i = [s_{i1}, s_{i2}, \cdots, s_{iN_c}]^T$. Let the spectral mask matrix $\mathbf{W} \in \mathbb{R}^{N_c \times K}$ be defined as $\mathbf{W} = [\mathbf{w}_1, \mathbf{w}_2, \cdots, \mathbf{w}_K]$, and the ith column vector $\mathbf{w}_i \in \mathbb{R}^{N_c \times 1}$ be defined as $\mathbf{w}_i = [w_{i1}, w_{i2}, \cdots, w_{iN_c}]^T$, which is the spectral mask vector for beam i and indicates which TDM carriers are allocated to beam i.

Let $\mathbf{A} = diag\{\alpha_1, \ldots, \alpha_K\}$ be the channel attenuation amplitude matrix over the user and $\mathbf{G}$ is the antenna gain matrix, which is defined as

$$\mathbf{G} = \begin{bmatrix} g_{11} & g_{12} & \cdots & g_{1K} \\ g_{21} & g_{22} & \cdots & g_{2K} \\ \vdots & \vdots & \ddots & \vdots \\ g_{K1} & g_{K2} & \cdots & g_{KK} \end{bmatrix}, \tag{2}$$

where g_{ij} is the square root of the gain of the j-beam on-board antenna towards the ith user. Let $\mathbf{H} = \mathbf{AG}$ be the overall channel matrix, and $\mathbf{W}_i = \mathrm{diag}\{\mathbf{w}_i\}$. Then, from the unified system model in equation (1), the received signal by all the N_c carriers for beam i, $\mathbf{y}_i \in \mathbb{C}^{N_c \times 1}$, can be expressed as

$$\mathbf{y}_i = h_{ii}\mathbf{x}_i + \sum_{k=1(k \neq i)}^{K} h_{ik}\mathbf{x}_k + \mathbf{n}_i, \tag{3}$$

where $\mathbf{x}_i$ is the spectral masked symbol vector for beam i, defined as $\mathbf{x}_i = \mathbf{W}_i\mathbf{s}_i$. The first term corresponds to the desired signals coming from the ith on-board antenna. The second term is the sum of interference signals from the other on-board antennas. $\mathbf{n}_i \in \mathbb{C}^{N_c \times 1}$ is a column vector of zero-mean complex circular Gaussian noise with variance σ^2 at beam i.

2.2 BF System in Time-Space Domain

For the BF SATCOM system, we are interested in the full frequency reuse pattern, therefore, the user streams are separated by different time-space beamforming directions, as opposed to frequency slot separation in the FLEX system.

The overall channel matrix $\mathbf{H}$ is the same as shown in Section 2.1, $\mathbf{F} \in \mathbb{C}^{K \times K}$ denotes the beamforming weight matrix be defined as $\mathbf{F} = [\mathbf{f}_1, \mathbf{f}_2, \cdots, \mathbf{f}_K]$, and $\mathbf{s} = [s_1, s_2, \cdots, s_K]^T$ is the symbol vector. s_i and $\mathbf{f}_i$ are the data symbol and the beamforming weight vector for beam i, respectively. We can reformulate equation (1) for the BF system with the desired signal and interference as (e.g. user i)

$$y_i = \mathbf{h}_i\mathbf{f}_i s_i + \sum_{k=1(k \neq i)}^{K} \mathbf{h}_i\mathbf{f}_k s_k + n_i, \tag{4}$$

where $\mathbf{h}_i$ and $\mathbf{f}_i$ are the ith row vector and ith column vector of $\mathbf{H}$ and $\mathbf{F}$, respectively.

3   Interference Management in Frequency Domain

In modern satellite networks, multibeam antenna technology is used because it can increase the total system capacity significantly [13]. However, each beam will compete with others for resources, e.g. power and bandwidth, to achieve satisfactory communications. This is due to the fact that the traffic demand among the beams of the coverage is potentially highly asymmetrical. The FLEX SATCOM system can minimize the co-channel interference by balancing the power and carrier allocation in frequency domain. In this section, we will address the problem (P1) of spectral mask vector design and power allocation for the FLEX SATCOM system.

3.1   SINR Formulation

For the FLEX SATCOM system in frequency domain, in order to derive the expression of SINR per beam from the system model, we can reformulated the spectral mask matrix as $\mathbf{W} = [\tilde{\mathbf{w}}_1^T, \tilde{\mathbf{w}}_2^T, \cdots, \tilde{\mathbf{w}}_{N_c}^T]^T$, where $\tilde{\mathbf{w}}_j = [w_{1j}, w_{2j}, \cdots, w_{Kj}]$, indicates which beams are allocated carrier j. Let the ith row of $\mathbf{H}$ be defined as $\mathbf{h}_i = [h_{i1}, h_{i2}, \cdots, h_{iK}]$ and $\tilde{\mathbf{h}}_i = \mathbf{h}_i|_{(h_{ii}=0)}$ is the channel of interference contribution. We assume that the amplitude of the transmitted symbols is normalized (i.e. $|x_{ij}|^2 = 1, \forall i, j$).

Then, the transmitted signal power of all the carriers for beam i can be given by the diagonal elements of the matrix $\mathbf{U}_i \in \mathbb{R}^{N_c \times N_c}$ as

$$\mathbf{U}_i = |h_{ii}|^2 \mathbf{W}_i \mathbf{W}_i^H. \tag{5}$$

And the co-channel interference power of all the carriers for beam i can also be given by the diagonal elements of the matrix $\mathbf{R}_i^{\text{int}} \in \mathbb{R}^{N_c \times N_c}$ as

$$\mathbf{R}_i^{\text{int}} = \text{diag}\left\{ \left[\tilde{\mathbf{h}}_i \tilde{\mathbf{w}}_j^H \tilde{\mathbf{w}}_j \tilde{\mathbf{h}}_i^H \right]_{j=1,2,\cdots,N_c} \right\}. \tag{6}$$

Thus, the interference power plus the noise matrix, $\mathbf{R}_i$, will be given as

$$\mathbf{R}_i = \mathbf{R}_i^{\text{int}} + \sigma^2 \mathbf{I}_{N_c}. \tag{7}$$

Consequently, the SINR for beam i, defined as $\mathbf{\Gamma}_i \in \mathbb{R}^{N_c \times N_c}$, can be expressed as

$$\mathbf{\Gamma}_i = \mathbf{U}_i (\mathbf{R}_i)^{-1}. \tag{8}$$

Note that in Section 3.2, the Rayleigh quotient optimization problem will be addressed based on the SINR formulation $\mathbf{\Gamma}_i$ in equation (8). Obviously, $\mathbf{\Gamma}_i$ is a diagonal matrix, because both $\mathbf{U}_i$ and $\mathbf{R}_i$ are diagonal matrixes. Thus, the SINR for the jth carrier used by beam i will be the jth diagonal element of the matrix $\mathbf{\Gamma}_i$. This means that for each carrier j of beam i, the SINR can be formulated as

$$\gamma_{ij} = \frac{|h_{ii} w_{ij}|^2}{\sum_{k=1(k \neq i)}^{K} |h_{ik} w_{kj}|^2 + \sigma^2}. \tag{9}$$

Consequently, the beam-level sum-rate can be expressed as

$$R_i = \sum_{j=1}^{N_c} \frac{B_{\text{tot}}}{N_c} \eta_{ij} = \sum_{j=1}^{N_c} B_c \eta_{ij}, \tag{10}$$

where $\eta_{ij} = f(\gamma_{ij})$ is the spectral efficiency, and $f(\gamma_{ij})$ equals to $\log_2(1 + \gamma_{ij})$ for Shannon limit with Gaussian coding, or can be a quasi-linear function in DVB-S2 [15] with respect to the SINR.

Note that the optimization problem with a SINR constraint is equivalent to the rate constraint. E.g., for the Gaussian coding case, if we consider that the rate required by the ith user is $\hat{R}_i$, the SINR requirement can be derived from $R_i = \log_2(1 + \gamma_i)$ as $\hat{\gamma}_i = 2^{\hat{R}_i} - 1$. Therefore, in the following sections, we focus on the optimization problem with a SINR constraint per user.

3.2   Resource Allocation Optimization with Interference Management

In order to best match offered and requested traffic per beam, we develop a methodology to jointly optimize power and carrier allocation. Existing results on the references [4, 5] on similar problems assume power limitation and the optimization is exclusively over the power allocation. However, we assume an additional degree of freedom: carrier allocation (spectral mask vector design). We propose to use Binary Power Allocation (BPA) ($|w_{i,j}|^2 = \{0, P_{\text{sat}}\}$) and quantized bandwidth allocation in order to decrease the complexity, where P_{sat} is the saturation power per carrier.

Optimization Problem Formulation. The problem we need to solve is both a problem of SINR balancing (as in [3,14]) and a problem of allocating the carriers (optimize the spectral mask matrix). We do not only balance the power allocation, but also optimize the strategy of carrier allocation (i.e. the structure of matrix $\mathbf{W}$). Therefore, the theory of SINR balancing is not applicable straightforwardly here. The problem can be formulated as

$$\max_{\mathbf{W}} \sum_{i=1}^{K} \frac{\gamma_i(\mathbf{W})}{\hat{\gamma}_i},$$
$$\text{subject to } \gamma_i \leq \hat{\gamma}_i, \tag{11}$$
$$\sum_{i=1}^{K} \mathbf{w}_i^H \mathbf{w}_i \leq P_{\text{tot}}; \text{ and } |w_{ij}|^2 = \{0, P_{\text{sat}}\}, \forall i, j.$$

where P_{tot} and P_{sat} are the total available satellite power and the saturation power per carrier, respectively, which are the constraints of satellite payload.

Algorithmic Solution. The general analytical solution of (11) is a complex problem due not only to the clear non-convexity but also to the need of preserving the geometry of the optimization model (i.e. $\mathbf{W}$). Therefore, we propose an iterative solution where each iteration is based on a two-step process as shown

Table 1. Algorithm solution for the flexible carrier allocation SATCOM system

> 1: Initialize: $R_i \Leftarrow 0, i = 1, 2, \cdots, K$.
> 2: Generating beam set $\mathcal{A}_s$:
> $$\mathcal{A}_s = \left\{ i_1, i_2, \cdots, i_N \Big| 0 \le \frac{R_{i_n}}{R_{i_n}} \le \frac{R_{i_n-1}}{R_{i_n-1}} < 1 \right\}.$$
> where $i_n \in \{1, 2, \cdots, K\}, n = 1, 2, \cdots, N$.
> 3: Repeat: $k = i_1$
> 4: Solve the Rayleigh quotient problem: $\max \dfrac{\mathbf{e}_j^H \mathbf{\Gamma}_k \mathbf{e}_j}{\mathbf{e}_j^H \mathbf{e}_j}$
> subject to $||\mathbf{e}_j||^2 = 1, \forall j$.
> 5: $w_{k,j} \Leftarrow \mathbf{e}_j^H \mathbf{e}_j (P_{\text{sat}})^{1/2}$.
> 6: Update $\mathbf{U}_k, \mathbf{R}_k, \mathbf{\Gamma}_k, \mathbf{W}$, and R_k.
> 7: go to step 3, until $k = i_N$.
> 8: go to step 2, until $\mathcal{A}_s$ is empty or $\sum_{i=1}^K \mathbf{w}_i^H \mathbf{w}_i \le P_{\text{tot}}$.

in Table 1. First, we obtain an analytical solution of the carrier allocation on a per-beam basis. Second, we obtain the power allocated to the selected carriers from the power constraint.

Note that, in Step 4, $\mathbf{e}_j \in \mathbb{R}^{N_c \times 1}$ is a unity column vector with only the jth element non-zero. Herein $\mathbf{e}_j$ is introduced to indicate which carrier is allocated. The solution of Rayleigh quotient problem shown in Step 4 is given as $\mathbf{e}_j = v_{\max}(\mathbf{\Gamma}_k)$. Hence, $w_{k,j}$ for jth carrier of beam k can be obtained with the solution of $\mathbf{e}_j$ as $w_{k,j} = \mathbf{e}_j^H \mathbf{e}_j (P_{\text{sat}})^{1/2}$.

In Step 6, the resource allocation matrix, SINR and data rate for the selected beam are updated. After the iterative algorithm convergence (the detailed study of convergence shown in [12, 2, 1]), we can obtain the resource allocation matrix (spectral mask matrix): $\mathbf{W}$.

4   Interference Cancellation in Time-Space Domain

As we indicated in Section 1, the interference cancellation can be realized by BF in the conventional system [2], where a subset of beams can be illuminated simultaneously by reusing the frequency bandwidth. In BF, each illuminated beam is multiplied by a beamforming weight vector ($\mathbf{f}_i$) for reducing/eliminating the mutual interference. This interference cancellation takes advantage of the spatial relative location between users in order to support multiple users simultaneously. As pointed out in [16,17], the sum-rate capacity optimization problem of MIMO BC using DPC and BF is discussed. However, in this paper, the problem is not maximize the sum-rate, but the beamforming weight vector design and power allocation for each user with individual SINR constraints (Problem P2). We will discuss this problem in this section.

4.1   SINR Formulation

Based on the unified system model presented in equation (1), we can define the pre-coded transmitted symbols, $\mathbf{x} \in \mathbb{C}^{K \times 1}$, as $\mathbf{x} = \mathbf{Fs}$, where $\|\mathbf{x}\|^2 \leq P_{\text{tot}}$, $\mathbf{F} \in \mathbb{C}^{K \times K}$ is the beamforming weight vectors (defined in Section 2.2) and $\mathbf{s} \in \mathbb{C}^{K \times 1}$ is the normalized symbol vector. Note that we need to normalize the symbol power after introducing the beamforming weight vector, we have that $p_i = \frac{\|s_i\|^2}{\varsigma_i}$ with $\varsigma_i = \|\mathbf{f}_i\|^2$.

Following the ith user received signal (in equation (4)), we can define the SINR as

$$\gamma_i = \frac{p_i \|\mathbf{h}_i \mathbf{f}_i\|^2}{\sum\limits_{k=1(k \neq i)}^{K} p_k \|\mathbf{h}_i \mathbf{f}_k\|^2 + \sigma_i^2} = \frac{\frac{\tilde{p}_i}{\varsigma_i} \|\mathbf{h}_i \mathbf{f}_i\|^2}{\sum\limits_{k=1(k \neq i)}^{K} \frac{\tilde{p}_k}{\varsigma_k} \|\mathbf{h}_i \mathbf{f}_k\|^2 + \sigma_i^2}. \tag{12}$$

where $\tilde{p}_i = \|s_i\|^2$ is the actually transmitted power towards the ith user. It can be observed that the downlink SINR is not a convex cost-function because of the coupling among the beamforming vectors in the denominator. This makes the optimization of the beamforming complex and computationally demanding.

4.2   Resource Allocation Optimization with Interference Cancellation

In the Zero-Forcing beamforming (ZFBF), weights are selected so as the co-channel interference is cancelled. Let $\mathcal{S} \subset \{1, \ldots, K\}$ denote the user subset from all the users in the co-channel beams that are selected for transmission. One easy choice of $\mathbf{F}(\mathcal{S})$ that gives zero-interference is the pseudo-inverse of $\mathbf{H}(\mathcal{S})$ (if channel matrix $\mathbf{H}$ is not full rank). Then, the ZFBF transmit beamformer matrix will be given by

$$\mathbf{F} = \mathbf{H}^\dagger = \mathbf{H}^H (\mathbf{H}\mathbf{H}^H)^{-1}, \tag{13}$$

thus $\|\mathbf{h}_i \mathbf{f}_i\|^2 = 1$, and $\|\mathbf{h}_i \mathbf{f}_k\|^2 = 0$ (if $k \neq i$).

Hence, assuming Gaussian codes the upper bound of the maximum throughput can be achieved by water-filling algorithm, and the water level is directly extracted from the overall payload power constraint. However, what we are interested is not the maximum throughput, but the resource allocation with individual QoS constraints (individual traffic requirement). This problem has been essentially solved on the terrestrial system in [14], and here we follow their approaches but with an instantaneous analysis instead of statistical.

With fixed beamforming matrix $\widetilde{\mathbf{F}}$, the downlink power allocation is a well-known SINR balancing problem (in [3,14]). The problem can be formulated as

$$\mathcal{C}(\widetilde{\mathbf{F}}, P_{\text{tot}}) = \max_{\mathbf{p}} \left(\min_{1 \leq i \leq K} \frac{\gamma_i(\widetilde{\mathbf{F}}, \mathbf{p})}{\hat{\gamma}_i} \right),$$

$$\text{subject to} \sum_{i=1}^{K} p_i \varsigma_i \leq P_{\text{tot}}. \tag{14}$$

where $\hat{\gamma}_i$ is the requested SINR, γ_i is the SINR with the optimized beamforming matrix $\widetilde{\mathbf{F}}$ and allocated power $\mathbf{p}$. $\mathbf{p} = [p_1, \cdots, p_K]$ is the power allocation vector for all the K users. The increased transmitted power compared to transmission without beamforming, quantified by precoding loss, which for square QAM constellations calculates to $\varrho_p^2 = \frac{M}{M-1}$. Even for moderate sizes M this loss is negligible and vanishes as M increases [10]. Therefore, we will not focus on this point in this paper.

The optimum of (14) can be achieved by an eigenvalue optimization problem (presented in [3]) as

$$\mathcal{C}(\widetilde{\mathbf{F}}, P_{\text{tot}}) = \frac{1}{\lambda_{\max}\left(\mathbf{\Psi}_E(\widetilde{\mathbf{F}}, P_{\text{tot}})\right)}. \tag{15}$$

where $\mathbf{\Psi}_E$ is the extended coupling matrix (defined in (12) of [3]). The optimal power vector $\mathbf{p}$ is obtained as the first K components of the dominant eigenvector of $\mathbf{\Psi}_E(\widetilde{\mathbf{F}}, P_{\text{tot}})$.

In [3], the authors also discussed how to jointly optimize the beamforming vector and power allocation. The problem can be written as

$$\mathcal{C}(P_{\text{tot}}) = \max_{\mathbf{F}, \mathbf{p}} \left(\min_{1 \leq i \leq K} \frac{\gamma_i(\mathbf{F}, \mathbf{p})}{\hat{\gamma}_i} \right),$$
$$\text{subject to} \sum_{i=1}^{K} p_i \varsigma_i \leq P_{\text{tot}}, \tag{16}$$
$$\|\mathbf{f}_i\|^2 = \varsigma_i, 1 \leq i \leq K.$$

where ς_i is the weighting factor in order to normalize the symbol power. The global optimum can be achieved by an eigenvalue optimization problem as

$$\mathcal{C}(P_{\text{tot}}) = \frac{1}{\min_{\mathbf{F}} \lambda_{\max}(\mathbf{\Psi}_E(\mathbf{F}, P_{\text{tot}}))}. \tag{17}$$

Therefore, the optimum beamforming weight matrix $\mathbf{F}$ is associated with the minimum of the maximal extended coupling matrix eigenvalue. And the optimal power vector $\mathbf{p}$ is obtained as the first K components of the dominant eigenvector of $\mathbf{\Psi}_E(\mathbf{F}, P_{\text{tot}})$.

5   Numerical Results

The simulations are carried out with the following assumptions

- Beams layout: the satellite system coverage is assumed to be regional (e.g. EU25 countries as presented in [2]), with 70 user beams for the whole coverage.

Table 2. Satellite payload parameters

Parameters	Value
Downlink frequency	19950Mhz
Available bandwith (B_{tot})	500Mhz
Output Back-Off (OBO)	4.5dB
Repeater loss (L_{repeater})	2.55dB
Antenna feed loss (L_{antenna})	1.17dB
Satellite Tx. antenna gain (G_{tx})	47.14dB
TWTA saturation power (P_{sat})	120Watts (if not otherwise stated)
Slope of the traffic distribution (β)	8×10^6bps (if not otherwise stated)
Propagation loss ($L_{\text{propagation}}$)	211.10dB
Ground terminal G/T ($G/T)_{gt}$	18.70 dB/K
Boltzmann constant (k_B)	$1.38 \times 10^{-23}\text{m}^2\text{kgs}^{-2}\text{K}^{-1}$

- Traffic distribution model: the distribution of traffic through the coverage area is highly unbalanced. As an example, we suppose a linear traffic requested distribution is defined as $\hat{R}_i = i\beta; i = 1, 2, \cdots, K$, β is the slope of the linear function.
- Antenna model: a tapered aperture antenna pattern is implemented in the simulation. And the extensive number of satellite payload parameters are list in Table 2.
- Performance metrics:

  - average spectral efficiency (η) is defined as the total useful allocated traffic with respect to the allocated bandwidth, i.e., $\eta = \dfrac{\sum_{k=1}^{K} \min\{R_k, \hat{R}_k\}}{\sum_{k}^{K} B_k}$, where B_k is the bandwidth allocated to beam k.
  - traffic Matching Ratio (MR) (ρ) is defined as the total useful allocated traffic with respect to the total requested traffic, i.e., $\rho = \dfrac{\sum_{k=1}^{K} \min\{R_k, \hat{R}_k\}}{\sum_{k=1}^{K} \hat{R}_k}$.

5.1 Simulation Results

The useful throughput (presented in Table 3) is defined as the traffic allocated to all the users with individual SINR constraint per user. We can see that the BF SATCOM system can achieve the best performance. About $15\% \sim 56\%$ improvement can be achieved by BF SATCOM system comparing with flexible one

Table 3. Useful throughput comparison

System scenario	FLEX (B_c =62.5MHz)				BF
f_R	1	3	4	7	1
Useful throughput (Gbps)	14.67	17.10	17.31	12.77	19.88

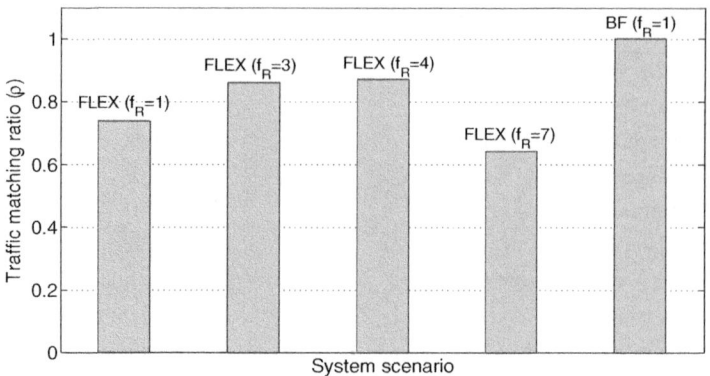

Fig. 1. Traffic matching ratio (ρ) of FLEX system and BF system ($B_c = 62.5$MHz)

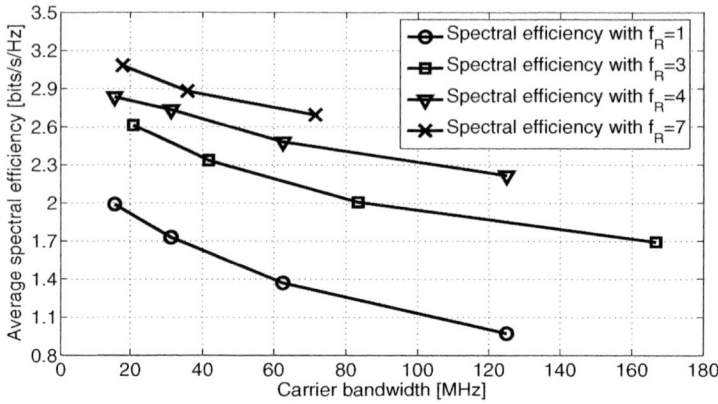

Fig. 2. Average spectral efficiency vs. B_c (for the FLEX system)

with different f_R. The reason is that the BF system reuse the whole bandwidth without co-channel interference. However, for the FLEX system, although more bandwidth is available by low frequency reuse factor, the co-channel interference will deteriorate the performance of the useful throughput.

In Fig. 1, the results show the same trend as the useful throughput in Table 3. We can see that the BF system with $\rho = 1$ can satisfy all the users' traffic requirement. The results of FLEX system show that the performance of ρ (also the useful throughput in Table 3) first increases and then decreases as the f_R increases. The reason is that, as we increase the f_R, the co-channel interference for the FLEX system will decrease, because the beams reusing the same frequency band are separated much farther from each other. However, the aggregated bandwidth will decrease as the f_R increases, since the total available bandwidth is fixed for each cluster (e.g. $B_{\text{tot}} = 500$MHz in our simulation).

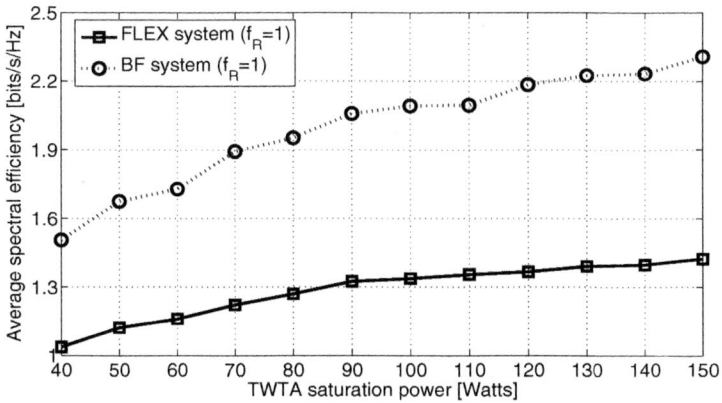

Fig. 3. Average spectral efficiency vs. saturation power P_{sat} ($B_c = 62.5\text{MHz}$, $f_R=1$)

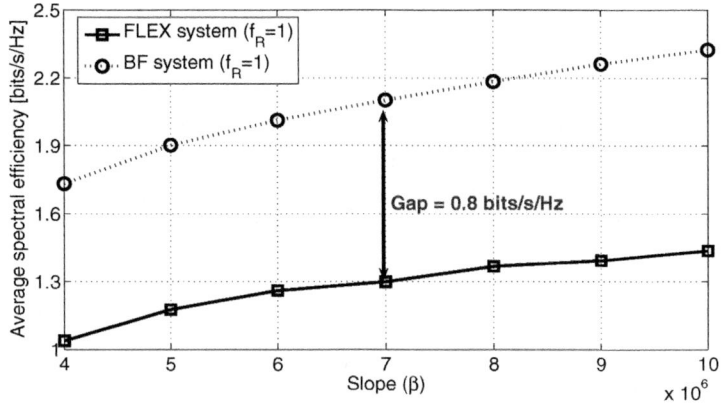

Fig. 4. Average spectral efficiency vs. slope (β) ($B_c = 62.5\text{MHz}$, $f_R=1$)

The performance of the average spectral efficiency as a function of the bandwidth granularity (B_c) of the FLEX system is shown in Fig. 2. The results show that the average spectral efficiency decreases as the B_c increases. The reason is that the greater the carrier bandwidth, the more the unused but offered traffic is, which implies greater difference between the total offered traffic and the total useful offered traffic. The results also indicate that, under a given bandwidth granularity, the lower the frequency reuse factor, the smaller the average spectral efficiency. Because we define the average spectral efficiency as the ratio of total useful allocated traffic to the aggregated bandwidth. Thus, although more aggregated bandwidth is available for the lower frequency reuse factor, the increased bandwidth is not efficiently used since the co-channel interference power is increased as we have indicated in Fig. 1. Conversely, for the case of higher frequency reuse factor, e.g., $f_R = 7$, it achieves the best average spectral efficiency because the bandwidth is used more efficient than the other cases.

We compare the average spectral efficiency of FLEX and BF systems in Fig. 3 by using the set of coding and modulation combinations in DVB-S2. This result is the expected one from the high average spectral efficiency that is obtained with the case of BF and high saturation power, along with higher available bandwidth, which allows perfect matching to heavily loaded beams.

Fig. 4 gives the simulation results for the average spectral efficiency with different linear traffic distribution slope. The results show that, for both FLEX and BF systems, the more unbalanced the distribution is, the larger average spectral efficiency can be achieved, because these two type of systems can take advantage of the nonuniformity of distribution by allocating resources more efficiently. This figure also shows that the spectral efficiency performance gap between FLEX and BF SATCOM system is about 0.8bits/s/Hz under the assumptions, i.e. $B_c = 62.5$MHz, f_R=1 and $P_{\text{sat}} = 120$Watt.

6   Conclusions

In this paper we have derived an analytical solution of the resource allocation for an interference-limited SATCOM system. We introduce two options for a multibeam payload: FLEX with resource allocation flexibility in the frequency domain, and BF conventional system. Two optimization problems are addressed with given individual constraints per-beam, P1: optimization of the spectral mask vector design and power allocation for FLEX system, and P2: beamforming weight vectors design and power allocation for BF system. The performance is evaluated based on the practical implementation, e.g. DVB-S2. The numerical results show that BF system with full frequency reuse ($f_R = 1$) provides the best performance, e.g. approx. 15% $\sim$ 56% improvement can be achieved by BF SATCOM system comparing with FLXE one with different f_R. The disadvantage of BF system is that the energy of the transmitted signal is increased, i.e. the precoding loss. But we have shown that this loss is negligible, especially for larger constellations modulation. For the FLEX system, the bandwidth can be utilized more efficiently with larger frequency reuse factor and smaller bandwidth granularity. We also evaluate the performance of FLEX and BF systems for the asymmetrical traffic distribution. For both systems, they can adapt well to the nonuniform traffic distribution, and BF system outperforms FLEX system about 0.8bits/s/Hz in terms of average spectral efficiency under the assumptions of $B_c = 62.5$MHz, f_R=1 and $P_{\text{sat}} = 120$Watt.

Acknowledgment. The authors gratefully acknowledge valuable discussions on the topic with the DLR and MDA. This work has been partly funded by the European Space Agency (ESA) under the project of "Study of Beam Hopping Techniques in Multi-Beam Satellite Systems".

References

1. Lei, J., Vázquez-Castro, M.A.: Frequency and Time-Space Duality Study for Multibeam Satellite Communications. In: Proc. IEEE Int. Conf. on Commun., Cape Town, South Africa, pp. 1–5 (May 2010)

2. Alberti, X., Cebrian, J.M., Del Bianco, A., Katona, Z., Lei, J., Vázquez-Castro, M.A., Zanus, A., Gilbert, L., Alagha, N.: System Capacity Optimization in Time and Frequency for Multibeam Multi-media Satellite Systems. In: 5th Advanced Satellite Multimedia Systems Conf., Cagliari, Italy (September 2010)
3. Schubert, M., Boche, H.: Solution of the Multiuser Downlink Beamforming Problem with Individual SINR Constraints. IEEE Trans. Veh. Technol. 53(1), 18–28 (2004)
4. Neely, M.J., Modiano, E., Rohrs, C.E.: Power Allocation and Routing in Multibeam Satellites with Time-Varying Channels. IEEE/ACM Trans. Netw. 11(1), 138–152 (2003)
5. Choi, J.P., Chan, V.W.S.: Optimum Power and Beam Allocation Based on Traffic Demands and Channel Conditions over Satellite Downlinks. IEEE Trans. on Wireless Commun. 4(6), 2983–2993 (2005)
6. Barcelo, J.E., Vázquez-Castro, M.A., Lei, J., Hjøungnes, A.: Distributed Power and Carrier Allocation in Multibeam Satellite Uplink with Individual SINR Constraints. In: IEEE Globecom, Honolulu, USA, pp. 1–6 (November–December 2009)
7. Zorba, N., Realp, M., Pérez-Neira, A.I.: An Improved Partial CSIT Random Beamforming for Multibeam Satellite Systems. In: 10th Int. Workshop on Signal Process. for Space Commun., Rhodes Island, Greece (October 6–8, 2008)
8. Yu, W., Cioffi, J.: Sum Capacity of Gaussian Vector Broadcast Channels. IEEE Trans. Inform. 52(2), 754–759 (2006)
9. Jindal, N., Jafar, S., Vishwanath, S., Goldsmith, A.: Sum Power Iterative Water-Filling for Multi-Antenna Gaussian Broadcast Channels. IEEE Trans. Inform. 51(4), 1570–1580 (2005)
10. Windpassinger, C., Fischer, R.F.H., Vencel, T., Huber, J.B.: Precoding in Multi-antenna and Multiuser Communications. IEEE Trans. on Wireless Commun. 3(4), 1305–1316 (2004)
11. Vázquez-Castro, M.A., Seco, G.: Cross-Layer Packet Scheduler Design of a Multi-beam Broadband Satellite System with Adaptive Coding and Modulation. IEEE Wireless Trans. 6(1), 248–258 (2007)
12. Lei, J., Vázquez-Castro, M.A.: Joint Power and Carrier Allocation for the Multi-beam Satellite Downlink with Individual SINR Constraints. In: IEEE Int. Conf. on Commun., Cape Town, South Africa, pp. 1–5 (May 2010)
13. Letzepis, N., Grant, A.J.: Capacity of the Multiple Spot Beam Satellite Channel with Rician Fading. IEEE Trans. Inform. 54(11), 5210–5222 (2008)
14. Weidong, Y., Guanghan, X.: Optimal Downlink Power Assignment for Smart Antenna Systems. In: IEEE Int. Conf. on Acoustics, Speech and Signal Process., Seattle, USA, vol. 6, pp. 3337–3340 (May 1998)
15. ETSI EN 302 307 v1.1.1, Digital Video Broadcasting (DVB): Second generation framing structure, channel coding and modulation system for Broadcasting, Interactive Services, News Gathering and other broadband satellite applications
16. Shariff, M., Hassibi, B.: A Comparison of Time-Sharing, DPC, and Beamforming for MIMO Broadcast Channels with Many Users. IEEE Trans. on Commun. 55(1), 11–15 (2007)
17. Taesang, Y., Goldsmith, A.: On the Optimality of Multiantenna Broadcast Scheduling Using Zero-Forcing Beamforming. IEEE Selected Areas in Comm. 24(3), 528–541 (2006)

Coordinated Multi-point Transmission Combined with Cyclic Delay Diversity in Mobile Satellite Communications[*]

Hee Wook Kim, Kunseok Kang, Bon-Jun Ku, and Do-Seob Ahn

Electronics and Telecommunications Research Institute,
305-350 Gajeong-dong Yuseong-gu Daejeon, Korea
{prince304,kskang,bjkoo}@etri.re.kr

Abstract. In OFDMA based MSS system, the beam planning with frequency reuse factor 1 may induce severe inter-beam interference due to the usage of the same subcarrier between user equipments of adjacent beams. In order to solve this problem, this paper shows a coordinated multi-point transmission scheme combined with cyclic delay diversity. The proposed scheme improves frequency usage efficiency in a beam boundary area and minimizes the interference between adjacent beams in an OFDMA based mobile satellite communication. In addition, Simulation results show that it makes the received signal to noise ratio increased due to diversity gain from cyclic delayed multi-point transmitted signals. This performance gain could be achieved without any modification of a conventional Single-Input and Single-Output tranceiver, differently from the space-frequency transmit diversity scheme.

Keywords: coordinated multi-point transmission, cyclic delay diversity, mobile satellite communications.

1 Introduction

Due to the increment of requirements to a high quality multimedia service, a mobile satellite service (MSS) system has been required to provide a broadband service. However, a very limited bandwidth has been allocated to the MSS. For example, a bandwidth of 30 MHz is allocated for the satellite component of IMT-2000 according to international mobile telecommunication union – radiocommunication sector (ITU-R). With this bandwidth, it is very difficult to realize a frequency reuse factor of three or seven because a wireless interface having a minimum bandwidth of 10 MHz is required to provide a broadband service. In practical, a frequency reuse factor of seven cannot be realized and the case of three requires that the entire frequency band has to be allocated to one operator. Therefore, it is essential to realize a MSS system having a frequency reuse factor of one to provide a broadband service.

---

[*] This work was supported by IT R&D program of MKE/KEIT, [KI001794, Development of satellite radio interface technology for IMT-Advanced].

G. Giambene and C. Sacchi (Eds.): PSATS 2011, LNICST 71, pp. 274–285, 2011.

Most of the candidate radio interfaces for the terrestrial components adopted multi-carrier transmission technologies such as orthogonal frequency division multiplexing (OFDM). Not much attention had been paid to the study on OFDM based satellite systems due to serious peak to average power ratio (PAPR) problems, especially for a high power amplifier in the satellite system. Nevertheless, resent results reported application of OFDM in satellite systems [1]-[4], not only to utilize the advantage of OFDM system such as capability of high speed transmission but also to keep commonalities between the terrestrial systems. For example, DVB-SH adopted OFDM transmission [7], which is the same signal format defined in DVB-H for terrestrial systems. The main reason for adopting OFDM stems from the fact that satellite and terrestrial transmitters from a single frequency network (SFN). Recently European Telecommunications Standardization Institute (ETSI) have started feasibility study on OFDM based scheme might provide better performance than the conventional wideband code division multiple access (WCDMA) based scheme [8].

In case of a CDMA based MSS system, a frequency reuse factor of one may be realized by using a different spreading code for each beam to reduce interference between adjacent beams. However, in case of an OFDMA based MSS system which has been considered as IMT-Advanced radio interface technology, it is not easy to realize a frequency reuse factor of one. In the OFDMA based MSS system, the beam planning with frequency reuse factor one can induces severe inter-beam interference due to the usage of the same subcarrier between user equipments (UEs) of adjacent beams, especially where a UE is located in a beam edge. As a solution to this interference problem, OFDMA based MSS systems applying fractional frequency reuse (FFR) was proposed [5][6]. In the FFR, the users at the beam edge operate with a fractional of whole subcarriers available. In addition, the edge users are not negligible differently in terrestrial systems. Relative spectral efficiency of the proposed FFR scheme compared to the conventional scheme with frequency reuse factor of seven was shown in the reference.

As more intelligent frequency reuse techniques, in this paper, we propose coordinated multi-point transmission and its combination with cyclic delay diversity in mobile satellite communications.

2   Fractional Frequency Reuse Technique in an OFDMA Based MSS System

Figure 1 shows a frequency reuse pattern with three subcarrier groups in an OFDMA based multi-beam MSS system. Each beam is partitioned into two regions and each frame is divided into two time sections, T1 and T2, in each beam. The first time section, T1 and T2, in each beam. The first time section, T1 is allocated to UEs in the beam centre with radius of R1, and all of the three subcarrier groups can be used for transmission during this time period. On the other hand, the second time section, T2 is allocated to UEs in the beam edge. During this time period, only a single subcarrier group out of three can be used. In other words, all the subcarriers are reused in the beam centre region during T1, while only a single group is allocated into each beam during T2. In this case, the allocated subcarrier group must satisfy the orthogonality condition. We note that the user signals in two regions cannot be transmitted simultaneously because we cannot ignore the interference from the adjacent beams.

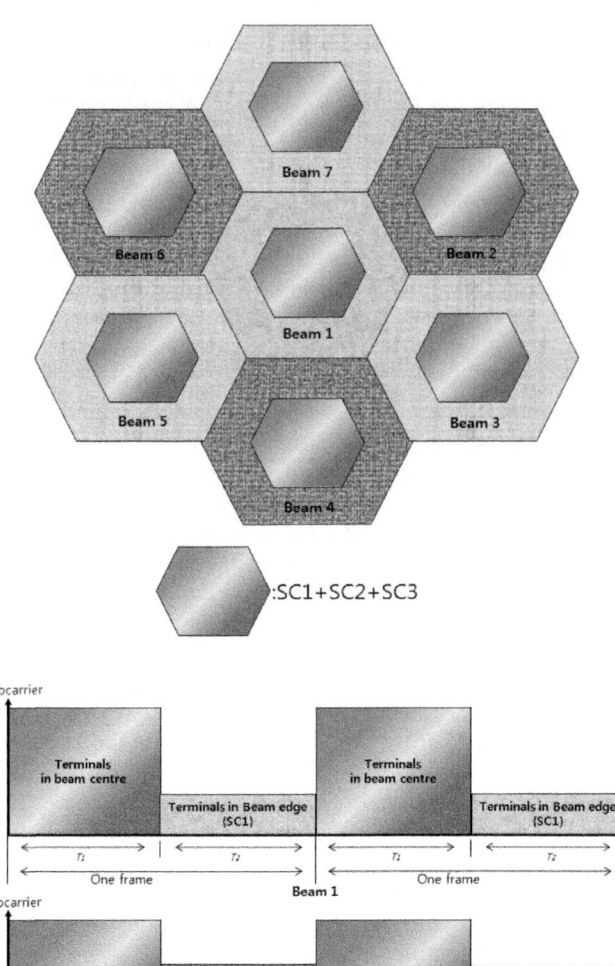

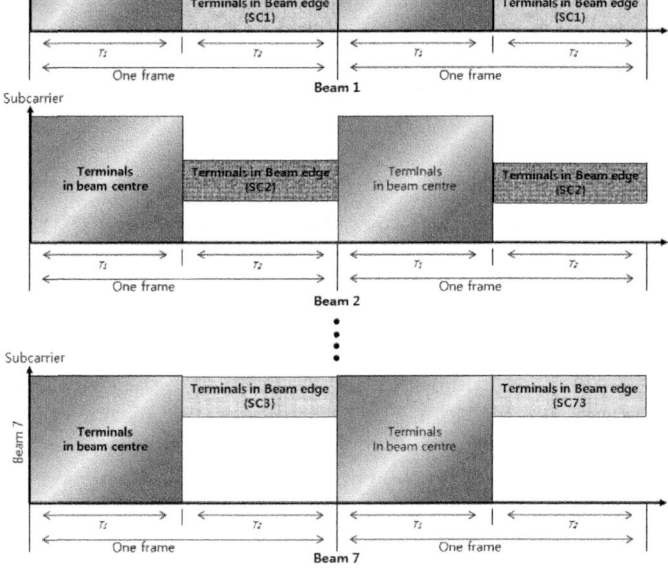

Fig. 1. Frequency reuse pattern for a multi-beam satellite system with three subcarrier groups

In case of using such a fractional frequency reuse scheme, maximum frequency usage efficiency of a beam boundary area is dropped to 1/3 compared to that of a beam centre area because a user in a beam boundary region uses only one of fractional subcarrier groups SC1 to SC3. Furthermore, a received Effective Isotropically Radiated Power (EIRP) from a satellite beam in a beam boundary region is lower than in a beam centre area. Accordingly, the UE capacity in a beam boundary is decreased and thus, digital divide between UEs in beam centre and edge regions is increased.

In order to solve these problems, we introduce a coordinated multi-point transmission for UEs located at a beam edge region. In the coordinated multi-point transmission scheme, beams cooperate with each other to provide a satellite communication service to a user rather than competing to each other. That is, the coordinated multi-point transmission scheme means a multi-beam transmission scheme that enables a signal from an adjacent beam to improve a communication service quality. The coordinated multi-point transmission will be described in detail though Sections 3 and 4.

3 Coordinated Multi-point Transmission in an OFDMA Based MSS System

The figure 2 shows a system using a coordinated multi-point transmission scheme. We assume that a satellite transmits signals for UE1 to UE3 through beam 1. UE1, UE2 and UE3 means terminals located at the beam centre region, a two beams-overlapped region, a three beams-overlapped region, respectively. In reference [5], users in a beam boundary area use different resources at adjacent beams in order to receive a signal from only one of overlapped beams without severe interference. On the other hands, in the coordinated multi-point transmission, a user in a beam boundary region receives its own signals through whole overlapped beams that enable to make coordinated multi-point transmission. For example, in reference [1], UE2 receives its own signal from only beam 1 and a signal from beam 3 overlapped with beam 1 may cause severe interference to UE2. Similarly, UE3 receives its own signal from only beam 1 and signals from beams 2 and 3 overlapped with beam 1 may cause higher interference to UE3 than UE2. On the other hands, in the coordinated multi-point transmission, the beam 3 signal does not cause interference to the UE1 signal but enhance it. Similarly, beams 1, 2 and 3 cooperate with each other in order to transmit the UE3 signal through the same resource. Accordingly, a reception performance of the UE2 and UE3 can be improved.

In conclusion, the coordinated multi-point transmission can improve a received signal to noise ratio (SNR) because a user receives a signal from adjacent multiple beams although the user is located at the beam edge region. Furthermore, interference can be avoided because an adjacent beam also transmits its own signal. If many users are located at a predetermined boundary region, the coordinated multi-point transmission scheme in an OFDMA based MSS system can flexibly allocate a large subcarrier region to the users at the predetermined boundary region. Thus, frequency efficiency can be improved.

Figure 3 illustrates a multi-beam MSS system consisting of one beam and six adjacent beams. In the multi-beam MSS system, a signal is transmitted during the same frequency band f1 from all beams to realize a frequency reuse factor 1. All beams are divided into a beam centre region and a beam boundary area similarly in reference [5]. In the case of beam 1, its boundary region is divided into six two-beam overlapped areas and six three-beam overlapped areas. In Fig. 3, whole subcarriers can be used in beam centre region while fractional subcarriers should be used in six two-beam overlapped areas and three-beam overlapped areas.

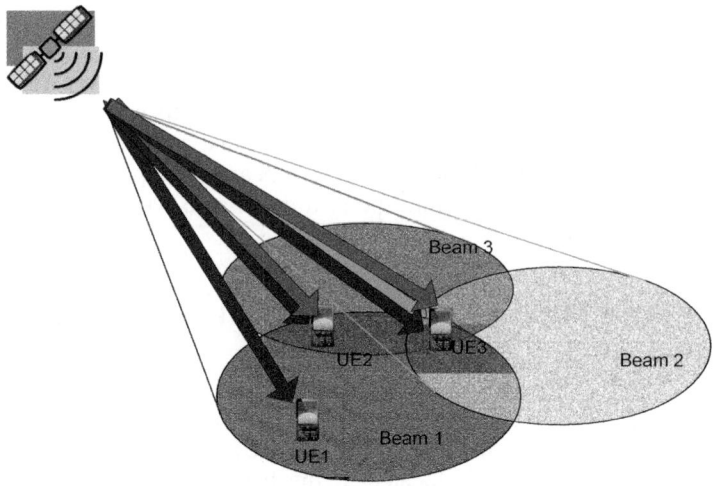

Fig. 2. Coordinated multi-point transmission concept in mobile satellite communication

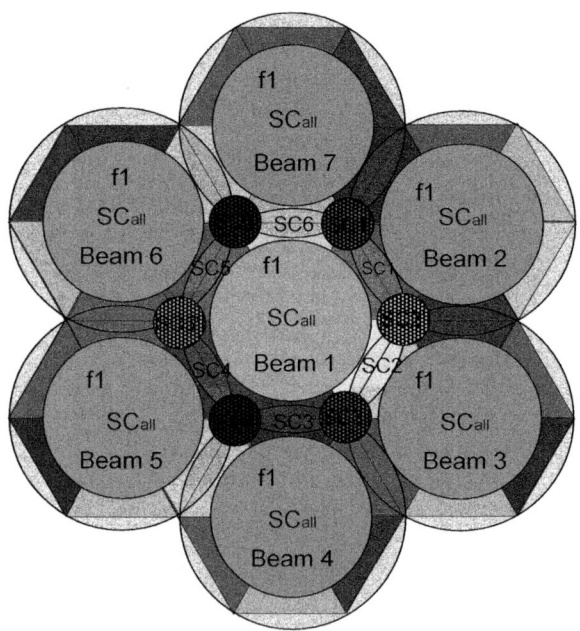

Fig. 3. Fractional frequency reuse pattern example for coordinated multi-point transmission

Figure 4 shows one example of frame structures to realize the fractional frequency reuse of Fig.3 for coordinated multi-point transmission in an OFDMA based MSS system. In this figure, one frame is divided into three transmission interval in time domain and six subcarrier groups in frequency domain. The first, second, and third transmission intervals, T1, T2, and T3 are allocated to a beam centre UE, a two-beam overlapped UE, and a three-beam overlapped UE, respectively. The beam centre UE can receive its own signal over whole subcarrier during T1 while the two-beam and three-beam overlapped UEs can have frequency resource over only predetermined fractional part of whole subcarriers such as subcarrier groups SC1 to SC6 and SC1' to SC6'. SC1 to SC6 and SC1' to SC6' are considered for six two-overlapped regions and six three-overlapped regions, respectively. The size of each subcarrier group can be flexibly decided depending on the traffic demand over each corresponding region. For example, in Fig. 4, we can know that the traffic demand in the two-beam overlapped regions would be high at the beam 1 edge region overlapped with beam. In the same principle, in case of the three-beam overlapped regions, the beam 1 edge region overlapped with both beam 3 and 4 or both beam 5 and 6 would require high traffic demand during T3. We can also control the duration of the time interval T1, T2 and T3, considering the required capacity from UEs in the beam centre, two-beam overlapped , and three-beam overlapped regions.

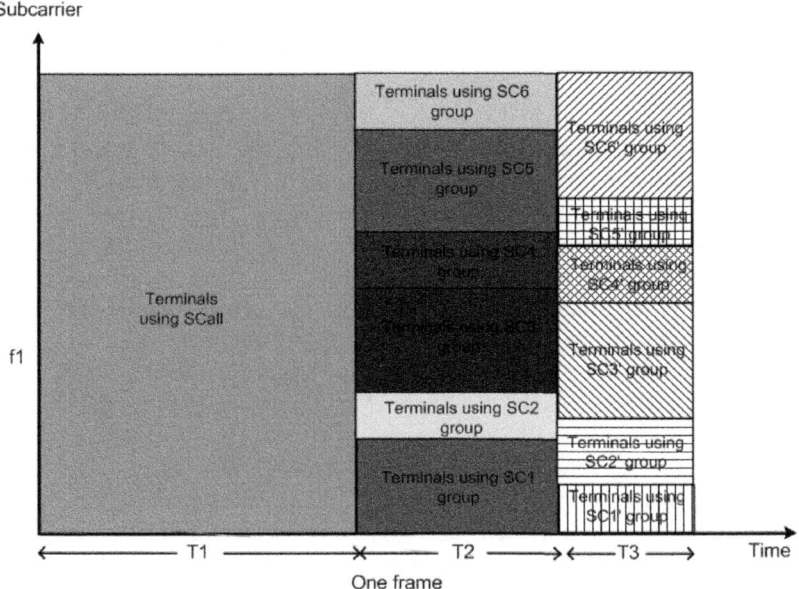

Fig. 4. Frame structure example for coordinated multi-point transmission

4   Coordinated Multi-point Transmission Combined with Cyclic Delay Diversity

As mentioned in Section3, the coordinated multi-point transmission in an OFDMA based MSS system makes the received SNR of beam edge UEs increased and the interference from adjacent beams reduced according to multiple reception of own signal over the same resource from the targeted and adjacent beams. However, because in this scheme an UE simply receives the multiple same signals, we cannot obtain any diversity gain. Therefore, in this Section, we consider the coordinated multi-point transmission applying cyclic delay diversity (CDD).

CDD is a diversity scheme used in OFDM based telecommunication systems, transforming spatial diversity into frequency diversity avoiding intersymbol interference. CDD involves transmitting the same set of OFDM symbols on the same set of OFDM subcarriers from multiple transmit antennas, with different delay on each antenna [9]. The delay is applied before cyclic prefix is added, thereby guaranteeing that the delay is cyclic over the Fast Fourier Transform (FFT) size.

Adding a time delay is identical to applying a phase shift in the frequency domain. As the same time delay is applied to all subcarriers, the phase shift will increase linearly across the subcarriers with increasing subcarrier frequency. Each subcarrier will therefore experience a different beamforming pattern as the non-delayed subcarrier from one beam interferes constructively or destructively with the delayed version from other coordinated beams. The diversity effect of CDD therefore arises from the fact that different subcarriers will pick our different spatial paths in the propagation channel, thus increasing the frequency-selectivity of the channel.

As a CDD for a case with two transmit antenna ports, we can express mathematically the received symbol r_k on the kth subcarrier as

$$r_k = h_{1k}x_k + h_{2k}e^{j\phi k}x_k \tag{1}$$

where hpk is the ricia fading channel from the pth beam, and $e^{j\phi k}$ is the phase shift on the kth subcarrier due to the delay operation. We can see clearly that on some subcarriers the symbols from the second transmit beam will added constructively, while on other subcarriers they will add destructively. Here, $\phi = 2\pi d_{cdd}/N$, where N is FFT size and d_{cdd} is the delay in samples.

The number of resulting peaks and troughs in the received signal spectrum pattern across the subcarriers therefore depends on the delay parameter dcdd: as dcdd is increased, the number of peak and troughs in the spectrum also increases. This help to illustrate how CDD enhances the channel coding gain by introducing frequency selectivity into a possibly flat fading channel.

Applying the CDD scheme to coordinated transmission, each coordinated multi-point transmitted signals from a satellite is applied to difference cyclic delay values in the same signal for a targeted user. This induces transmit diversity between coordinated multi-point transmitted signals since different cyclic delay values makes different frequency selectivity in the coordinated transmission signals.

5   Simulation Results

In order to assess the performance enhancement of the proposed diversity scheme, we apply a simple simulation model. We assume the signals for beam 1 and beam 2 are modulated using the QPSK scheme and, in a case of coded transmission, coded in the convolutional coding scheme with generation polynomial of $[172\ 133]_8$, code rate of

1/2, constraint length of 7, and the length of a codeword with 288. The decoding of a received signal is done by Viterbi algorithm. We also assume that the amplitude of fading from the satellite beams to the receiver is uncorrelated or fully correlated Rician distributed with factor K of 3 dB. As in a normal satellite channel, we assume that fading is constant across whole subcarriers in one OFDM symbol and the receiver has perfect knowledge of the channel. Finally, we consider that the signal power from each beam of the satellite put on equal power. This case corresponds to a UE in a boundary region between two beams.

Figure 5 and 6 show bit error rate (BER) performance of the coordinated multi-pointed transmission schemes over an uncorrelated Rician distributed channel in cases of uncoded and coded transmission, respectively. We assume that the number of total subcarriers in one OFDM symbol is 288 corresponding to the length of one convolutional coding codeword.

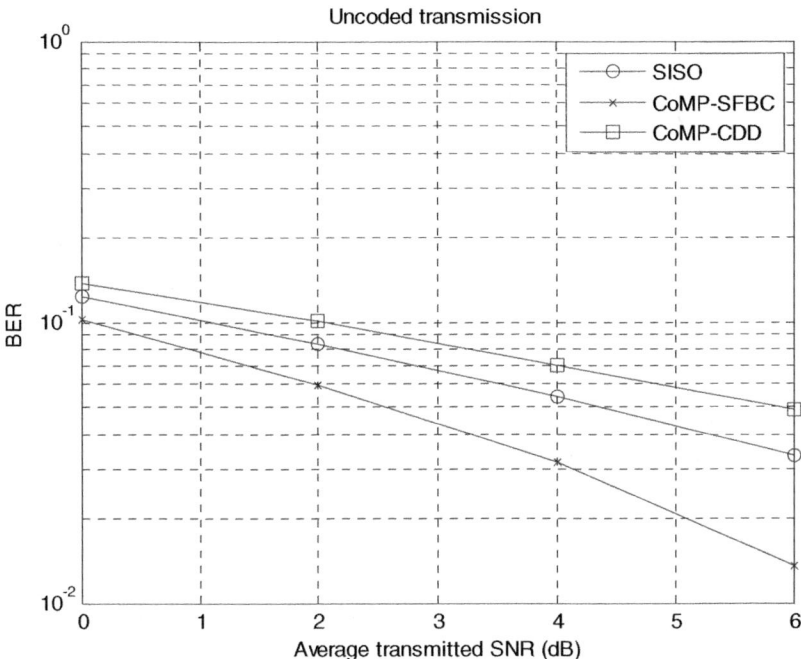

Fig. 5. Uncoded transmission in multi-beam mobile satellite communication

In Fig. 5, it is noted in uncoded transmission that the coordinated transmission with CDD has no advantages and performance degradation due to loss of the received SNR while the coordinated transmission with space frequency block code (SFBC) applying to Alamouti code can make performance improved due to space diversity gain.

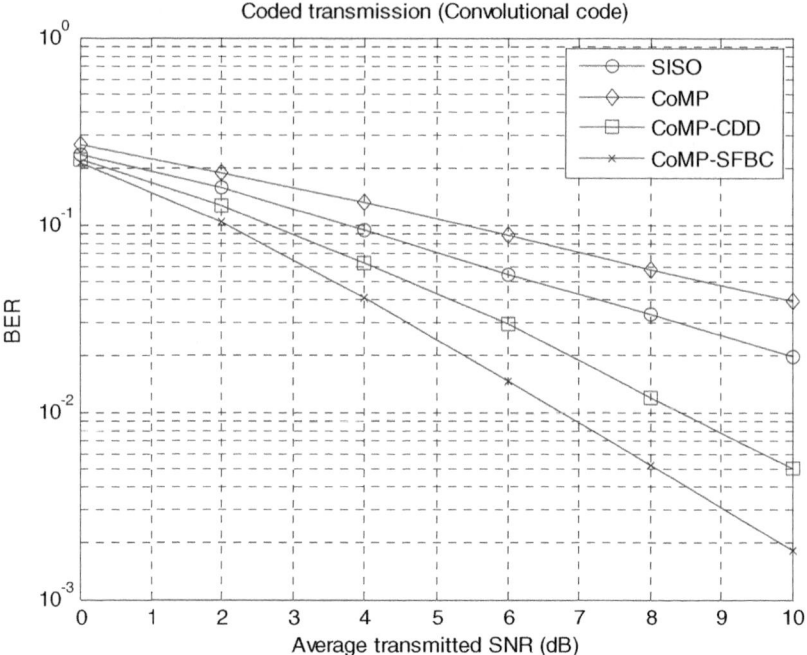

Fig. 6. Coded transmission in multi-beam mobile satellite communication

In Fig.6, it is noted in coded transmission that the coordinated transmission should be combined with one of possible diversity schemes for performance enhancement. In the conventional single transmission, whole coded bits in one codeword suffer from the same channel fading due to the frequency flat fading characteristic of satellite communication channel, and thus coding performance is not good. By the way, in the coordinated multipoint transmission with CDD, a cyclic delayed beam signal increase frequency selectivity on the coordinated transmission channel, and thus we can get more coding gain over one codeword corresponding to the length of whole subcarriers. Clearly, the coordinated multipoint transmission with SFBC has best performance due to perfect diversity gain.

Figure 7 shows BER performance of the coded coordinated multi-pointed transmission over an uncorrelated Rician distributed channel according to various cyclic delay shift values. As seen in the figure, a degree of spatial diversity gain in the CoMP-CDD scheme depends on the value of cyclic delay shift. From the figure, we know that this scheme get best diversity gain when the value of cyclic delay shift is 1/16 of the length of whole subcarriers. It comes from the fact that the power variance of the CoMP-CDD signal is largest in that value. In addition, it is noted in the figure that performance of the CoMP CDD scheme is degraded in the cyclic delay shifts less than 1/16 of the length of whole subcarriers as well as corresponding to the half of the length of total subcarriers.

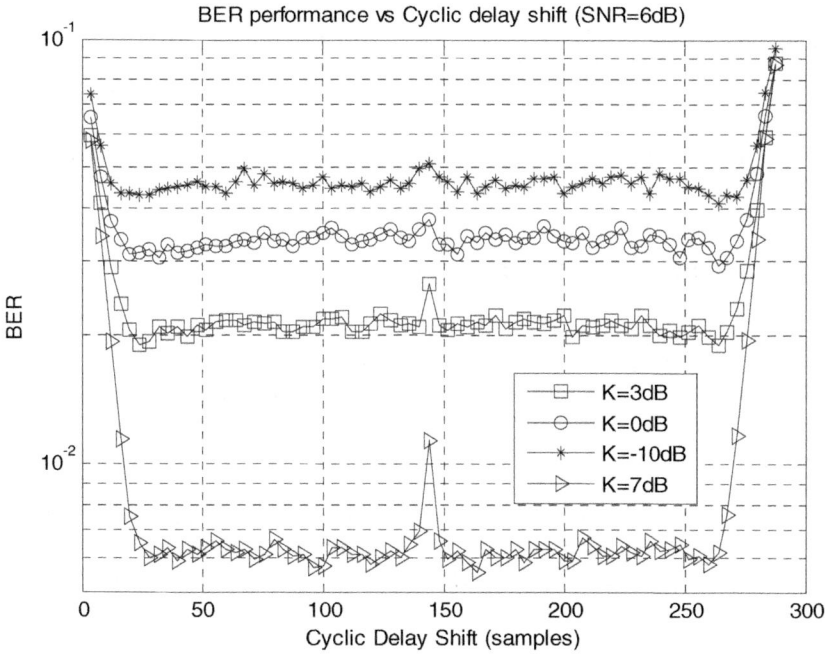

Fig. 7. BER performance gain according to various cyclic delay shifts

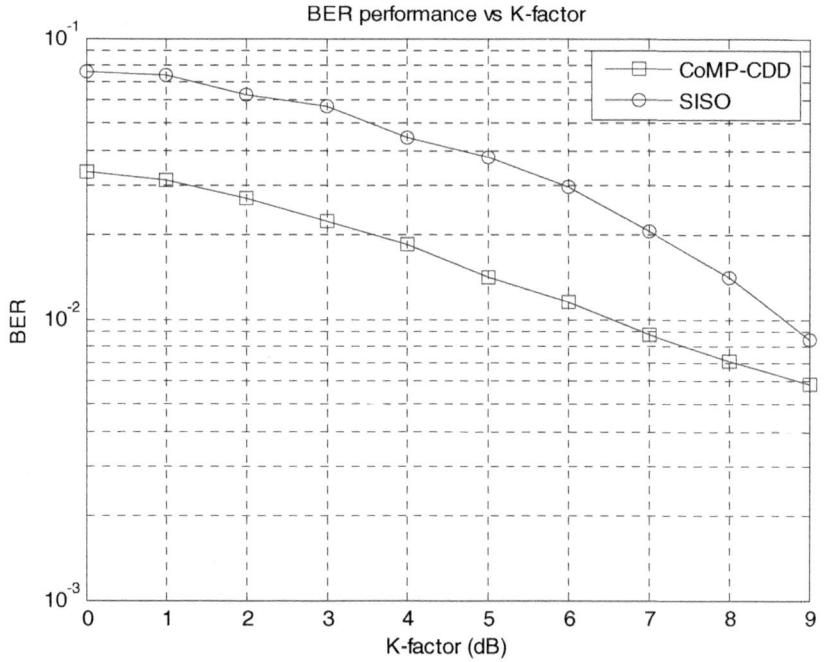

Fig. 8. BER performance gain according to various K-factors

Finally, Fig. 8 represents BER performance of the coded transmission according to various K-factors. As seen in the figure, performance enhancement due to diversity gain is kept below K-factor of 6 dB while BER performance is degraded as K-factor is increased over 8 dB. It comes from the reason that the dominant factor for performance improvement would be Rician fading from a strong line-of-sight component rather than diversity gain from the CDD scheme. It is noted in coded transmission.

6   Conclusion and Further Works

In this paper, we proposed a coordinated multi-pointed transmission and its combination with cyclic delay diversity in a mobile satellite system. We introduced the basic concept and frame structure of the proposed cooperative transmission and its combination example with the cyclic delay diversity scheme for performance enhancements. Multiple beam signals generate signals in the way that the multiple received signals from several beams can be combined to produce spatial diversity gain. We demonstrated from the simulation results that the proposed method can provide stable improved performance in an OFDM based multi-beam mobile satellite system for a beam edge region and coded transmission. We note that this performance gain could be achieved without any modification of a conventional SISO receiver, differently from the space-time transmit diversity scheme. By the way, in the scheme, the frequency reuse efficiency may be reduced because adjacent beams have to cooperate with each other to communicate with only one user. However, I believe that such a possibility can be overcome through performance enhancement in a edge region with the proposed scheme. Detail analysis on overall system capacity considering these both reduced frequency reuse efficiency and improved performance will be addressed in the future.

References

1. Cioni, S., Corazza, G.E., Neri, M., Vanelli-Coralli, A.: On the use of OFDM radio interface for satellite digital multimedia broadcasting systems. International Journal of Satellite Communications and Networking 24(2), 153–167 (2006)
2. Wand, L., Jezek, B.: OFDM modulation schemes for military satellite communications. In: IEEE Military Communications Conference 2008, pp. 1–7 (November 2008)
3. Kim, H.W., Ku, B., Kang, K., Ahn, D.S.: Efficient time and frequency offset estimation for OFDM based MSS systems. In: IEEE Vehicular Technology Conference 2006, pp. 1–5 (September 2006)
4. Ham, W., Erdogan, A.T., Arslan, T., Hasan, M.: High-performance low power FFR cores. ETRI Journal 30(3) (June 2008)
5. Kim, H.W., Kang, K., Ahn, D.S.: OFDMA based mobile satellite communication applying fractional frequency reuse technique. In: AIAA ICSSC 2009 (May 2009)
6. Park, J.M., Ahn, D.S., Lee, H.J., Park, D.C.: Feasibility of coexistence of mobile satellite service and mobile service in co-frequency bands. ETRI Journal 32(2), 255–264 (2010)

7. ETSI EN 302 583 v.1.1.1.: Digital Video Broadcasting (DVB); Framing structure, channel coding and modulation for satellite services to handheld devices (SH) below 3 GHz (March 2008)
8. ETSI TR 102 443: Satellite Earth Stations and Systems (SES); Satellite component of UMTS/IMT-2000; Evaluation of the OFDM as a satellite radio interface (August. 2008)
9. Sesia, S., Toufik, I., Baker, M.: LTE; the UMTS long term evolution from theory to practice. John Wiley & Sons Ltd., Chichester (2009)

The SANDRA Communications Concept – Integration of Radios

Jim Baddoo, Peter Gillick, Rex Morrey, and Aleister Smith

Thales Research and Technology (UK)
Worton Drive, Worton Grange, Reading, RG2 0SB, UK
Thales Avionics
Manor Royal, Crawley, RH10 9HA, UK
{jim.baddoo,peter.gillick,rex.morrey}@thalesgroup.com
aleister.smith@uk.thalesgroup.com

Abstract. The Single European Sky Air Traffic Management (ATM) research programme SESAR has identified continued growth in demand for aircraft communications as air traffic increases and communications become more network centric. Alongside existing systems such as VHF Data Link Mode 2, new systems such as the L-band Digital Aeronautical Communication System (LDACS) and the Aeronautical Mobile Airport Communications System (AeroMACS) are being proposed. This growth is likely to increase the size, weight and cost of avionics radio communication equipment, so there is a need to examine new radio architectures which will help limit these increases. SANDRA is a European Commission programme which aims to design and demonstrate an integrated communications system using software defined radio techniques. The concepts behind the integrated communications system are described, including improved modularity using high-speed digital links, security, redundancy and certification. The specific requirements of the SANDRA programme and details of the proof-of-concept demonstrator are outlined.

Keywords: integrated modular radio, integrated communications system, aeronautical communications system, software defined radio, avionics.

1 Introduction

Over the years, systems for aeronautical communications have grown and evolved in order to meet increasing demands and support new technologies. Increasing demand has arisen from the continued growth in air traffic and the need to exchange increasing amounts of data. New technologies have included the transition from analogue to digital systems and the more recent preference for Internet Protocol (IP) - centric services.

These trends are set to continue for the foreseeable future. Air traffic in Europe is projected to nearly double by 2025, increasing from 9 million flights per year in 2005 to 17 million in 2025 [1]. In Europe, new developments are being coordinated through the Single European Sky ATM (Air Traffic Management) Research Joint Undertaking

G. Giambene and C. Sacchi (Eds.): PSATS 2011, LNICST 71, pp. 286–299, 2011.

(SESAR JU) programme, whose founding members are the European Commission and EUROCONTROL [2].

Typical application areas for aeronautical communications include Air Traffic Management (ATM), Aircraft Operational Control (AOC), Airline Administrative Communication (AAC) and Airline Passenger Communications (APC). In order to support growth in these areas, additional radio systems have been proposed, including:

- EUROCONTROL L-band Digital Aeronautical Communication System (LDACS) to augment the VHF systems
- European Space Agency Iris Programme for Air Traffic Management (ATM)
- Aeronautical Mobile Airport Communications System (AeroMACS) [3], based on Worldwide Interoperability for Microwave Access (WiMAX)
- High-bandwidth satellite communications systems based on Digital Video Broadcasting (DVB) standards.

However, supporting these new systems represents a considerable extra burden of size, weight, complexity and cost in aircraft avionics equipment, should the new radio systems be implemented in stand-alone equipment as has been traditionally the case. Moreover, although it has been suggested that the new systems will eventually replace the legacy communications systems, the likelihood is that there will be a lengthy period where aircraft will need to be fitted with all of the systems for global interoperability. This is the forecast expressed by SESAR, and the additional airborne equipment required during this transition phase severely threatens the realisation of the future communications vision.

A different approach aiming at a broader level of integration is therefore needed in order to achieve the required increase of capacity, safety, security and efficiency of air transportation operations while at the same time keeping the complexity and cost of on-board networks and equipment at a sustainable level.

In October 2009, the European Commission (EC) launched a Seventh Framework Programme called SANDRA [4] to examine "Seamless Aeronautical Networking through integration of Data links, Radios, and Antennas". SANDRA aims to design, specify and develop an integrated aircraft radio communication architecture to improve efficiency and cost-effectiveness by ensuring a high degree of flexibility, scalability, modularity and reconfigurability. The programme will examine a number of integration possibilities, including:

- the integration of communication service provision at the network layer using the Internet Protocol (IP) as a unification technology
- the integration of radios which typically cover the realisation of the physical layers, data link layers and network layers of specific radio communications waveforms
- the integration of an L-Band antenna with a Ku-Band antenna.

SANDRA aims to provide a truly integrated modular approach for a global aeronautical network and communication architecture. The overall programme is designed to lead to performance evaluation of the system in a laboratory environment, followed by flight trials.

A number of concepts relating to the integration of radio systems are described, covering issues such as system boundaries, separation of processing from transceivers, security, segregation, redundancy and certification. The SANDRA functional architecture is briefly described, splitting it into the Integrated Router (IR) and the Integrated Modular Radio (IMR). Special attention is then paid to the SANDRA Proof-of-Concept IMR, looking at some of the design choices that are being made and highlighting how the programme will exercise IMR concepts.

2   Integrated Communication System Concepts

2.1   System Boundaries

The system boundary for an integrated communications system includes all avionics radio sub-systems and is illustrated in Figure 1. Items such as Subscriber Identity Module (SIM) cards, which are required for certain waveforms, are considered to be part of the system.

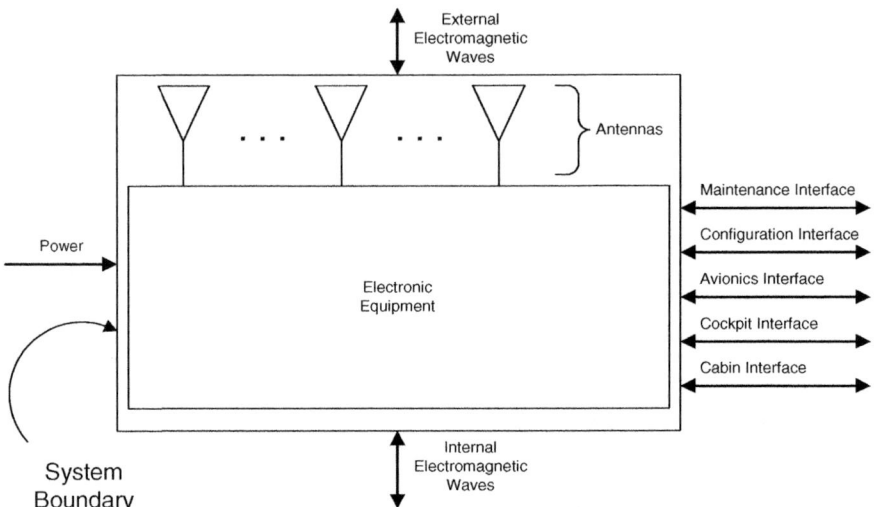

Fig. 1. System Boundary for Integrated Communications System

The boundary for the maximum system covers the following interfaces:

- *Maintenance interface* – this allows the system to be installed and maintained, and includes the ability to upgrade software and firmware
- *Configuration interface* – this allows the system to be configured, for example, setting certain internal IP addresses
- *Avionics interface* – this provides access to avionic systems outside the IMR, for example, to obtain or provide navigation information
- *Cockpit interface* – this provides voice and data services to the cockpit

- *Cabin interface* – this provides voice and data services to the cabin, for example, through the In Flight Entertainment (IFE) system
- *Electromagnetic wave interfaces for radio services* - the Radio Frequency (RF) electromagnetic wave interfaces at the antennas. The actual antennas are within the system boundary
- *Power input* - this provides power to the system.

2.2 Candidate Waveform Examples

A number of candidate example waveforms are listed in Table 1. This provides an illustration of the range of different waveforms that may need to be supported, but is not a comprehensive list. The tables include the currently anticipated Design Assurance Level for each system, but this may change in the future.

Table 1. Candidate Waveform Examples

System	Tx/Rx	Frequency Range (MHz)		Modes	Expected Design Assurance Level
HF	Tx/Rx	2.8	24	Analogue Voice Data (HFDL ACARS)	D
VHF	Tx/Rx	118	137	Analogue Voice Data (VDL Mode 2) Data (VDL Mode 4)	C
Aircell CDMA	Rx Tx	849 894	851 896		E
L-DACS	Tx/Rx	960	1215		C
Inmarsat	Rx Tx	1530 1626.5	1559 1660.5	Aero C Aero H/H+ Aero I Aero L Mini M Aero	D
				Swift64 SwiftBroadband	E
Iris	Rx Tx	1545 1646.5	1555 1656.5		C
Iridium	Tx/Rx	1616	1626.5		D/C
Iridium NEXT	Tx/Rx	1616	1626.5		D/C
AeroMACS	Tx/Rx	5091	5150		C
DVB-S2	Rx Tx	10700 14000	12750 14500		E

2.3 Separation of Processing from Transceivers

Key to the integrated communications system approach is the partitioning of the radio functionality in order to identify common functionality which can be combined or hosted more efficiently. With this in mind, radio functionality may be split into the following major areas:

- *Front-end Functionality* – This covers antennas and other items that may need to be located close to the antennas, such as antenna matching units, diplexers, low noise amplifiers and high power amplifiers.
- *Transceiver Functionality* – This broadly covers the analogue aspects of radio technology, typically operating in the Intermediate Frequency (IF) and Radio Frequency (RF) domains. It usually includes amplifiers, mixers and filters to provide up and down conversions between an IF representation of and an RF representation of signals. It also covers the conversion between analogue and digital representation of signals.
- *Processing Functionality* – This broadly covers the digital aspects of radio technology, where processing is typically carried out in Field Programmable Gate Arrays (FPGAs) and processors. The later have traditionally been split into Digital Signal Processors (DSPs) and General Purpose Processors (GPPs), but this distinction is less important today since some processors can carry out both digital signal processing and general purpose processing efficiently.

An example partitioning of radio functionality is illustrated in Fig. 2. There is some flexibility in the location of certain functions. For example digital up/down conversion

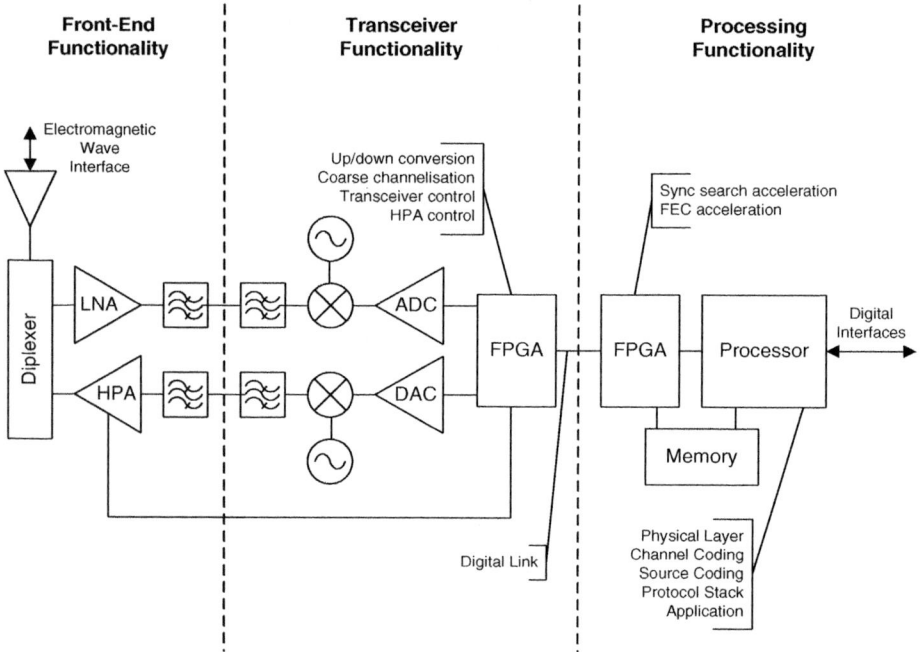

Fig. 2. Example Partitioning of Radio Functionality

and coarse channelisation may be viewed as part of transceiver functionality or processing functionality. Alternatively, it may be split between the two.

Avionics radio equipment has traditionally been based on a federated architecture where a dedicated radio is provided for each type of radio system (e.g. VHF). The above functional partitioning illustrates an opportunity for an integrated communications system where the processing capability is common across different radio systems and can either be reused or shared. This approach is similar to the Integrated Modular Avionics (IMA) [5] approach which moved the provision of other types of avionics equipment from a federated architecture to an integrated architecture. In fact, the integration of the communications system within the IMA is also being examined within the SANDRA programme.

2.4 Digital Serial Links

The partitioning of radio functionality described above supports an architecture where a common baseband processing solution is used in combination with transceiver functionality that is designed for different systems or area of the spectrum. This gives the potential for cost saving by maximising reuse of the design, isolating the transceiver functionality if required and physically separating the location of transceiver functionality and processing functionality. It also provides architectural flexibility since in some systems, it is desirable for the transceiver functionality to be located close to the antenna to reduce cable losses.

Such an architecture is enabled by modern high-speed digital links which can be placed between the processing units and transceiver units. Recent technology trends have moved from multi-drop parallel buses to point-to-point serial links, examples being Serial RapidIO (SRIO), Peripheral Computer Interconnect Express (PCIExpress), Ethernet and the Common Public Radio Interface (CPRI) [6].

Point-to-point links are attractive since they are faster, simpler and more reliable than multi-drop links. Modern high-speed serial links include equalisers at the receivers to improve signal integrity. The higher speeds are also achieved because the clock is embedded in the signal, and there is no need to keep several electrical lines synchronised as is required with parallel buses.

2.5 Security

Security is an important consideration in the provision of avionics equipment and can be viewed as covering confidentiality, integrity and availability of systems and data.

Confidentiality problems can be caused, for example, by eavesdropping on communications and gaining unauthorised access to systems through loopholes in the access control measures.

In terms of integrity, attackers may, for example, attempt to modify data, inject false data or simply selectively delete data and thereby cause the aircraft systems to make incorrect decisions due to erroneous or missing information. This could be achieved by intercepting communications, for example, or by introducing malicious software onto a processing platform.

Availability problems can be caused by attackers blocking communications or causing vital systems to fail through, for example, exploitation of software bugs in the applications or their underlying operating systems.

There are several groups and organisations that are looking at various aspects of security for avionics. The activities of these groups and organisations are summarised below.

Airlines Electronic Engineering Committee (AEEC) is an airlines and airframe manufacturers standards group that co-ordinates several security activities for avionics. The standards and working groups of interest are:

- ARINC 811 – Aircraft Information Security Process. This is a model for assessing security risk, implementing appropriate countermeasures, operating these countermeasures and feeding results back to risk assessment.
- ARINC Manager of Air-Ground Interface Connection (MAGIC) – This group is looking at integrated communications systems on board aircraft, initially for airline and passenger systems but with the ability to extend this to ATM and AOC systems later on.

Radio Technical Commission for Aeronautics (RTCA) Inc is a US based organisation that develops aviation standards. A similar organisation in Europe is the *European Organisation for Civil Aviation Equipment* (EUROCAE). The RTCA SC-216 group is working on security issues, and so is the EUROCAE WG-72 group. Both groups are collaborating to develop the following security related documents:

- Minimum Aviation System Performance Standards (MASPS) for Aeronautical Electronic and Networked Systems Security
- Security Assurance and Assessment Processes and Methods for Safety-related Aircraft Systems.

Air Transport Association of America, Inc (ATA) is an airline trade association based in the US. The following working group and specification cover security matters:

- Digital Security Working Group – this group provides a forum for addressing the application of digital security technologies and establishing best practice and conventions.
- Spec 42: Aviation Industry Standards for Digital Information Security – this specifies identity management solutions based on Public Key Infrastructure (PKI) technology.

EUROCONTROL is the European organisation for the safety of air navigation, and the *Federal Aviation Administration (FAA)* is an agency of the US government that regulates and oversees all aspects of civil aviation in the US. The following study has been carried out by EUROCONTROL and the FAA:

- Communications Operating Concept and Requirements for the Future Radio System (COCR) [7] – This focuses on air traffic control and air operations, and includes a security risk assessment in these areas. Confidentiality is rated as of low importance except for business needs, but integrity and availability are rated as potentially of very high importance across a wide range of systems. The main recommendation is the use of cryptographic communications security as the main form of protection.

The above specifications and studies for security are likely to be relevant to the integrated communications system approach.

2.6 Segregation

Segregation applies at two levels in the context of the integrated communications system approach. First of all there is often a need to segregate the provision of services, for example, segregating cockpit services used by the crew from the cabin services used by passengers. Secondly, where software applications share the same resources (e.g. processor and memory) there is a need to segregate the software applications so that they cannot interfere with each other.

The more traditional federated approach has seen an application performed by dedicated resources, i.e. a software application would run on its own processor with dedicated memory and other peripheral resources. The segregation between the various applications was clear and generally defined in hardware such that a fault occurring in that unit could easily be prevented from affecting other applications. However, the obvious drawback with the federated approach is the use of more resources than necessary, which often requires duplication for fault tolerance, further exacerbating the problem.

IMA has emerged as a design approach challenging the federated architecture by reducing the costs associated with acquisition, space, power, weight, cooling, installation and maintenance. IMA uses a single computer system (with replication for fault tolerance), to provide a common computing resource for several applications. This is achieved by employing an operating system that provides space (i.e. memory) and time partitioning for example, an ARINC 653 compliant operating system.

Similar issues apply in the integrated communications system approach. Cockpit and cabin services segregation can be achieved through software partitioning, or by using different processing platforms for the different services. Software partitioning can also be provided through an appropriate operating system. However, it should be noted that the tolerable latencies in software defined radio applications are generally less than those in current IMA applications. Latency issues limit the number of waveform applications that can share the same processing platforms using time partitioning.

2.7 Redundancy

Authorities, such as the FAA, specify a minimum level of redundancy of equipment that must be operational before an aircraft is allowed to take off. This redundancy is specified so that a single failure does not cause the loss of any vital communication channel.

In the federated approach, redundancy is provided by replicating dedicated equipment, taking up space and adding weight. In the integrated communications system approach, there are more opportunities for supporting redundancy, including the ability to switch from one processing unit to another in the case of failure.

2.8 Certification

Certification is one of the most difficult areas in the development of avionics systems. It needs, for good reasons, to be done thoroughly, and tends to err on the side of caution to ensure that safety requirements are fully satisfied. The subject is large and complex and addressed in various documents, including the following:

- SAE ARP4754 Guidelines for Development of Civil Aircraft and Systems [8]
- RTCA DO-178B / ED-12B: Software Considerations in Airborne Systems and Equipment Certification [9]
- RTCA DO-248B: Final Report for Clarification of DO-178B [10]
- RTCA DO-254 / ED-80: Design Assurance Guidance for Airborne Electronic Hardware [11]
- RTCA DO-297: Integrated Modular Avionics (IMA) Development Guidance and Certification Considerations [12].

A careful examination is required between the certification impact for dedicated radio equipment as in the federated approach, and the certification impact where common processor platforms are employed, as in the integrated communications system approach.

3 SANDRA Overview and Functional Architecture

The SANDRA programme aims to design and demonstrate an integrated communications system for civil aviation. The network architecture is illustrated in Fig. 3 and supports the following waveforms:

- Analogue VHF and Voice Data Link Mode 2 (VDL2) in the VHF band
- Inmarsat Broadband Global Access Network (BGAN) in L-Band
- Aeronautical Mobile Airport Communications System (AeroMACS) in C-Band
- Digital Video Broadcasting - Satellite - Second Generation (DVB-S2) in Ku-band (receive only).

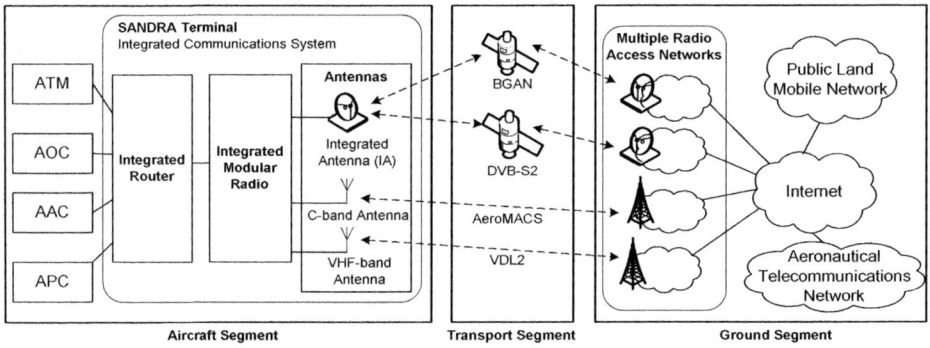

Fig. 3. SANDRA Network Architecture

The SANDRA Terminal in Fig. 3 is an integrated communications system made up of an Integrated Router (IR), an Integrated Modular Radio (IMR) and a number of antennas. The IR carries out high level functions such as routing, security, Quality of Service (QoS) provision and mobility. The IMR carries out low level functions such as the radio communication layers, radio resource allocation and QoS mapping. The IR/IMR split is in fact an arbitrary but pragmatic partitioning within the SANDRA programme that has been necessary to allow the scope of the work packages to be more easily delimited.

The realisation of the IMR within the programme will now be described, illustrating how it will demonstrate some of the key concepts in the integrated communications system approach.

4 SANDRA Proof-of-Concept Integrated Modular Radio

4.1 High-Speed Digital Serial Link Selection

The IMR architecture employs a high-speed digital serial link between transceiver functionality and the processing functionality illustrated in Fig. 2. This link mainly carries baseband samples of the receive signal and baseband samples or symbols of the transmit signal. In addition, it carries control and status signals. An important requirement for this link is the required bit rate. The estimated bit rate for each system is given in Table 2, based on 16 bit samples and two times oversampling with reference to the minimum Nyquist requirement for the bandwidth of interest.

Table 2. Bit Rate Requirement for High-Speed Digital Links

	VHF	BGAN	AeroMACS	DVB-S2
Max channel bandwidth (MHz)	0.025	0.20	20[1]	36[1]
Bit rate for 16 bit samples with 2x oversampling (Mbit/s)	1.6	12.8	1280	2304
Estimated total bit rate including controls and status (Mbit/s)	1.8	13.0	1300	2400
Note 1: The proof-of-concept demonstrator covers lower bandwidths				

Latency is an important issue in radio equipment, and is relevant to the high-speed serial link.

At the protocol level, radio communication systems are often based on frames which are about 10 ms or more in duration. Protocols typically allow a period of a frame or more in between a radio receiving a frame and generating a response. This period allows items such as physical layer receive completion, block error correction, protocol handling and transmit preparation to be carried out after a frame has been received. For this purpose, a useful scheme is to deliver receive sample blocks and take out transmit sample blocks every 0.5 ms in order to realise a suitable degree of block processing granularity. A link latency of the order of 100 μs ensures that the round trip delay over the link is less than the granularity of 0.5 ms and so reduces the impact of the link delay.

In radio communications applications, other controls such as fast Automatic Gain Control (AGC) or inner loop power control can require significantly shorter response times. For example, the maximum round trip delay in the CPRI specification (excluding cable delays) is 5 μs, in order to support the UTRA-FDD (UMTS Terrestrial Radio Access - Frequency Division Duplexing) inner loop power control.

The required latency of the link depends on the types of applications that need to be supported and the desired location of functionality at each end of the link. A round trip delay of the order of 100 μs means that fast responses such as fast AGC or inner loop power control will need to be implemented at the *transceiver* unit. A round trip delay with the CPRI figure of 5 μs means that fast responses such as fast AGC or inner loop power control can be implemented at the *processing* unit.

The required alignment of a transmit burst to a receive signal varies from system to system. One method of achieving alignment is to drive the Analogue to Digital Converter (ADC) and Digital to Analogue Converter (DAC) from the same clock source, thus providing a fixed relationship between transmit samples and receive samples. A transmit function in software or firmware can then set up a transmission aligned to the timing defined by the ADC and DAC pair. In this scenario, the only impact of the performance of the link (i.e. its latency) will be on the amount of memory required in the transceiver unit to buffer receive and transmit samples.

Additional requirements include environmental requirements and support for locating transceiver functionality and processing functionality at different ends of the aircraft, for example through the use of optical fibres.

The following candidate high-speed serial interfaces have been examined:

- PCI Express (PCIe)
- Serial RapidIO (SRIO)
- Common Public Radio Interface (CPRI).

CPRI has been selected for use in the SANDRA demonstrator since it has the following desirable characteristics:

- specifically designed for links between transceiver functionality and processing functionality
- supports line rates of up to 6144 Mbit/s, well in excess of the requirements of Table 2
- supports electrical and optical media, with the latter supporting distances of over 10 km
- supports the distribution of frequency references, should this be desirable
- flexible enough to support a variety of radio waveforms
- round trip delay is guaranteed to be as low as 5 μs since unlike PCIe and SRIO, it is framed based rather than packet based
- supports a number of topologies such as chains, trees and rings.

4.2 Processing Card and FPGA Card Selection

The processing platform for the SANDRA demonstrator is hosted in a compact PCI chassis. A Commercial Off-The-Shelf (COTS) card has been selected from Concurrent Technologies, with an Intel i7 processor. It also provides Express Mezzanine Card (XMC) sites connected to the processor via PCIExpress.

A COTS XMC card with a Xilinx SX95T FPGA and Small Form-factor Pluggable (SFP) connectors has been selected from Innovative Integration, to provide an FPGA capability and support CPRI connectivity.

The above choices have been made to support rapid development and demonstration rather than target a product.

4.3   Real-Time Operating System Selection

The QNX Real-Time Operating System (RTOS) has been selected for use on the processing platform since it has a strong micro-kernel architecture, supports the Portable Operating System Interface for Unix (POSIX), supports multi-core processors, supports time partitioning and has very capable trace and debug facilities.

Once again, the choice has been made to support rapid development and demonstration rather than target a product since, for the latter, the choice of operating system would need to be assessed in terms of DO-178B certification.

4.4   SANDRA IMR Demonstrator

The SANDRA IMR demonstrator is illustrated in Fig. 4, showing two common processing platforms, optional frequency references, three transceivers and one receiver. The processing platforms are connected to the transceivers and receiver by CPRI. The processing platforms also connect to each other and to the external Integrated Router (IR) via Ethernet.

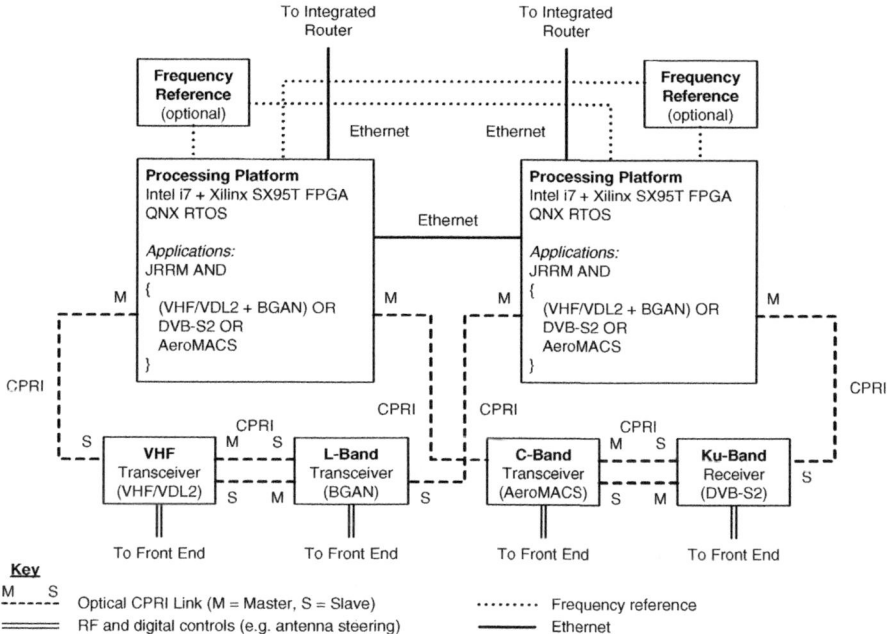

Fig. 4. SANDRA IMR Demonstrator

The processing platforms run Joint Radio Resource Management (JRRM) applications, with one acting as a master and the other as a slave. They interface to the IR for the provision of communication services and decide which waveforms should run on each platform. The platforms are intended to run VHF/VDL2 and BGAN waveforms at the same time, or an AeroMACS waveform or a DVB-S2 waveform.

The IMR demonstrator supports redundancy in terms of the high-speed links and processing platforms. If a processing platform running a high priority waveform goes down, it will be possible to disable the lower priority waveform on the second platform and start running the high priority waveform on that platform.

A full IMR is likely to require CPRI based switches to more easily configure links between processing platforms and transceivers.

5 Conclusions

The SANDRA proof-of-concept demonstrator will allow a number of concepts relating to the Integrated Modular Radio approach for avionics to be examined. These include the separation of processing functionality from transceiver functionality through the use of high-speed digital links, the use of a common processing platform for a variety of waveforms, dynamic radio resource management, support for asymmetric data links through the joint use of BGAN and DVB-S2, support for legacy radio links such as VHF and some support for redundancy.

Acknowledgement. The research leading to these results has been partially funded by the European Community's Seventh Framework Programme (FP7/2007-2013) under Grant Agreement n° 233679. The SANDRA project is a Large Scale Integrating Project for the FP7 Topic AAT.2008.4.4.2 (Integrated approach to network centric aircraft communications for global aircraft operations). The project has 31 partners and started on 1st October 2009.

References

1. EUROCONTROL: Strategic Guidance in Support of the Execution of the European ATM Master Plan. Edition 1.0 (2009)
2. SESAR Joint Undertaking, http://www.sesarju.eu
3. EUROCONTROL: IEEE 802.16E System Profile Analysis for FCI'S Airport Surface Operation. Edition 1.3 (2009)
4. SANDRA (Seamless Aeronautical Networking through integration of Data links Radios and Antennas), http://www.sandra.aero
5. RTCA: Integrated Modular Avionics (IMA) Development Guidance and Certification Considerations, DO-297 (2005)
6. CPRI: Common Public Radio Interface (CPRI); Interface Specification. V4.1 (2009)
7. EUROCONTROL/FAA: Communications Operating Concept and Requirements for the Future Radio System, Version 2.0
8. SAE: Guidelines for Development of Civil Aircraft and Systems, ARP 4754 (2010)

9. RTCA: Software Considerations in Airborne Systems and Equipment Certification, DO 178B (1992)
10. RTCA: Final Report for Clarification of DO-178B, DO-248B (2001)
11. RTCA: Design Assurance Guidance for Airborne Electronic Hardware, DO-254 (2000)
12. RTCA: Integrated Modular Avionics (IMA) Development Guidance and Certification Considerations, DO-297 (2005)

The SANDRA Communications Concept - Future Aeronautical Communications by Seamless Networking

Simon Plass[1], Christian Kissling[1], Tomaso de Cola[1], and Nicolas Van Wambeke[2]

[1] German Aerospace Center (DLR)
Institute of Communications and Navigation
82234 Wessling - Oberpfaffenhofen, Germany
{simon.plass,christian.kissling,tomaso.decola}@dlr.de
[2] Thales Alenia Space
Business Segment Navigation & Communications
Toulouse, France
Nicolas-van-wambeke@thalesaleniaspace.com

Abstract. The EU SANDRA (Seamless Aeronautical Networking through integration of Data-Links, Radios and Antennas) project [1] aims to design, specify and develop an integrated aircraft communication system to improve efficiency and cost-effectiveness by ensuring a high degree of flexibility, scalability, modularity and reconfigurability. SANDRA aims at the definition of an access to an open system resulting in a collection of communications technologies targeted at specific operational settings. Within the paper the main ideas of this envisaged communications concept are addressed. Furthermore, quality of service management strategies are assessed with respect to their applicability and efficiency in the ATM (Air Traffic Management) context. Additionally, techniques for the selection of links for data transmission and the interaction between technology independent and technology dependent components in the networking architecture are covered by means of standardized communication protocols such as IEEE 802.21 and ETSI BSM extensions.

Keywords: Seamless networking, IPv6, QoS, aeronautical communications.

1 Introduction

The vision of ACARE (Advisory Council for Aeronautics Research in Europe) for 2020 [2] shows that Europe has to create a seamless system of air traffic management that copes with up to three times more aircraft movements than today by using airspace and airports intensively and safely. The development of sophisticated ground and satellite-based communication, navigation and surveillance systems will make this possible. Furthermore, goal is to reduce significantly the noise nuisance, and therefore, large airports can operate around the clock. Finally, this will ensure flying safely in all weathers and aircraft are running on schedule 99% of the time.

G. Giambene and C. Sacchi (Eds.): PSATS 2011, LNICST 71, pp. 300–313, 2011.

There is a need for this new approach in order to achieve a broader level of integration for the required increase of capacity, safety, security, and efficiency of air transportation operations which keeps at the same time the complexity and cost of on-board networks and equipments within a sustainable level. For these goals the SANDRA integrated network concept with IPv6 as final unification point (target 2025 and beyond) has to be developed. Recently, ICAO (International Civil Aviation Organization) adopted IPv6 for use within its future IP-based aeronautical telecommunications network (ATN) [3].

In the following the approach towards seamless networking integration is described. Details are given in the structure of the SANDRA communications concept and its working structure. Furthermore, quality of service management strategies are assessed with respect to their applicability and efficiency in the ATM context. In particular addressing the service demands of ATM communication, such as strict latency and loss limitations are considered herein. This also covers techniques for the selection of links for data transmission and the interaction between technology independent and technology dependent components in the networking architecture by means of standardized communication protocols such as IEEE 802.21 and ETSI BSM extensions.

2   The SANDRA Programme for Seamless Networking

The vision of SANDRA [4] is the integration of aeronautical communications systems using well-proved industry standards to enable a cost-efficient global provision of distributed services. SANDRA system is considered as a 'system of systems' addressing five levels of integration: Service integration of a full range of applications and services (ATS, AOC, AAC, APC); Network integration of different radio access technologies through a common IP-based aeronautical network and interoperability of network technologies (ACARS, ATN/OSI, ATN/IPS); Radio integration of radio technologies in an Integrated Modular Radio (IMR) platform [5], [6]; Antenna integration of an asymmetric high data rate downlink by development of an hybrid Ku/L band SatCom antenna; WiMAX adaption for integrated multi-domain airport connectivity. i.e., AeroMACS.

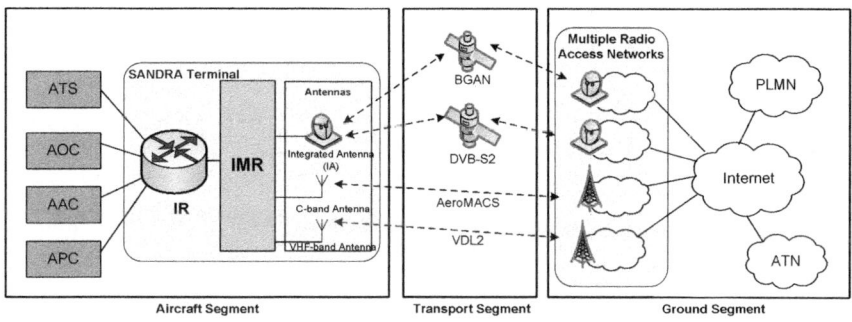

Fig. 1. SANDRA Network Architecture [6]

Considering the communications network, SANDRA spans across three segments, i.e., aircraft segment, transport segment (resp. wireless access segment) and the ground segment, as shown in Fig. 1. The aircraft segment contains the main functional components: the integrated router (IR), the integrated modular radio (IMR) and the antennas consisting of a hybrid Ku/L band integrated antenna (IA), a VHF band antenna and a C-band antenna..

The integration of different service domains with very heterogeneous requirements through a cost-effective and flexible avionic architecture is thus one of the main challenges addressed by SANDRA. Under this perspective, the SANDRA communications system presents a key to enable the global provision of distributed services for common decision making based on the System Wide Information Management (SWIM) concept [7], and to meet the high market demand for broadband passenger and enhanced cabin communications services.

2.1 Aeronautical Seamless Networking Environment of SANDRA

The following sections and Fig. 2 give an overview of the main tasks if the seamless networking aspects in the SANDRA project, namely: interworking of different data link technologies (ground-based, satellite-based, airport systems as main streamline for validation, and air-to-air MANET as long term extension), interoperability of network and transport technologies (ACARS, ATN/OSI, IPv4, IPv6 networks), and integration of operational domains (ATS, AOC, AAC, APC).

Additionally, a large effort is spent in the validation and testing of the SANDRA integrated airborne network design [8]. Capacity limits and overall system performance on future air traffic scenarios, services, and applications are assessed. This gives input to the development of a prototype implementation and its testing. In 2013, the overall SANDRA concept will be validated and demonstrated within a test bed and during flight trials. The planned lab environment is illustrated in Fig. 2.

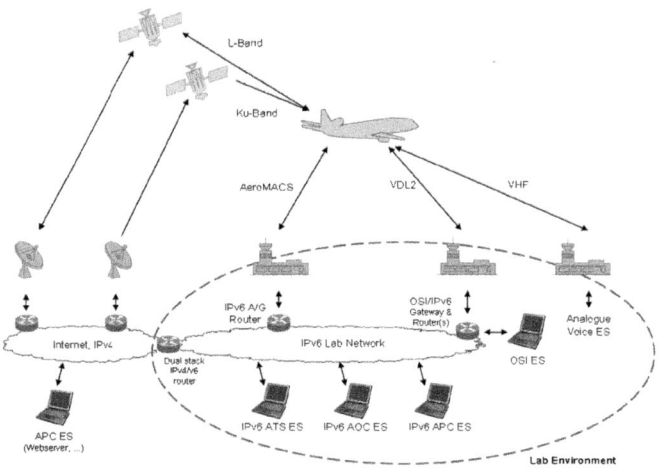

Fig. 2. Aeronautical seamless networking environment of SANDRA

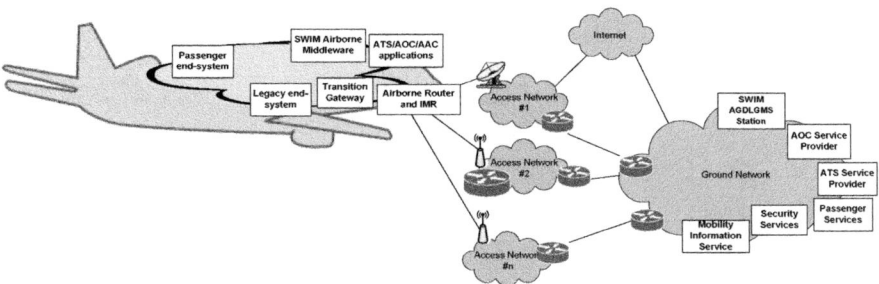

Fig. 3. High Level overview of the SANDRA system from a functional point of view [9]

2.2   Network Architecture and Interoperability

Topology Level: The aim of the SANDRA network architecture is to allow for onboard end systems to communicate with other end systems located on the ground through potentially more than one radio link at a given time (e.g. combining higher throughput satellite with lower latency AeroMACS). From a topology point of view, the functional architecture can be illustrated as shown on Fig. 3.

On the airborne side of the network, several functional entities are represented, like the passenger end systems which will mainly use the SANDRA network architecture in order to access the Internet and specific ATS/AOC applications which will communicate with ATS/AOC service providers on ground through the use of multiple access networks technologies (e.g., AeroMACS, satellite communications, etc.). On the ground side of the network, the counterparts to several of the airborne side functional entities are presented. In order to provide efficient service, the SANDRA network architecture relies on various functionalities provided on the ground like the Mobility Information Services and the Security Services. The SANDRA system supports the SWIM architecture and the related airborne and ground components of the SWIM architecture are also shown in Fig: 3. The SWIM based ATS/AOC applications will interface with the SWIM airborne middleware on the airborne side and the SWIM Air-Ground Datalink Ground Management System (AGDLGMS) stations on the ground. In order to support multiple data links with variable characteristics and constraints (e.g. local coverage of AeroMACS vs. global coverage of the satellite, variable data rate and latency), the onboard network and the various access networks are interconnected by the airborne router and the IMR. It is at this level that all the functionalities related to Quality of Service, Resource Management, Packet Scheduling and Link Selection take place. Furthermore, mobility and security functions are also strongly linked to the Packet Processing that takes place in the airborne router and IMR. Finally, in order to provide interoperability with legacy end systems which might be using protocols not natively supported by the SANDRA system, the airborne architecture includes a transition gateway. This gateway implements transition mechanisms required in order to adapt legacy protocols to the SANDRA network architecture (these include but might not be limited to tunnelling, protocol translation, higher layer proxying, etc.) [9].

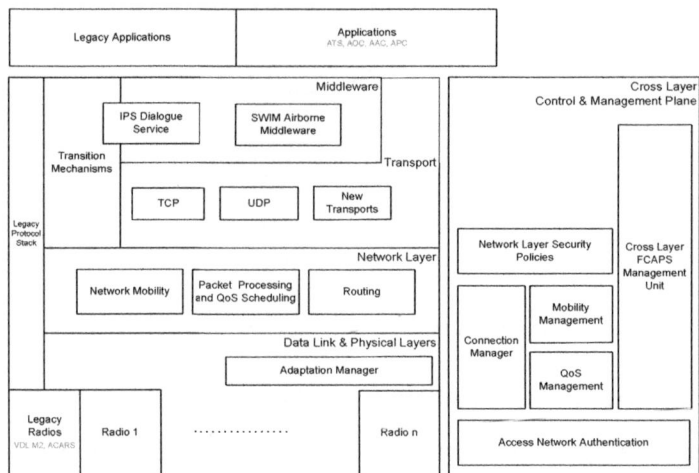

Fig. 4. High level overview of the Airborne Functional Architecture

Airborne Level: The main focus of the SANDRA project is on the airborne aspects of the functional architecture. However, some of the functions to be supported by the system may require the presence of a ground located counterpart. In this case, the study of these ground elements will be performed for completeness purposes.

Fig. 4 presents a high level overview of the functional architecture focusing on the airborne level. In order to simplify the description, the presentation is made following a layered architecture (similar to the OSI model) to which several cross-layer extensions are provided following the principles described in [10].

A clear separation between the user (data) and control planes has been performed as presented on Fig. 4. The user (data) plane includes all the functional blocks that are directly related with the transport of data while the control plane assures the control of this transport as well as the management of the SANDRA system. At the topmost layer of the architecture, the Applications, both legacy and non-legacy applications are to be supported by the SANDRA system. For legacy applications, three possibilities exist:

1. These applications can use the legacy protocol stack and legacy radio, which corresponds to them operating as they would without the SANDRA system.
2. These applications can use transition mechanisms in order for the traffic they generate to be transported using the SANDRA system
3. Or, these applications can be adapted in order for them to directly interface with a middleware such as the dialogue service or SWIM through the use of a SWIM adapter.

Non-legacy applications are considered to be using one of the supported middleware layers or to directly use any of the transport layer protocols supported by the architecture. At the network layer, several functionalities such as packet processing and QoS scheduling are implemented in addition to functionalities related to network mobility and security. The network layer can either directly make use of radios if these radios implement the correct interface or it can be connected to the IMR.

The IMR implements the data links and physical layers of the protocol stack and also provides an abstraction layer between the radios and the network layer packet processing. The Adaptation Manager block in the user plane at the Data Link and Physical layers acts as this abstraction layer. The Adaptation Manager is responsible for interfacing the multiple radios to the network layer in a common and standard way even if these radios do not provide similar interfaces to be connected to the network layer.

In parallel to the data plane (in which user data is processed), the functional network architecture presented on Fig. 4 includes a cross-layer control and management plane. The SANDRA system requires a close integration of the network, data link and physical layers in order to support complex networking scenarios such as seamless mobility, handover, QoS management and security. In order to perform this, several functional entities coordinating the interactions between the functions implemented at the various layers of the stack have been identified. These functionalities range from the overall management of the SANDRA system elements (based on the FCAPS model [11],[12]) to the control and management of connection in terms of QoS and Link Selection. Additionally, security and mobility related functional entities are also present. [9].

2.3 Network Design

Since IPv6 is the unification point in the SANDRA network, there is the need of the design and adaptation to an aeronautical internet. Main focus within this task is the handling of the network management and also of the resource management. Additionally, effort is spent for the development of new and efficient handover and mobility management algorithms and concepts, respectively. Also an IPv6 based naming and addressing architecture will be provided. Due to the high degree of mobility on a global scale and the heterogeneous network environment (i.e. short-range and long-range terrestrial as well as satellite access technologies), work on a network mobility (NEMO) based IPv6 protocol started in contrast to the ICAO chosen Mobile IPv6 protocol supporting only host mobility.

3   Quality of Service Management and Interoperability

3.1 QoS Definition for Aeronautical Networks

The term Quality of Service (QoS) is used in a variety of different ways and often depends also on the context that it is used in. One notion of QoS denotes the performance of a service from the users view. A measure for the grade of QoS is how good the performance attributes of a service match with the demands made on it. The kind of attributes which are relevant and need to be fulfilled depend thus naturally on the context of the service. While for many other services perceived or qualitative QoS measures are applicable, the ATM communication environment envisaged here makes high and precise demands on different attributes, presented later on in detail. [14] provides a good overview and summary of different aspects of QoS provision in the context of heterogeneous networks, such as present in the ATM environment considered in SANDRA.

The provision of QoS in an operational and safety critical aeronautical environment is however considerably different from the applications and demands in the Internet. Service parameters such as defined in [13] thus cannot be directly applied here. The most intuitive reason for this is that a violation of QoS attributes in Internet applications results in a reduced service quality, which is naturally undesirable and bothersome for users, but has not necessarily implications on operational events and safety of life. In the aeronautical domain, for the management of air traffic this is decisively different. Late arrival of e.g. directive commands issued by the controller for the pilot can have catastrophic effects. Also corrupted messages or multiple receptions of messages can have such serious consequences, affecting the safety of the airplane and the passengers. For this reason it is not sufficient if the QoS mechanisms for ATM communication try to achieve the requirements as far as possible but it is necessary that the requirements are definitely met. In a joint study of Eurocontrol and the Federal Aviation Administration (FAA), potential future communication technologies which are suitable to provide the necessary safety and regularity of flight have been investigated and requirements for the future application services have been derived. The results of this study have been published in the so called "Communications Operating Concept and Requirements for the Future Radio System (COCR)" [15]. Within this study the concepts of ATM have been analyzed from an operational perspective and the expected technical requirements have been formulated, also for services which are not yet deployed but are expected to be deployed in the future. The results in the COCR provide information for all operational services with respect to their periodicity, volume and technical requirements. The main QoS requirements for the services can be categorized into transmission delay, expiration time and continuity. The transmission delay (denoted $TD_{95,FRS}$) hereby sets the maximum transmission latency until successful reception at the receiver within which 95% of all messages must have arrived. For messages which are not fulfilling this requirement, e.g. due to a packet loss requiring retransmission or buffering delays, the fraction of messages specified by the continuity requirement must have arrived within the expiration time. The COCR specifies the QoS requirements per service, but also for aggregated Classes of Service (CoS).

Within the SANDRA QoS activity, the problem is addressed how different communication links can be integrated into a seamless network and which mechanisms and approaches are suitable to allow provision of the required QoS. SANDRA hereby focuses on the network layer QoS mechanisms mainly. Fig. 5 illustrates the general approach. One requirement for the layer 3 QoS mechanisms is that they must be interoperable and independent of the type of used link. Going beyond this, also the uniform interfaces (denoted Service Access Points, SAP in the following) to the technology dependent L2 are in the scope of SANDRA and discussed hereafter in more detail.

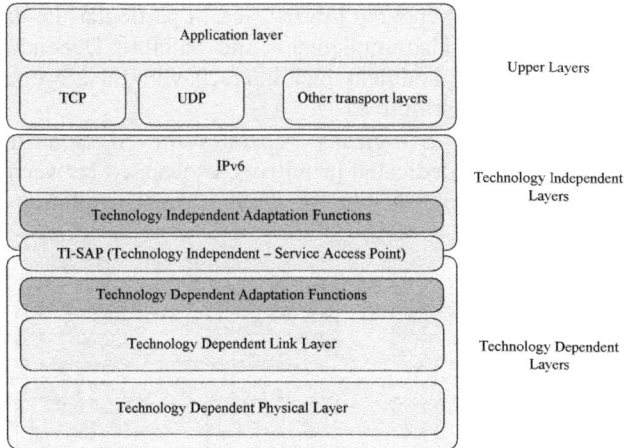

Fig. 5. Functional interaction between technology independent higher layers and technology dependent lower layers

3.2 QoS Mapping in the SANDRA Architecture

As straightforward from the considerations drawn in the previous section, the necessity for the SANDRA architecture is to simultaneously manage different QoS traffic profiles and transmission technologies over which different services have to be handled, translate into a QoS mapping problem. Beside the technical challenges that arise in selecting the L2 queues to which the traffic has to be forwarded depending on the QoS requirements (scheduling and QoS mapping problem), a particular attention has to be reserved to the characteristics of the QoS architecture, being embedded in the SANDRA's. Apart from the specific QoS model being adopted (IntServ or DiffServ as sketched in the following sections), some attention has to be addressed to how L3 and L2 intercommunicate, by preserving the QoS requirements specified in the SLSes of the specific traffic service. In this respect, different approaches can be applied. Ad-hoc solutions can be deployed, by extending for instance the functionalities and the related primitives already available from the ISO/OSI protocol stack. Given the scope of the SANDRA framework, it is instead better to have a model in line with architectures currently or going to be standardised. In this perspective, the features offered by the ETSI BSM protocol architecture are worth being considered. The main peculiarity consists in the definition of the SI-SAP interface, virtually separating the upper layer (Satellite Independent, SI) from the lower layers (Satellite Dependent, SD) and providing dedicated primitives to efficiently manage QoS, Address Resolution and Multicast functionalities over satellite.

The overall ETSI BSM protocol architecture is depicted in the following picture, where the main components are:

- SI layer: it implements the upper layer and in particular the IP protocol (versions 4 or 6). It also incorporates the Satellite Independent Adaptation Function (SIAF) module, which is responsible for adapting the SI functions to the characteristics of the lower layer specification, through dedicated primitives.

- SD layer: it implements the lower layer, in particular the datalink and the physical ones. It also implements the Satellite Dependent Adaptation Functions (SDAF) module, which interacts with the aforementioned SIAF through dedicated primitives.
- SI-SAP interface: it logically separates the SI from the SD layers, providing a set of dedicated primitives, exchanged between the SIAF and SDAF modules, responsible for QoS, address resolution and multicast functionalities.

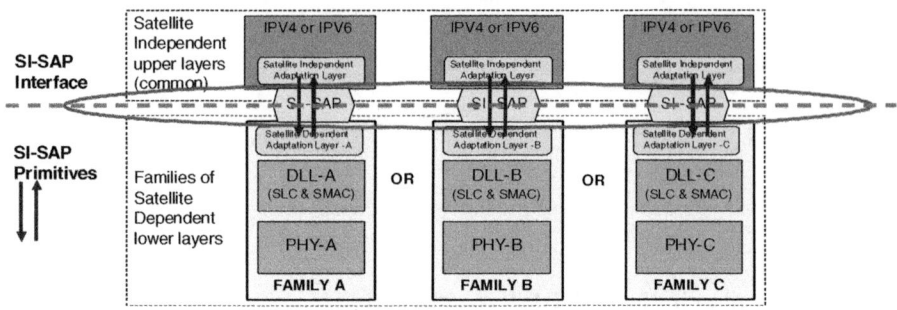

Fig. 6. ETSI BSM protocol architecture and SI-SAP interface definition

In this light, it is reasonable to extend the principles of the ETSI BSM protocol architecture for application in the SANDRA framework, to particularly address the QoS requirements of aeronautical networks.

In fact, two main "ingredients" of the SI-SAP interface can be re-used and properly extended to match the requirements of the SANDRA functional architecture: the Queue Identifier (QID) and the QoS primitives. The former is defined in the ETSI BSM protocol architecture as identifier of the L2 physical queues, so to allow an efficient QoS mapping between L3 and L2 queues, through the dedicated QoS primitives. The latter, in turn, allows actually implementing the QoS mapping algorithms and offering the essential tool to perform the resource allocation, based on the requests coming from the upper layers. The QoS problem in the SANDRA network involves not only resource allocation issues but also transmission technology selection, thus requiring the extension of the current SI-SAP interface functionalities along with the use of the IEEE 802.21 architecture in terms of the Media Independent Handover (MIH) functions. In practice, the QID has to be conceptually extended in a way that it incorporate both queue and link identifiers. Besides, the integration and the interaction of the ETSI BSM and the IEEE 802.21 architecture is of primary importance to perform the communication of the link selection to the upper layer and perform the resource allocation based on the requirements notified from the higher layers (e.g., application protocol or management plane). To this end, the SI-C-QUEUE primitives will be conveniently extended in their scope so to also include the new functionalities, thus allowing the different components to interwork properly according to the SANDRA network characteristics.

At this point, the final point to be addressed is the way the described protocol architecture integration (ETSI BSM and IEEE 802.21 namely) can be finally embedded in the real architecture of the SANDRA network. In this respect, a particular attention has to be reserved to the IR and IMR interaction. Although the SI-SAP interface has been conceived to logically separate the upper from the lower layers within a satellite terminal, it can be easily extended to physically separate two different components, by distributing the implementation of the primitives. This can be done by re-thinking the SI-SAP interface as the separating IR and IMR; these, in turn, will implement the related QoS primitives, thus working as the SIAF and SDAF modules in the original ETSI BSM architecture.

The overall system function can be then summarised in the following operations:

- In case the QoS requirements are constrained to a specific link by the upper layer, the IR will signal the selected transmission technology along with QoS request in a dedicated QID to the IMR, which in turn will forward the forthcoming data traffic to the specified transmission link. The availability of the transmission link is known after the start-up phase, which is accomplished by suitably combining the SI-C-QUEUE-open primitives with the MIH functionalities.

- In case no link-constrained request is performed by the upper layer, the IR simply signals the IMR about the QoS requests. In turn, the IMR will be responsible for running the link selection algorithm to identify the transmission technology most appropriate to match the received QoS requests. Also in this case the signalling is performed through real exchange of the SI-SAP primitives; in particular, in this case the QID will basically contain an identifier for the QoS request and a default value of the transmission technology, being it not explicitly selected by the upper layers.

- In case a link was no longer available or its availability was reduced (upon notification through the specific MIH functions), the IMR would in turn notify it to the IR through the corresponding enhanced SI-C-QUEUE primitives to trigger a new resource allocation. The IR in turn will run a new resource allocation request to match the new link configuration, by modifying or demanding the assignment of a new QID.

The overall interaction between the SANDRA components is represented in the following Fig. 7.

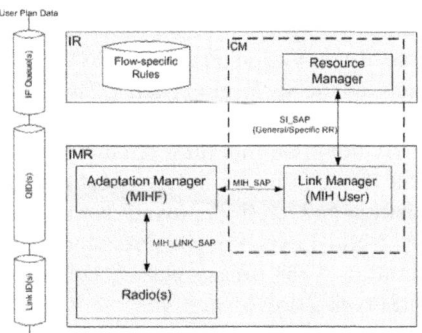

Fig. 7. Interaction between IR and IMR modules within the SANDRA network

3.3 QoS Management Architecture

In contrast to QoS architectures which are deployed in the internet, the QoS design here has to comply with a range of security and safety requirements which limit the freedom of choice for a QoS architecture considerably. The selected QoS management architecture should also rely on well established and standardized solutions. From today's perspective, one of the major design constraints is the strict separation of operational (ATS and AOC) and non-operational (AAC and APC) services within the network due to safety. While this separation is a real requirement nowadays, in SANDRA an all-integrated, seamless network is envisaged for the far future, which integrates also operational and non-operational services and provides the required safety at the same time. Naturally this has also an impact on the QoS architecture. One major impact of this different design is on the Connection Admission Control (CAC) and Congestion Control (CC) functionalities. In a purely operational network for instance, neither a rejection of a communication request of a pilot is acceptable, nor is delaying the transmission of the message to a later point in time to avoid congestion, if this would mean violating the maximum latency of the message. Overload situations which lead to such events such has to be prevented by proper network dimensioning and CAC/CC mechanisms are not strictly applicable. When integrating operational (op) and non-operational (non-op) services the situation changes however. Here it must be ensured that the op services always get the resources and priority they need and may not be affected by the non-op services, such as passenger communication. For the QoS architecture it is thus necessary to deploy methods for prioritization of the op-services and to have mechanisms ready which allow flow shaping, CC and CAC for the non-op services. The detailed functional QoS management and the interaction of the scheduling, CC and CAC functionalities is going beyond the scope of this paper, so the focus here is put on the suitability analysis of different QoS architectures, namely DiffServ and IntServ in an aeronautical operational environment.

3.3.1 IntServ QoS Approach
The IntServ architecture [16], [17] was developed for supporting specific QoS for end-to-end sessions across networks. In this approach, single flows (representing a stream of packets) are identified and treated individually. Every packet is checked for the resources it is entitled to receive. For this purpose the state of all flows in the network has to be periodically signalled among the routers in the end-to-end path of each flow. The Resource ReSerVation Protocol (RSVP) [17] was designed for this purpose. IntServ also has connection admission control mechanisms as an integral part of its functionality which admits new traffic to the network only if sufficient resources are available. By doing all this IntServ can guarantee hard upper bounds for packet delays and packet loss caused by buffer overflow. Moreover IntServ can rely with RSVP on an existing and well deployed signalling protocol. The per-flow treatment also allows Multi-Level-Priority-Preemption (MLPP) which can be beneficial to differentiate ATM messages according to their priority and urgency.While these IntServ features match very well with the QoS requirements in the ATM environment, the application of IntServ would have several major drawbacks. As is the case for all IntServ architectures, the main drawback is the

scalability of the system and the signalling overhead. The traffic profile of ATM message exchange as predicted in the COCR consists of mainly small messages in the order few bytes, reaching at maximum several kilobytes in single cases. In the downlink for instance (i.e. aircraft to ground in ATM terminology) the maximum message size is 2763 bytes for the FLIPINT service. Estimations on the traffic profile have shown that the maximum message arrival rate hereby is slightly below 1 msg/s per aircraft at maximum, having an average of less than 0.1 msg/s per aircraft. This means in practice that either for every message a dedicated IntServ flow would have to be initiated and signalled, or an IntServ flow needs to be setup and kept alive for a longer time without being used most of the time, and accepting the overhead caused by the periodic keepalive messages necessary for this. Besides the volume overhead of the IntServ signalling also the time required for session initiation is an important overhead, considering that some messages have latency requirements as low as 0.74 s (DG-B) and 1.4 s (DG-C). For GEO satellite links already the session initiation would consume a considerable fraction of the maximum latency. Finally the heterogeneous and highly mobile environment, consisting of different link technologies and the belonging different access networks and the need for intra- and inter-technology handovers causes path changes. A change in the end-to-end path would then result also in additional IntServ session re-establishment overheads.

3.3.2 Differentiated Services (DiffServ)
DiffServ [18],[19] is the second well known QoS architecture specified by the IETF. In contrast to IntServ no individual flows can be distinguished but only different aggregated classes of traffic. Instead of a guaranteed forwarding behaviour for every flow, DiffServ defines the per-hop forwarding behaviour for the aggregate classes. For identification of the aggregate, the Traffic-Class field in the IPv6 headers are used. Since in DiffServ only traffic aggregates are treated instead of per-flows, no hard guarantees for the availability of resources and the end-to-end QoS performance can be given. An overdimensioning of resources is thus necessary here in order to meet the QoS requirements. The overdimensioning affects for instance the buffer sizes in the schedulers to avoid packet drops due to buffer overflow but also the available datarates on the links. While in theory the definition of one DiffServ aggregate per COCR CoS would be possible (resulting in 12 aggregates), in practice a smaller number of DiffServ aggregates improves the scalability and reduces the complexity. In this case the application CoS need to be mapped by a classifier into the suitable DiffServ aggregates. Since all COCR CoS have different demands for maximum latency, an aggregation into fewer DiffServ aggregates implies also an increase of the required bandwidth, since the latency of the most demanding service in a DiffServ aggregate has to be met since DiffServ is not distinguishing within an aggregate. In other words services which could tolerate a longer latency need to be transmitted in fewer time (i.e. the time of the most demanding service) what results in a higher demand in terms of data rate. For a DiffServ QoS approach also appropriate estimation and dimensioning of the network capacities is essential and requires a good model for the prediction of the amount of traffic to be transported including an additional buffer for unexpected traffic bursts. Such an (over)dimensioning on the other hand can also mean a waste of resources if capacity is strictly allocated per aggregate class and cannot be shared among different aggregates and considering the

highly bursty traffic profile. On the other hand a DiffServ architecture has significant advantages over an IntServ approach which outweigh the aforementioned drawbacks. Most important of all the issues with scalability do not exist here since only aggregates have to be treated instead of single flows. DiffServ is such much more suitable for the highly populated global ATM network under consideration with respect to this. Moreover a change of the end-to-end path, as can happen due to intra- and inter-technology handovers is not an issue here since no re-establishment of the RSVP tunnels is required anymore. Also the signalling overhead of IntServ for session initiation and keepalive can be saved while saving also the time for flow establishment which is beneficial for the overall delay profile.

3.4   Conclusions on QoS Architecture

A flow-oriented architecture such as IntServ would have the feature of guaranteeing a certain end-to-end behaviour, but is not suitable w.r.t. the bursty traffic profile, having only spurious transmission of single messages which have also only small size. The signalling overhead is considerable w.r.t. the small message payloads and also the additional time demand for a session initiation is considerable w.r.t. the latency requirements. A flow-oriented QoS architecture such as IntServ is thus no preferable solution for application in an ATM. The alternative QoS architecture matching better with the given scenario is thus DiffServ. For deployment of a DiffServ QoS architecture several design parameters have to be kept in mind, in particular the correct dimensioning of the resource trunks, mapping of application CoS into aggregate classes and priority scheduling. The main benefits here are the scalability also for a large and global ATM network. Also a change in the network point of attachment, e.g. due to a handover are not an issue here. The data volume and signalling delay overheads of IntServ can be saved here as well. For an integration of operational with non-operational services in the same network, however further specification of the mechanisms ensuring a safe separation of these two domains and appropriate mechanisms for CC, CAC and flow control of the non-op services need to be specified. This work is currently under definition within SANDRA.

4   Summary

In this perspective paper, the demands on an integrated aircraft communications system were laid out. Within the EU Project SANDRA a concept for a functional network architecture is developed which satisfies these requirements. The architecture was presented from a topology level and then there was a focus on the airborne architecture which is the core of the SANDRA network functional architecture. Furthermore, QoS mechanisms were discussed and summed up in detail.

Acknowledgments. The research leading to these results has been partially funded by the European Community's Seventh Framework Programme (FP7/2007-2013) under Grant Agreement n°233679. The SANDRA project is a Large Scale Integrating Project for the FP7 Topic AAT.2008.4.4.2 (Integrated approach to network centric aircraft communications for global aircraft operations). The project has 31 partners and started on 1st October 2009.

References

1. SANDRA (Seamless Aeronautical Networking through integration of Data links Radios and Antennas), http://www.sandra.aero
2. European Aeronautics: A Vision for 2020, ISBN 92-894-0559-7
3. International Civil Aviation Organization: Manual for the ATN using IPS Standards and Protocols (Doc 9896), 1st (ed.), Unedited Advance version (February 2009)
4. Barba, A., Battisti, F.: SESAR and SANDRA: new paradigms for aeronautical communications. In: Third International ICST Conference on Personal Satellite Services (PSATS 2011), Malaga, Spain (February 2011)
5. Baddoo, J., Gillick, P., Morrey, R., Smith, A.: The SANDRA Communications Concept - Integration of Radios. In: Giambene, G., Sacchi, C. (eds.) PSATS 2011. LNICST, vol. 71, pp. 291–304. Springer, Heidelberg (2011)
6. Ali, M., Xu, K., Pillai, P., Hu, Y.F.: Common RRM in Satellite-Terrestrial based Aeronautical Communication Networks. In: Giambene, G., Sacchi, C. (eds.) PSATS 2011. LNICST, vol. 71, pp. 333–346. Springer, Heidelberg (2011)
7. SWIM, http://www.swim-suit.aero
8. Fazli, E.H., Luecke, O.: Design Aspects of Networking Testbed for IPv6-Based Future Air Traffic Management (ATM) Network. In: Giambene, G., Sacchi, C. (eds.) PSATS 2011. LNICST, vol. 71, pp. 319–332. Springer, Heidelberg (2011)
9. SANDRA – FP7 233679: D3.2.1 Consolidated SANDRA network and interoperability Architecture (July 2010)
10. Srivastava, V., Motani, M.: Cross-layer design: a survey and the road ahead. IEEE Communications Magazine 43(12), 112–119 (2005)
11. ITU-T Recommendation M.3010: Principles for a Telecommunication management network (1996)
12. ITU-T Recommendation M.3400: TMN management functions (2000)
13. International Telecommunication Union, ITU-T G.1010 End-user multimedia QoS categories (November 2001)
14. Marchese, M.: QoS over Heterogeneous Networks. John Wiley & Sons Ltd., Chichester (2007)
15. Eurocontrol, FAA, Communications Operating Concept and Requirements for the Future Radio System (COCR), Eurocontrol/FAA (2007)
16. Wroclawsky, J., Braden, R. (eds.): RFC 2210: The Use of the Resource Reservation Protocol with Integrated Services. IETF (1997)
17. Zhang, L., Berson, S., Herzog, S., Jamin, S., Braden, R. (eds.): RFC 2205: Resource ReSerVation Protocol (RSVP) — Version 1 Functional Specification. IETF (1997)
18. Nichols, K., Blake, S., Baker, F., Black, D.: RFC 2474: Definition of the Differentiated Services Field (DS Field) in the IPv4 and IPv6 Headers (1998)
19. Blake, S., Black, D., Carlson, M., Davies, E., Wang, Z., Weiss, W.: An Architecture for Differentiated Services (1998)

A Networking Testbed for IPv6-Based Future Air Traffic Management (ATM) Network

Oliver Lücke and Eriza Hafid Fazli

TriaGnoSys GmbH, Weßling-Oberpfaffenhofen, Germany
{oliver.luecke,eriza.fazli}@triagnosys.com

Abstract. This paper presents a networking testbed, which is developed within the EC FP7 project SANDRA[1] to test and validate IPv6-based protocols for future Air Traffic Management (ATM) network. The development was originally initiated within the EC FP6 project NEWSKY which aimed at developing a concept for a global, heterogeneous communication network for aeronautical communications, based on IPv6 protocol stack. The SANDRA project aims at designing and implementing an integrated aeronautical communication system and validating it through a testbed and, further, in-flight trials on an Airbus A320. Central design paradigm is the improvement of efficiency and cost-effectiveness by ensuring a high degree of flexibility, scalability, modularity and re-configurability. Whereas the NEWSKY testbed is considered to be a *proof-of-concept*, the SANDRA testbed will represent a *prototype* aircraft communication system, integrating prototypes developed and implemented in SANDRA, comprising AeroMACS, Integrated Modular Radio (IMR), Integrated Router (IR), and a novel Ku-band antenna.

Keywords: NEWSKY, SANDRA, ATM, ATC, COCR, ATS, AOC, AAC, APC, IPv6, testbed, in-flight trials, aeronautical communication, RoHC, RoHCoIPsec, NeXT, NEMO, MIPv6, Integrated Modular Radio, Integrated Router.

1 Introduction

Several European research activities are being undertaken with the goal to develop improved communication technologies for aeronautical communication. These activities comprise ground-based, satellite-based, aircraft-to-aircraft and airport communication for all different application classes, both safety and non-safety, namely air-traffic services (ATS), airline operational and administrative communication (AOC, AAC), and aeronautical passenger communication (APC).

[1] The research presented in this paper has been partially funded by the European Community's Seventh Framework Programme (FP7/2007-2013) under Grant Agreement n° 233679. The SANDRA project is a Large Scale Integrating Project for the FP7 Topic AAT.2008.4.4.2 (Integrated approach to network centric aircraft communications for global aircraft operations). The project started on 1st October 2009 and the consortium consists of 31 multinational partners from 11 countries and from industry including several SMEs, research institutions and public bodies.

G. Giambene and C. Sacchi (Eds.): PSATS 2011, LNICST 71, pp. 314–327, 2011.
© Institute for Computer Sciences, Social Informatics and Telecommunications Engineering 2011

In this paper, the development of a networking testbed to test and validate IPv6-based protocols for future Air Traffic Management (ATM) network is presented.

The development was originally initiated within the EC FP6 project NEWSKY (Networking the Sky for Aeronautical Communications), which aimed at developing a concept for a global, heterogeneous communication network for aeronautical communications, based on IPv6 protocol stack. The NEWSKY network integrates different applications (ATS, AOC, AAC, and APC) and different data link technologies (legacy and future long range terrestrial radio, satellite, airport data link, etc.) using a common IPv6 network layer. For proof-of-concept, the NEWSKY testbed has successfully implemented NEWSKY network mobility, handover, and quality of service solutions, and tested and demonstrated them over real satellite link [1].

Also the EC FP7 project SANDRA (Seamless Aeronautical Networking through integration of Data-Links, Radios and Antennas) [2] aims at designing and implementing an integrated aeronautical communication system and validating it through a testbed and, further, in-flight trials on an A320. Central design paradigm is the improvement of efficiency and cost-effectiveness by ensuring a high degree of flexibility, scalability, modularity and re-configurability.

Whereas the NEWSKY testbed is considered to be a *proof of concept*, the SANDRA testbed will represent a *prototype* aircraft communication system, integrating prototypes developed and implemented in SANDRA, comprising AeroMACS, Integrated Modular Radio (IMR), Integrated Router (IR), and a novel Ku-band antenna.

The design of the SANDRA testbed network was already finalised and the SANDRA testbed detailed interfaces, software, and hardware specifications were started recently and are ongoing.

SANDRA focuses on the air-to-ground communication and on the development of the on-board airborne SANDRA terminal. The SANDRA terminal, in particular the IR and the IMR have to jointly implement the capabilities of resource allocation among heterogeneous link access technologies and link reconfigurability (e.g. when new links become available or previously available become unavailable, including handover between links). The required technology-dependent functions (such as control of the heterogeneous link technologies) reside in the IMR, whereas technology-independent functions are implemented in the IR, while using IP to achieve convergence and interoperability between the different link access technologies. Although the focus is on the airborne terminal, the testbed of course also has to implement a ground network side.

The main aspects of the testbed design and the testbed core concepts and components are presented in this paper, namely the security and QoS (quality of service) provision concepts for segregation of safety and non-safety domains in an integrated system, the Integrated Router, IPv6 and IPv4 internetworking.

The paper is organised as follows.

First, in Sec. 2 the NEWSKY project will be shortly described, including the study objectives and the architecture of an IPv6 network testbed for NEWSKY demonstration is also described.

The following Sec. 3 contains a short overview of the SANDRA main goals and testbed purpose. The main differences of the SANDRA testbed from the NEWSKY testbed will become evident. This includes improvements in the protocols and configuration taking into account lessons learnt in NEWSKY, additional features such as automatic modem and bandwidth control, Integrated Modular Radio (IMR), and the additional requirement that the testbed shall be integrated in an aircraft for in-flight test.

Sec. 4 focuses on presenting the overall SANDRA testbed network architecture. This includes a reference network architecture, which is independent of restrictions specific to the testbed implementation, and the actual testbed network architecture which has to take into account various constraints, e.g. unavailability of IPv6 satellite access networks.

Sec. 5 presents in some more detail the selected topics related to the SANDRA testbed which are specifically interesting aspects of implementation and investigation; IPv6 over IPv4 network traversal and header compression.

Finally, Sec. 6 gives a short summary and presents the next steps and schedule for the testbed implementation and laboratory and flight trials.

2   NEWSKY Testbed Overview

The main objective of the laboratory testbed development in NEWSKY was to implement some basic functionalities of the NEWSKY networking design, in particular with respect to mobility, security, and quality of service (QoS) aspects. However, due to the constraints within the project, not all design aspects are implemented in the testbed. This section describes the NEWSKY testbed architecture, components, and points out the differences between the design and the implementation wherever applicable.

The laboratory testbed consists of two main subnetworks: the *airborne mobile network* representing an aircraft, and the *ground network* representing a connected ATC and airline network, and the public Internet. The two main subnetworks are connected by two different data links. The network architecture is depicted in Fig.1.

The airborne network consists of the Mobile Network Node (MNN) and the Mobile Router (MR). The MNN represents the airborne user terminal, which could belong either to the cockpit (e.g. an interface for pilot-ATC communication) or to the cabin (e.g. passenger's laptop). As part of the NEMO (Network Mobility) protocol (see description below), a Home Agent (HA) is installed in the ground network. The Corresponding Node (CN) acts as the other end point of the communication, e.g. it could be the air traffic controller, or the airline. An interface to the public Internet is also implemented to emulate passenger Internet connectivity.

Further testbed components are described in the following:

Network Mobility Protocol. Mobile IPv6 (MIPv6)/NEMO [5,6] and its extensions have been selected by the International Civil Aviation Organization (ICAO) Aeronautical Communications Panel Working Group I (ACP WG-I) to be the solution for global network mobility in future IPv6-based aeronautical telecommunication network (ATN). NEWSKY took the same approach by specifying MIPv6 as its solution for mobility. Network Mobility (NEMO) protocol is introduced to extend MIPv6, to enable the mobility of a complete network instead of just a single host. The testbed uses Linux based NEMO protocol from the Nautilus6 project [14].

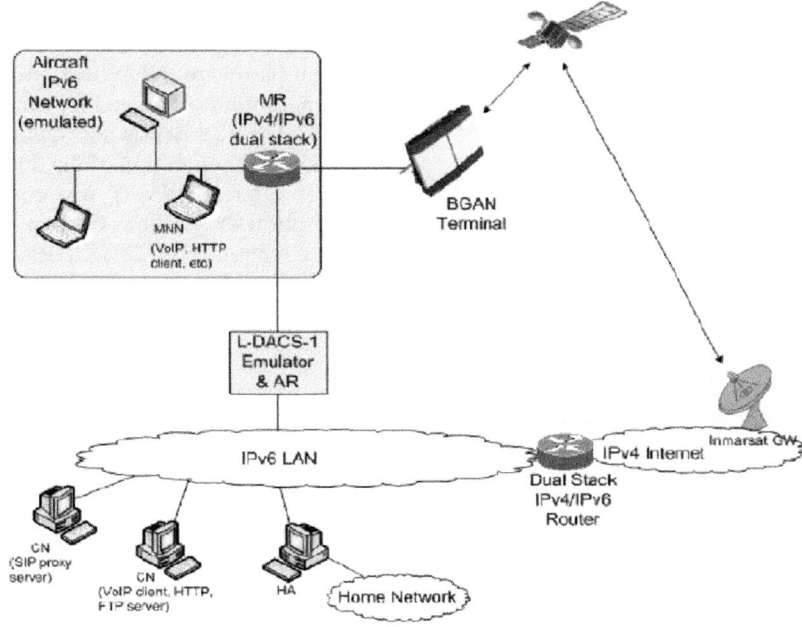

Fig. 1. NEWSKY laboratory testbed network architecture

Data Links. Two data link technologies are integrated in the testbed, namely the Broadband Global Area Network (BGAN) system from Inmarsat, and an emulated terrestrial link of a potential future air-ground data link technology, L-DACS-1 (L-band digital aeronautical communications system). The satellite link is a real physical link through the Inmarsat I4 satellites, whereas the emulator is software which adds delays to and limits the data rate of an Ethernet link, representing the physical and data-link layer of L-DACS-1.

IPv4 Network Traversal. Although NEWSKY focuses only on IPv6, in reality the Inmarsat BGAN system is based on IPv4. In the testbed, we implemented the Network Crossing via Translation protocol (NeXT) to efficiently traverse the satellite IPv4 network. NeXT isan overhead optimised NAPT-PT (Network Address and Port Translation - Protocol Translation) mechanism [10].

Applications. Future ATM communications will rely more on data services, and the usage of voice as in the current VHF-based radio system will be greatly reduced. Therefore in the testbed we implemented some representative IP-based applications, e.g. VoIP and FTP data transfer.

QoS Management. NEWSKY specifies DiffServ to guarantee the prioritisation of message coming from different applications. In the testbed this is implemented using Linux based *tc* tool to direct IPv6 packet flows into different queue classes, instead of putting traffic class marks directly to the packet as in DiffServ.

Domain Separation. For safety and security reasons, the on-board aircraft mobile network is divided into safety-critical subnetwork for ATS/AOC applications, and non-safety-critical subnetwork for APC applications. In the current airborne network this separation is achieved physically with different hardware. NEWSKY proposed a future airborne network architecture, where the separation is achieved logically using IPsec security tunnels. At the time of the project, however, this was not implemented, and a firewall is configured at the MR to enable a simpler separation of the domains.

Improvement of the testbed was continued even after NEWSKY was completed. Some of the new improvements include the integration of Multiple Care-of Address (MCoA) extension for NEMO [7], and the implementation of NEWSKY IPsec based domain separation. The testbed served as an initial proof-of-concept of the NEWSKY's future ATM networking design, and is to be further developed into a prototype within the SANDRA project.

3 SANDRA Key Elements

A key contribution of the SANDRA project to the definition of future aeronautical communication system is the integration on different system levels to overcome the drawbacks (such as limited coverage of direct A/G data links, high delay of satellite systems, etc.) of existing aeronautical communications systems based on single radio technologies and individual radio access systems:

- Integration of services from the safety and non-safety domains (ATS, AOC, AAC, APC).
- Integration of different networks, based on interworking of different radio access technologies through a common IP-based aeronautical network whilst maintaining support for existing network technologies (ACARS, ATN/OSI, ATN/IPS, IPv4, IPv6).
- Integration of different radio technologies in an Integrated Modular Radio (IMR), applying concepts for Integrated Modular Avionics (IMA) to communication avionics.
- Integration of Ku- and L-band antennas in a single hybrid Ku/L band satellite communication antenna providing an asymmetric broadband link.
- Design and implementation of AeroMACS (similar to WiMAX) for integrated multi-domain airport connectivity.

To prove all of the concepts developed in the SANDRA global system architecture the IMR will be capable of interfacing with a sufficient number of bearers, comprising Inmarsat SwiftBroadband (SBB) class 6/7, Ku DVB-S2 (2nd generation Digital video broadcast, receive only), AeroMACS, VHF Voice and VDL Mode 2 (VDL2).

Closely related to the key contributions, the SANDRA partners confirmed in Dec. 2009[2] that key requirements for SANDRA comprise

- Support of data and digital voice services for ATS, AOC, AAC and APC.
- Support of interfacing with legacy and future data links.

---

[2] At the kick-off of the SANDRA sub-projects "SP2 - Requirements and Global System Architecture" and "SP3 - Seamless Networking"

- Ensuring network interoperability comprising ACARS, ATN/OSI, IPv4, IPv6.
- Deploying IPv6 as unification.
- Compatibility with ATN/IPS.
- Performing link selection based on information from applications, policies, and link layer information.

A further key element of the SANDRA project is the testbed, which has the purpose to implement and demonstrate as many of these SANDRA key elements as possible for validation of the overall SANDRA concept and architecture up to Technology Readiness Level (TRL) 5-6[3].

The SANDRA testbed design and implementation will also take into account more specific and comprehensive requirements definitions prepared within SANDRA in various Work Packages (WP). SANDRA activities being relevant for the testbed specification are "WP3.1 - Detailed network requirements", "WP3.2 - Network architecture", "WP3.5 - IPS network design", "WP4.2 - IMR interface", "WP7.1 - identification of specific hardware available for the testbed", "WP7.2 - Use cases and scenarios to be covered by the testbed", "WP7.3 - Identification and Implementation of Applications".

According to the envisaged TRL, a central goal of the SANDRA testbed is the development of a SANDRA *prototype* (in contrast to NEWSKY's testbed for *proof-of-concept*), integrating the prototypes implemented in SANDRA (Integrated Router (IR), Integrated Modular Radio (IMR), novel Ku-band antenna, and AeroMACS) for validation&verification but also demonstration

a) in a laboratory set-up of a high-fidelity representation of the target SANDRA system, including the airborne and the ground networks, and also comprising a realistic system environment for the airborne network (i.e. an aircraft) and
b) during flight trials (planned for 2nd quarter 2013).

In addition to the SANDRA prototypes all network elements will be implemented and integrated in the testbed that are necessary to create the high-fidelity representation of the SANDRA target system (e.g., global geographically distributed networks, mobility via NEMO including extensions, security provisions, techniques for improving efficiency (e.g. Robust Header Compression/RoHC, etc.), including the 4 domains (ATS, AOC, AAC, and APC) for the ES which will host the applications and all other entities in the airborne and in the ground networks (cf. Sec. 4 and 5for details).

The detailed testbed requirements specification and implementation phase was started in Oct. 2010 and is expected to be on-going until end of 2nd quarter 2012. The subsequent integration of all components and testbed trials will continue until 1st quarter 2013.

In addition to the above stated main purpose of the testbed, further suggestions and new algorithms from SANDRA activities running in parallel to the testbed design and implementation will be considered for implementation in the testbed.

---

[3] TRL 5: components and/or breadboard validation in *relevant* environment; thus, flight trials will be performed within SANDRA. TRL 6: system & subsystem model or prototype demonstration in a relevant environment (ground or space). The highest TRL (TRL 9) corresponds to the actual system "flight proven" through successful mission operations.

Thus this paper shows work in progress and cannot provide all details of the SANDRA testbed, in particular presentation of results will have to wait until the testbed implementation is finalised and trials have been performed.

Finally, as stated above, the SANDRA testbed will be deployed in a laboratory environment and in an aircraft for flight trials.The testbed set-up for the flight trials will be subject to some restrictions, e.g. no Ku-band link will be available for the flight trials because of the too high costs for the installation and certification of the Ku-band antenna on the aircraft fuselage. However, we will not further address in the following the differences between the two testbed set-ups (laboratory and flight trials) and will not address in more detail the restrictions for the flight trials.

4 SANDRA Testbed Network Architecture Overview

4.1 Testbed Reference Network Architecture

In Oct. 2010, the SANDRA testbed network architecture was finalised in a Software Requirements Document (SRD). In this SRD, the SANDRA testbed reference network architecture is defined, which initially is independent of the restrictions specific to the testbed (e.g. no IPv6 satellite network); a schematic overview is shown in Fig.2.[4]ATC (air traffic control, safety related) ground networks and (public) Internet (non-safety related) and CN for the 4 domains are also available in other geographical regions (region 2 and 3 in reference architecture diagram)but not shown to keep the figure simpler.

Depending on geographical location, the MR connects to a respective AR. ATC regional ground subnetworks are interconnected to implement global connectivity and the same holds for the Internet and its regional subnetworks.

The ES (MNN and CN) are not addressed in detail this paper, which only deals with the IP network up to the interfaces where the ES are connected to.

Although integration of legacy data links and interoperability with ACARS, ATN/OSI, IPv4 are considered essential for SANDRA (cf. Sec. 3), in the testbed the focus is on IP networking (IPv6 and IPv4).However, integration of legacy ES and applications (VDL2, ACARS, ATN/OSI, analog voice) into the testbed is under investigation in terms of use cases/scenarios, requirements, and implementation but also effort and may be included in the testbed.

4.2 Testbed Network Architecture

In contrast to the reference SANDRA testbed network architecture presented in the previous section, the actual testbed architecture and implementation will have to take into account various constraints, such as the necessity to include IPv4 access and ground networks also for ATS/AOC because of the current unavailability of IPv6 satellite networks.

---

[4] Because NEMO is used for mobility support, the reference architecture uses the nomenclature of [6], comprising Mobile Network Nodes (MNN), Mobile Router (MR), Home Agent (HA), and Correspondent Node (CN).

The testbed ground-network will mainly be set-up at TriaGnoSys (TGS) premises. Also the airborne side will initially be set-up at TGS premises as well.

Due to limitations for the flight trials that do not apply to the laboratory set-up (e.g. no Ku-band antenna on aircraft), only a subset of the laboratory equipment of the airborne network will be used in the flight trials and installed in the aircraft.

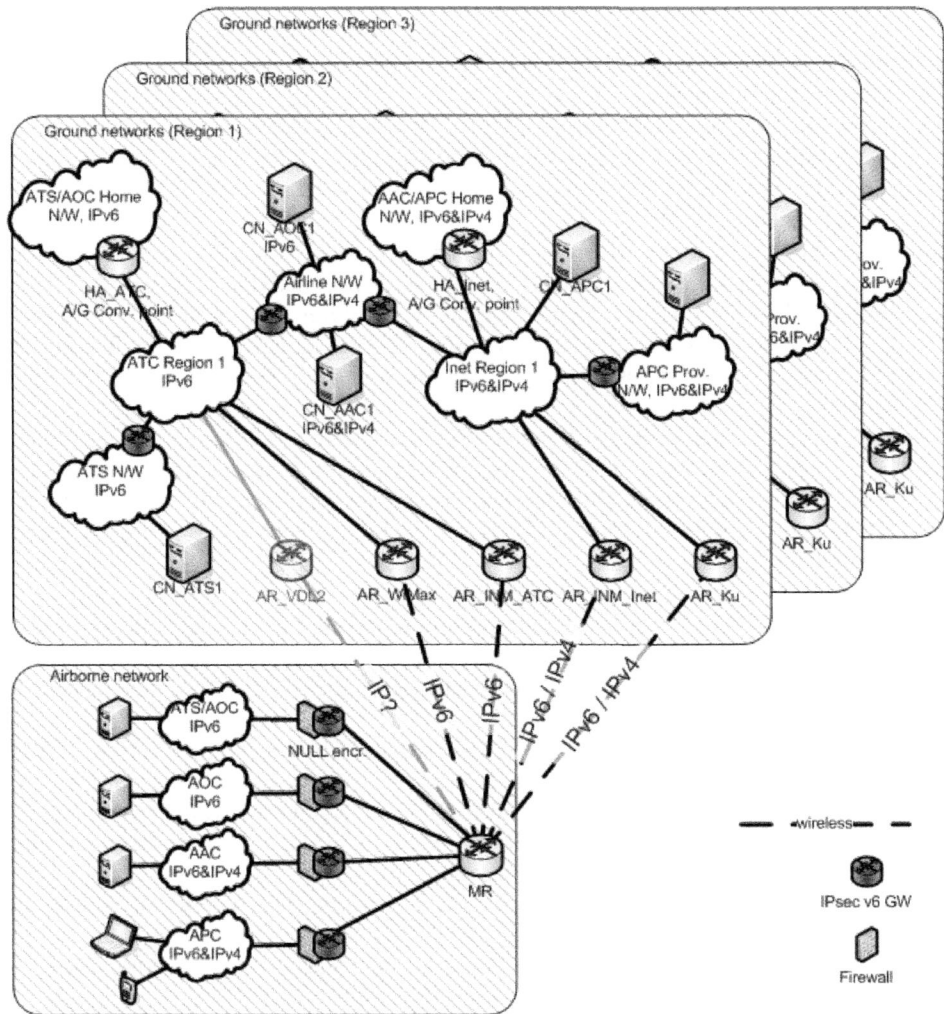

Fig. 2. SANDRA reference network architecture

Note that, as already stated above, the actual testbed network architecture will comprise IPv4 access and ground networks between the IPv6 only ATS/AOC airborne and ground networks due to the lack of support of IPv6 in the satellite networks available for the testbed implementation.

In contrast to ATS/AOC, for AAC/APC the IPv4 access and ground networks are a desired element of the testbed, because support of IPv4 hosts and access networks is required for AAC/APC.

The network nodes in the airborne network are:

- Airborne mobile network nodes (MNN) in the domains ATS, AOC, AAC, and APC; these are hosting the applications. Note that middleware and/or transition gateways may need to be deployed to integrate end-nodes that do not provide IPv6 interfaces. Note that these MNN and any middleware or transition gateways are not addressed in this paper.
- Firewalls at the entry/exits points of the 4 operational and non-operational domains to inhibit entry/exit of packets from malicious hosts (intentional or unintentional). Firewalls will not be implemented in the testbed because no penetration tests will be performed.
- IPsec (IPv6) gateways to provide authentication and integrity, with or without encryption, for safety (ATS/AOC) and non-safety related data traffic (AAC/APC). Further, IPsec enforces in the airborne network segregation of the safety domains from the non-safety domains. Also VLANs, port or tag based, are possible options to further enforce segregation but, like the firewalls, will not be implemented in the testbed.
- The Mobile Router (MR, IPv4 and IPv6) (also referred to as the Integrated Router, IR), is responsible for controlling the modems (for the fall-back solution[5]) or the IMR, resource allocation, QoS provision at IP packet level, etc. The NEMO related functionalities of the MR are specified in [6] and [7], including, e.g., registration of multiple CoAs at the HA.
- The network nodes in the ground network are:
- Access routers (AR); at least one AR is deployed for each access technology (different provider may deploy different ARs or several geographically distributed ARs are deployed together with multiple radio access stations). Access routers provide CoA to the MR.
- Home Agents (HA); at least one HA is deployed. The functionality of the HA are again specified in [6] and [7]. An HA further represents an air-ground routing convergence point, being aware of all routing paths (multiple CoAs) to the aircraft MR and forwarding traffic by policy routing to the MR on a particular routing path, e.g. based on QoS requirements.
- IPsec gateways being the counterparts of the airborne IPsec gateways securing the ATS, AOC, AAC, and APC domains.
- Correspondent Nodes (CN) in the domains ATS, AOC, AAC, APC; these are hosting the respective applications. Like the MNN, the CN are not addressed in detail in this paper.

---

[5] Already included in the SANDRA proposal is a fall-back solution *without* the IMR, in which case the IR directly interfaces with respective COTS (Commercial off- the-shelf) modems (SBB, DVB-S2, WiMAX).The fall-back solution is included to mitigate a potential risk of delay in the IMR implementation due to the high complexity of this prototype development. In this case, the IR has to implement some functionalities of the IMR, such as modem control and handover management.

In an aeronautical communication system that is deployed world-wide, MNN/MR, AR, HA, and CN are generally geographically distributed. Latencies between nodes that are close to each other are typically small, whereas latencies between nodes that are far away from each other may be significantly larger[15]. To reflect this, the testbed ground network is separated in three regions (three regions are considered sufficient for most scenarios, e.g. that AR, HA, and CN are located in at most three different regions); latencies between nodes in the same region are small, whereas latencies between nodes in different regions may be significantly larger.

Further, it is currently not foreseen to implement separate ATC and public Internet ground networks (cf. reference network architecture from previous section). However, if later definition of use cases, scenarios, and applications, not known currently, will require this, then separate ground networks will be implemented by deployment of additional routers (possibly virtual).

The resulting SANDRA testbed network architecture is shown in Fig.3.

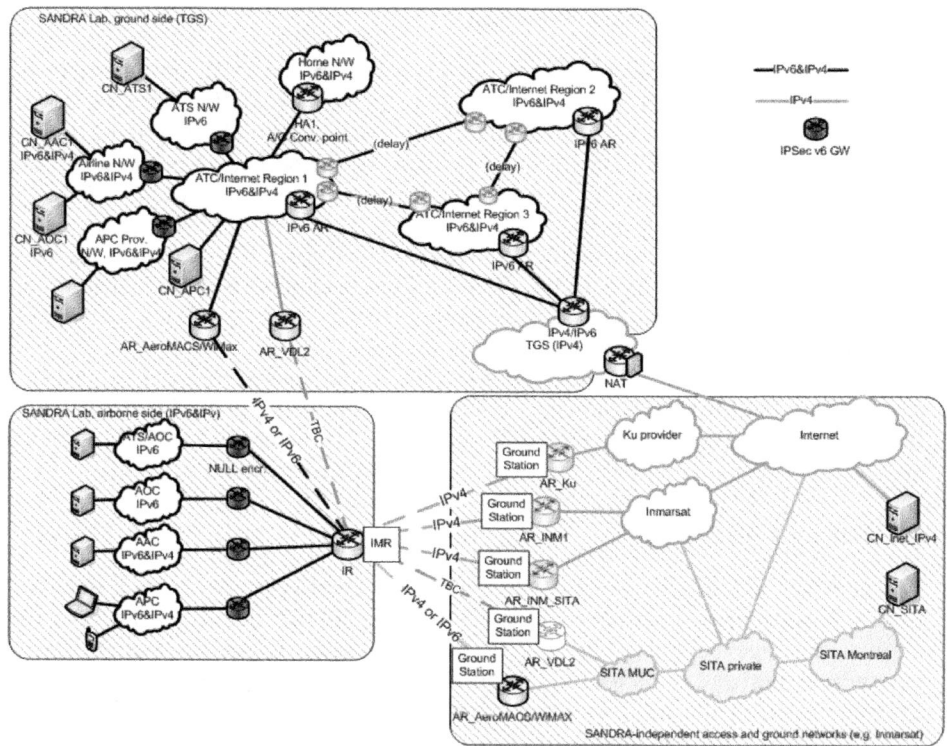

Fig. 3. Testbed network architecture overview, including IPv4 access and ground networks

As for NEWSKY, the mobility solution for the SANDRA testbed is NEMO [6], including the possibility to register multiple Care-of-Addresses [7].NEMO Route Optimisation (RO) will be considered (e.g. Global HAHA, Correspondent Router , etc.), however, implementation in the testbed will depend on an assessment of benefit and effort of the implementation, in particular if open source implementations are not yet available.

Use of multiple CoAs requires synchronisation of policy routing rules at the MR and the HA. This is addressed in a recent IETF draft [8], which also will be considered in the SANDRA testbed.

Finally, NEMO Basic Mobility Support as specified in [6] does not foresee IPv4 access networks. Two options allowing deployment of NEMO over IPv4 access networks will be considered for implementation in the testbed; first, RFC 5555 [9] and, second, a NAPT-PTbased solution [10] that was already successfully deployed in the NEWSKY testbed.

5   Selected Topics in the SANDRA Testbed

Due to the limited number of pages available for this paper we have to restrict a more detailed description to few selected topics in the SANDRA testbed.

5.1   IPv6 Over IPv4 Network Traversal

For each connected access network, the IR obtains from the IMR (or directly from a modem in case of the fall-back solution) one or more IPv6 or IPv4 CoAs (e.g. SBB allocates an IPv4 address for each primary PDP context), which are provided by the respective AR (by which protocol this is specifically done may depend on the IMR interface to the IR, the specific modem, the access network etc., which are not yet defined in the necessary detail).

In case RFC 5555 is deployed for IPv4 network traversal, IPv4 CoA are registered at the HA via a binding update [9]. IPv6 packets, including NEMO signalling, are tunnelled between MR and HA in IPv4 and a UDP header is added in case that NAT (Network Address Translation) is present on the path.

When NeXT is deployed, traversal of the IPv4 access network for IPv6 packets is achieved via IPv4/IPv6 translation. A context is set-up between NeXT master (in the IR) and slave (e.g. located at the border between IPv4 Internet and IPv6 SANDRA ground networks), mapping IPv4 CoA to an IPv6 CoA for NAPT-PT.

5.2   Header Compression

The network architecture which is driven by the security, mobility, and quality of service requirements, imposes the addition of a significant amount of header to the original message payload. Fig. 4 shows the structure of packets sent to the air interface at the egress of the IR. A quick observation shows that the amount of overhead justifies the need for implementing a header compression mechanism in particular for low bandwidth links.

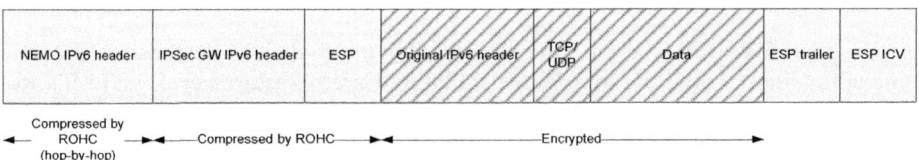

Fig. 4. Packet structure going out from airborne Integrated Router

Plenty of header compression schemes currently exist and are standardised in various IETF RFCs. For the SANDRA testbed, Robust Header Compression (RoHC) [4] is chosen because of its flexibility and the well-known good performance also over long-delay, error prone links, such as satellite links. Further, RoHC is the header compression scheme considered in the ICAO ATN IPS SARPS (ICAO Standards and Recommended Practices) for optional implementation in IP nodes [16].

RoHC could generally be deployed between several instances[6]

- between the IR and the HA to reduce overhead of the IPv4 and IPv6 packets tunnelled by NEMO. Note: compression gains, respectively, implementation complexity is significantly affected by IPsec deployment, in particular if encryption is used (cf. next bullet)
- if IPsec is deployed the relatively new RoHCoIPsec related RFCs could be implemented [11,12,13] and deployed in the respective IPsec gateways,
- between the IR and the AR, e.g. within a PPP context, if supported by the AR.

RoHC requires access to IP and TCP/UDP header for compressing packets. When these headers are encrypted (e.g. when using IPsec ESP) the compression is not possible. RoHC has defined a profile for ESP, but it only enables the compression of IP and the 8-Byte ESP header. Moreover, the compression works only in a hop-by-hop basis. The encrypted inner header remains untouched, as indicated in Fig. 4.

A framework for integrating RoHC over IPsec (RoHCoIPsec) is defined in [11]. It defines the modification required in IPsec protocol stack to allow the compression of IPsec encrypted packets.

Further investigations will be performed in the frame of the testbed development to assess in detail whether the expected compression gain obtained by implementing RoHCoIPSec justifies the implementation effort. Also alternatives to RoHCoIPsec will be investigated, such as Multilayer-IPsec [17].

Due to regulatory constraints, ANSPs (Air Navigation Service Providers) do not allow data traffic in their network to be encrypted. This puts ATS, AOC traffic in a particular situation, where IPsec is used for ensuring domain separation, authenticity, and integrity of the packets, but not their confidentiality. As the inner IPv6 and transport layer header are visible, a dedicated RoHC profile could be defined to compress all the headers except the outermost one, as it is needed for routing.

5.3 ATS and AOC Data Traffic

Performance assessment in the testbed, not only for the assessment of RoHCoIPsec, requires realistic ATS, AOC, AAC, and APC data traffic.

For ATS and AOC, the services defined in the COCR (Communications Operating Concept and Requirements for the Future Radio System) will be used[3]. A challenge is to obtain a communication traffic profile per aircraft (i.e. packet length, and frequency of occurrence), imposed by the COCR messages to the network. A realistic estimate needs to take into account the COCR service instance, which states, for example, that a message exchange shall take place once per sector, once per domain in airport, etc.

---

[6] However, nesting RoHC is not recommended and it needs to be investigated how this issue can be addressed in the SANDRA testbed.

Our COCR traffic estimate is obtained from the methodology and tool developed in the framework of the EC project ANASTASIA (Airborne New and Advanced Satellite Techniques & Technologies in A System Integrated Approach) and ESA Iris programmes. The methodology uses information from a global flight route simulation. The obtained information, e.g. average flight duration, average flight duration in a domain, etc., is used to approximate and normalise the COCR service instances, thus simplifying the complex multi-service COCR traffic.

The methodology has been presented to the (satellite) ATM community several times, including Eurocontrol and NexSAT meetings in particular, and received comments have been continuously used to improve methodology and tool. Theaverage traffic intensity of COCR messages (in bit-per-second per aircraft) will be used, e.g. in the context of deriving RoHCoIPSec compression gain.

6 Summary and Next Steps

This paper has presented a detailed overview over the SANDRA testbed, including a comparison with its precursor, the NEWSKY testbed.

It was stated that the main difference between the two testbeds is that the SANDRA testbed will represent a *prototype* of a future aeronautical communications system, whereas the NEWSKY testbed resembled a *proof-of-concept.*

This has quite far reaching consequences in terms of complexity of the testbed design and implementation efforts and also for verification&validation and trials, which need to be performed in a relevant environment, comprising integration on an aircraft and performing in-flight trials.

In contrast to the NEWSKY proof-of-concept, the SANDRA prototype needs a high degree of automatism for QoS provision, dynamic link selection, and resource allocation, etc. These functionalities are jointly implemented in SANDRA by the Integrated Router and the Integrated Modular Radio, which both were not present in NEWSKY.

Detailed design and implementation of the SANDRA testbed components including the SANDRA prototypes will continue until 2nd quarter 2012. Subsequently, all testbed components will be integrated, followed by extensive laboratory trials (until 1st quarter 2013) to verify&validate the SANDRA testbed and to investigate the performance of the deployed prototypes and protocols.

After successful integration of the testbed, flight trials on an Airbus A320 will be performed in the 2nd quarter 2013 to assess performance in a real-world scenario.

Of course, we intend to publish the final testbed design and trials results in future publications.

References

1. Fazli, E.H., Via, A., Duflot, S., Werner, M.: Demonstration of IPv6 Network Mobility in Aeronautical Communications Network. In: ICC 2009, Dresden, Germany, http://newsky-fp6.eu/pdf_papers/proceedings/ ICC2009_NEWSKY.pdf (last visited January 11, 2011)
2. http://www.sandra.aero (last visited January 11, 2011)

3. EUROCONTROL/FAA, Communications operating concept and requirements for the future radio system, COCR Version 2.0 (May 2007)
4. Pelletier, G., Sandlund, K.: RObust Header Compression Version 2 (RoHCv2): Profiles for RTP, UDP, IP, ESP and UDP-Lite. RFC 5225 (April 2008)
5. RFC 3775, Mobility Support in IPv6 (June 2004)
6. RFC 3963, Network Mobility (NEMO) Basic Support Protocol (January 2005)
7. RFC 5648, Multiple Care-of Addresses Registration (October 2009)
8. IETF Internet Draft Flow Bindings in Mobile IPv6 and NEMO Basic Support draft-ietf-mext-flow-binding-10.txt (September 14, 2010)
9. RFC 5555, Mobile IPv6 Support for Dual Stack Hosts and Routers (June 2009)
10. Via, A., Fazli, E., Duflot, S., Riera, N., Werner, M.: IP Overhead Comparison in a Test-bed for Air Traffic Management Services. In: Proceedings of the Vehicular Technology Conference 2009, VTC 2009 (2009)
11. RFC 5856, Integration of Robust Header Compression over IPsec Security Associations
12. RFC 5857, IKEv2 Extensions to Support Robust Header Compression over IPsec
13. RFC 5858, IPsec Extensions to Support Robust Header Compression over IPsec
14. http://www.nautilus6.org/ (last visited January 11, 2011)
15. Ayaz, S., Bauer, C., Arnal, F.: Minimizing End-to-End Delay in Global HAHA Networks Considering Aeronautical Scenarios. In: Proceedings of the 7th ACM International Symposium on Mobility Management and Wireless Access (MobiWAC 2009), Tenerife, Canary Islands, Spain, pp. 42–49 (2009)
16. ICAO Doc 9896, Manual on the Aeronautical Telecommunication Network (ATN) using Internet Protocol Suite (IPS) Standards and Protocols (1st ed.) (2010)
17. Zhang, Y.: A Multilayer IP Security Protocol for TCP Performance Enhancement in Wireless Networks. IEEE Journal on Selected Areas in Communications 22(4), 767–776 (2004)

Common RRM in Satellite-Terrestrial Based Aeronautical Communication Networks

Muhammad Ali, Kai Xu, Prashant Pillai, and Yim Fun Hu

Schools of Engineering Design and Technology, University of Bradford
Bradford, UK
{m.ali28,k.j.xu,p.pillai,y.f.hu}@bradford.ac.uk

Abstract. This paper presents a collaborative radio resource management (CRRM) scheme to support seamless aeronautical communications using satellite and terrestrial access technologies. The CRRM adopts and extends the IEEE 802.21 Media Independent Handover (MIH) framework and the ETSI Broadband Satellite Multimedia (BSM) SI-SAP concept to split the CRRM functions between the upper layers (layer 3 and above) and the lower layer (link layer and physical layer) of an aircraft terminal. Upper layer functions are managed by an integrated router (IR) on-board the aircraft and lower layer functions are provided by an on-board integrated modular radio (IMR) consisting of heterogeneous radio access technologies. A joint radio resource manager (JRRM) provides the abstraction layer for mapping higher layer functions into lower layer functions to enable collaboration. The CRRM scheme and its associated general signaling procedures are described in detail. Through the CRRM scheme, the connection establishment functions and seamless handovers between different radio technologies are performed by combining MIH primitives and BSM primitives. Analytical time-delay analysis is carried out to evaluate the signaling delay for connection establishment and handover procedures.

Keywords: Aeronautical networking, BSM, MIH, CRRM/JRRM, Handover, AeroMACS, DVB-S2.

1 Introduction

The EU Project SANDRA (Seamless Aeronautical Networking through integration of Data-Links, Radios and Antennas) [1] aims to design, specify and develop an integrated aircraft communication system primarily for air traffic management to improve efficiency and cost-effectiveness in service provision by ensuring a high degree of flexibility, scalability, modularity and reconfigurability.

The SANDRA system is a 'system of systems' addressing four levels of integration: Service Integration, Network Integration, Radio Integration and Antenna Integration. From the communications network point of view, SANDRA spans across three segments, namely, the Aircraft segment, the Transport segment and the Ground segment, as shown in Fig. 1. The Aircraft segment consists of three main physical components: the Integrated Router (IR), the Integrated Modular Radio (IMR) and the

G. Giambene and C. Sacchi (Eds.): PSATS 2011, LNICST 71, pp. 328–341, 2011.

Antennas. These three components form the SANDRA terminal. The IR is responsible for upper layer functionalities, such as routing, security, QoS and mobility. The IMR takes care of lower layer radio stacks and functions including radio resource allocation, QoS mapping and adaptation functions. Through Software Defined Radio (SDR) [2] the IMR supports dynamic reconfigurability of operations on a specific radio link at any time and provides the flexibility for accommodation of future communication waveforms and protocols by means of software change only. The physical separation between the IR and the IMR has the advantage of increased modularity and identifying distinct management roles and functions for higher layer and lower layer components with IP providing the convergence. The Antennas include a hybrid Ku/L band Integrated Antenna (IA), a VHF antenna and a C-band antenna. The IA is a hybrid Ku/L band SatCom antenna to enable an asymmetric broadband link. The various end-systems i.e. Air Traffic Service (ATS), Airline Operation Centre (AOC), Airline Administrative communication (AAC) and Aeronautical Passenger Communications (APC) [3] are all connected to the IR.

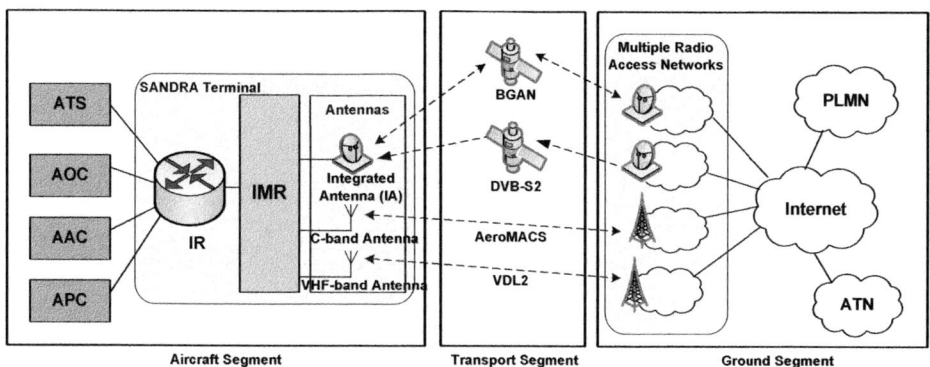

Fig. 1. SANDRA Network Architecture

In the Transport segment, four radio transport technologies are considered, namely, VDL mode 2 [4] in VHF band, BGAN [5] in L-band, DVB-S2 [6] in Ku-band and AeroMACS - a WiMAX [7] equivalence for aeronautics communications - in C-band.

The Ground segment consists of multiple operators; multiple Radio Access Networks (RANs) and their corresponding core networks, the Aeronautical Telecommunication Network (ATN), the Internet and possibly the Public Land Mobile Network (PLMN, for passenger communications). The RANs can also be connected directly to the ATN and the PLMN on the ground. In order to provide mobility and security services for aeronautical communications, functional components such as the mobility server, security and authentication server are required in the ground segment to provide corresponding mobility and security information services. These components will be provided by the ATS/AOC/AAC and APC service providers of the ATN on ground.

This paper presents a functional architecture of the SANDRA terminal for radio resource management (RRM) and an approach to partition the functional entities between the IR and IMR for the configuration and reconfiguration of radio links during the connection establishment and handover.

2   The SANDRA RRM Architecture

Radio resource management includes network based functions and connection based functions. Network based functions encompass admission control, load control, packet scheduling and radio bearer allocation. Connection based functions include handover control and power control. Handover control function handles handover initiation, decision and execution processes to ensure call/session continuity.

The SANDRA project concentrates on radio link selection decision among the four heterogeneous radio access technologies during admission control, packet scheduling and handover processes. Radio bearer allocation functions for individual radio link access technologies will not be considered as this is outside the scope of the project. More specifically, this paper describes a RRM architecture framework where radio link selection decisions are made collaboratively between the IR and the IMR.

To provide efficient RRM among heterogeneous networks, QoS and routing related parameters have to be considered along with the real-time link conditions and characteristics. A collaborative RRM mechanism is derived to monitor and manage the resources available in the different radio links to ensure that the application QoS/SLA requirements are met during connection establishment and handover processes. Fig. 2 shows the collaborative RRM functional and protocol architecture of the SANDRA terminal.

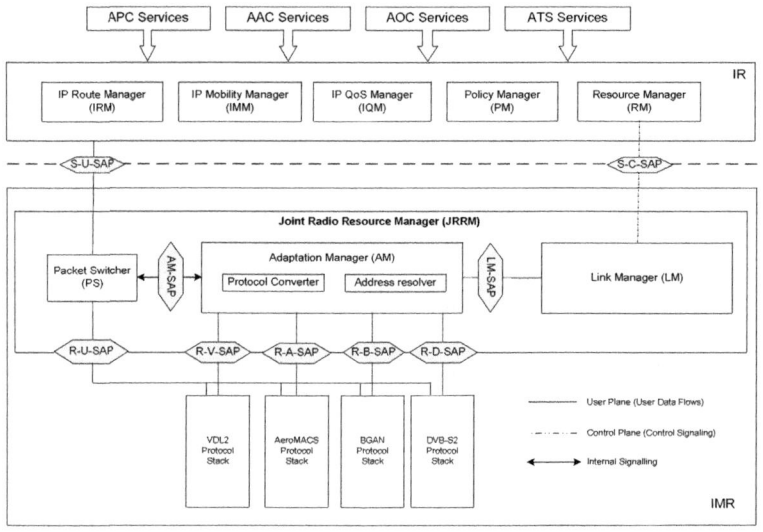

Fig. 2. RRM Functional Architecture

From the protocol stack point of view, the application layer supports APC, AAC, AOC and ATS services. The IR represents the network layer of the OSI stack and contains the following functional entities:

- IP Route Manager (IRM): This entity provides IP routing and IP address management functions. It interacts with the IP Mobility Manager (IMM) to enable mobility management functions to be carried out. For NEMO [8] support, the IRM will manage the IPv6 and IPv4 prefixes assigned to the mobile network. It is also responsible for setting up Mobile IP tunnels to the correspondent node and reverse tunnels to the Home Agent using IPv6 encapsulation.

- IP QoS Manager (IQM): This entity maps application layer QoS parameters onto to those in the IP layer. It also performs packet classification, packet buffering and packet scheduling functions.
- IP Mobility Manager (IMM): This entity enables the SANDRA terminal to roam from one network domain to another while maintaining session connection during handover when there is a change in the point of attachment of the terminal. Specifically, the IMM is responsible for neighbor discovery, router advertisement, mobility header management and binding update processing. It is also responsible for handling multihoming and network mobility (NEMO) functionalities.
- Policy Manager (PM): The PM manages and maintains a database of the flow-specific policies or rules that specify the traffic flow characteristics (e.g. QoS requirement, cost, security level, etc.) that the radio links may need to meet. These policies may also contain preferences of radio links for individual flows. These rules will facilitate the link selection decision for traffic flow with policy constraints.
- Resource Manager (RM): This entity monitors the availability of the different radio links, any reserved resources on them and maintains a view on the different IP traffic flows. It identifies whether there is a need for more resources based on the type of session requests or on the current status of the IP queues.

The IMR representing the data-link and the physical layer of the OSI stack consists of four different radio protocol stacks. A SANDRA specific Joint Radio Resource Manager (JRRM) is also located here. It is responsible for managing and controlling the resources made available to the underlying radios in a uniform and consistent manner and provides a single common interface between the IR and the IMR. Functional entities included in the JRRM are the Adaptation Manager (AM), the Link Manager (LM) and the Packet Switcher (PS). Both the AM and the LM are control plane entities whereas the PS is a user plane entity.

The AM supports protocol mapping for protocol conversion between the IR and the IMR and address resolution functions for mapping between network layer identities onto link specific identities. In addition, it also carries out a switching function equivalent to the MIH Function (MIHF) for handover to support handover services including the Event Service (ES), Information Service (IS), and Command Service (CS), through service access points (SAPs) defined by the IEEE 802.21 MIH [9] working group.

The LM performs link selection and link configuration functions and together with the RM in the IR forms the Connection Manager (CM). The CM as a whole acts as a MIH user in the MIH framework.

The PS is responsible for switching data packets received from the IR in the user plane to the destined radio modules according to a packet switching table generated and passed by the address resolver in the AM during connection establishment. The packet switching table essentially contains the mapping of the QIDs [10] defined in the BSM SI-SAP concept onto different radio link identifiers (Link IDs). As a result, each data packet can be switched directly to the radio modules without passing through the AM in the user plane.

3   SANDRA Collaborative RRM Mechanism

The collaborative RRM mechanism defined in SANDRA considers collaborative connection management, collaborative QoS management and collaborative mobility management in relation to admission control, packet scheduling and handover functions as described in the previous section through cross-layer collaboration between functional entities in the IR and the IMR and using the session concept. While the IR is responsible for managing the network layer connections, the IMR is responsible for the link layer connections. The collaborating entities, the RM and the LM, are grouped into a single cross-layer entity - the CM. The relationships of the CM with the PM, IQM, IMM and AM is depicted in Fig. 3.

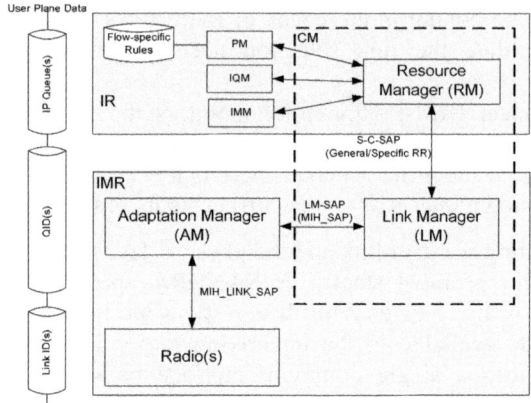

Fig. 3. Cross-layer collaborative connection management

3.1   The SANDRA Sessions Concept

The SANDRA Sessions mechanism is used for connection management and connection bindings between the IR and the IMR in order to provide efficient RRM and to meet the QoS requirements. Sessions are used to map corresponding data queues within the IP layer and the link layer.

In SANDRA, the term "session" corresponding to a single connection is between a given IP queue and a link-layer queue. As shown in Fig. 4, there is a one-to-one mapping between the IP queues and the sessions. In BSM [10], the QID is used for mapping IP queues to BSM queues. This QID concept has been adopted and extended within SANDRA to identify the IP queues in the IR and the corresponding session with the IMR. Hence, the sessions will be represented by a QID, which is unique for every session.

Every session between the IR and the IMR will have a certain QoS profile based on the desired application requirements that are being carried by the session. It is important that the QoS parameters of the IP queue are mapped to the session QoS and then further mapped to the QoS of the link-layer queue.

Different sessions are required for the different support radio links within the SANDRA system. Hence, if two active radio links are present, then at least two sessions are required to carry data over the two radio links, i.e. one for each radio. In other words, the same session cannot be used for carrying data over two radio links.

Radio technologies like VDL2 and DVB-S2 only provide single type of radio bearers and hence limited RRM functionalities. Only a single type of radio connection may be established between the terminal and the ground infrastructure of these radio technologies at any given time. Therefore within SANDRA, only a single session is required between the IR and the IMR for these radio technologies. Newer radio technologies like BGAN and AeroMACS support multiple radio bearers of different QoS simultaneously. As such, multiple sessions may be present between the IR and the BGAN and AeroMACS radio stacks.

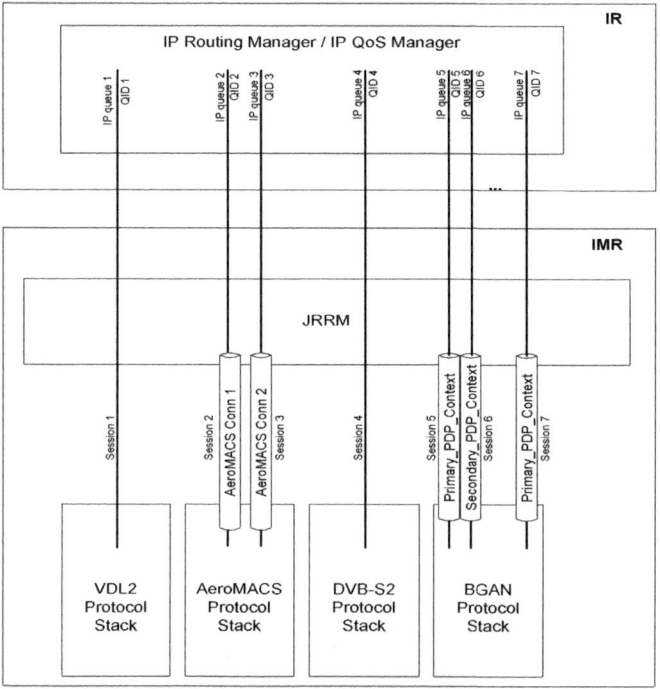

Fig. 4. The SANDRA Sessions concept

A session can be general or dedicated, depending on whether the radio request (RR) is a general RR or a specific RR. A general session means that a session is established without specific restrictions apart from satisfying the QoS requirements. In this case, the LM can make decisions on link selection - to select the most suitable link available which can satisfy the requirements specified in the RR. In addition, user traffic from multiple applications can be transmitted on the same session. The RM in the IR will make the decisions on how to use these general sessions to transmit user data. Thus, the IR and IMR make collaborative decisions on the session establishment. The former decides on the way how user data from multiple applications will be transmitted on different sessions whereas the latter decides on the most suitable link access technology for the requested session. A dedicated session is established for a specific application; only data from this specific application can be transmitted on the requested session. If a RR from the IR includes policy based routing decision, for example a dedicated radio link should be used for this specific RR due to regulatory or geographical constraints, the LM in the JRRM of the IMR should try to set up the session on the specified radio link. In this case, the IR makes decision on which link access technology should be used and informs the IMR through a specific RR so that the IMR can collaborate by allocating resource on the specified radio link for the requested session.

3.2 Collaborative Connection Management

Connection management functions including connection establishment, connection termination and connection modifications are carried out by the CM that enables decisions on network layer and link layer connections to be established by the RM in the IR and the LM in the IMR collaboratively.

As shown in Fig. 3, the PM is connected to a database of the application flow-specific rules and policies. These policies may govern the decision on which radio links can be used for different applications. Such policies may be based on the type of applications, type of planes, the location of the plane and flight path, etc. It may also be based on security and regulatory requirements that may restrict the applications to be transported over a specific radio link. Applications that do not have strict requirements specified by the policy manager may be transported over any one of the available links that satisfies the application QoS.

Upon reception of a session request specifying the application QoS requirement, the RM will decide whether it is a general session request or a dedicated session request by checking with the PM for any flow-specific rules or policies with which the requested applications must comply. Such policies may restrict the choice of radio links for its transportation. There may be some applications that do not explicitly send a request to the IR specifying the required QoS but may start directly sending the application data to the IR. The Resource Manager would be responsible for managing the resources required for such traffic also. As such, the RM carries out the first level of decision making. It is responsible for deciding when new resources are required, when resources are released, etc. It will also perform link selection decision upon receiving dedicated session requests.

The LM responsible for controlling the radio links and performs the second level of RRM related decision making for connection establishment. In the case of general session request, the LM performs suitable link selection by mapping the application QoS requirements onto the resource availability and the quality of the available links. The radio link that can most satisfy the QoS requirements will be selected and a session between the IR and the selected radio link will be established.

3.3 Collaborative QoS Management

In relation to satisfying the QoS requirements upon a service request, the IQM in the IR will control and manage the IP Queues. On receiving data from the higher layers, the IP QoS manager performs packet classification based on the type of application and perform packet marking using Diffserv codepoints. Codes corresponding to the QoS requirements are added to the IP header of each packet before sending it to the IMR. The IR also performs packet level scheduling of all incoming application packets based on their QoS requirements. The IR sees the different sessions between the IR and the IMR as different data tunnels though which different data needs to be sent. Application data as a result of a dedicated session request will be sent over a dedicated session, otherwise they can be sent over any available sessions that may satisfy its QoS requirements.

The IMR needs to be able to also setup appropriate link-layer connections that meet the desired QoS that is requested by the IR. This requires mapping the higher layer QoS parameters to the link-specific QoS parameters. If the radio link network cannot meet the desired QoS then another suitable link may be selected that could satisfy the QoS. If none of the available radio links is able to meet the desired QoS then the session request is rejected. The IR may then re-issue the resource request with the modified QoS parameters.

The IP QoS Manager in the IR is responsible for monitoring the IP queues to make sure that there are no packet drops within the system. The Packet Switcher in the IMR is also responsible to monitor any packet drops. These performance metrics need to be reported to the management Unit in the IR via the management plane.

When the existing sessions are not able to satisfy the QoS needs of the application, then new session may be setup or additional resources may be requested on the existing radio links. This would require QoS re-negotiation with the ground networks.

3.4 Collaborative Mobility Management

The SANDRA system supports multihoming where the IR can be connected to multiple ground networks via different radio links at any given time. Due to location constraints, handover support across different radios is required. For example, the AeroMACS radio technology would be primarily available only at the airports during taxiing, taking-off and landing whereas satellites will be the primary means for communications when the airplanes are at cruising attitude. In addition, an airplane may move out of coverage of a given satellite link and then enter into another. The fast movement of the airplanes presents another complexity for mobility management in terms of handover.

In SANDRA, NEMO will be used by the IR for providing local and global mobility solutions and seamless mobility across the different networks. The IR and the IMR work in a collaborative manner to provide a cross layer mobility management solution. The IR may request the IMR to handover sessions from one radio link to another if there are some rules that dictate that different links may be used by an application during different phases of the flight. The IMR will also periodically monitor the link conditions and if it detects that a given link is no longer available then it will initiate the handover procedure. In the case of a general session, the Link Manager will select another suitable active link that may already be active for other general sessions. The Link Manager will then handover the old link to the new link and informs the IR about the handovers. The IR may then initiate the NEMO/Mobile IP binding updates to the ground networks. In case a special session is already active on this link, the LM will inform the IR about this session so that the Resource Manager may perform suitable link selection for this session.

4   RRM Signaling Procedures

Attempts have been made to construct the message sequence charts (Fig. 5 and Fig. 6) to demonstrate how MIH primitives can incorporate BSM SI-SAP primitives for general session establishment (link selection by LM) and mobile controlled handover. From the figures, the BSM SI-SAP primitives are shown as the signaling messages carried over the interface between the IR and IMR. These SI-SAP primitives will trigger a sequence of MIH link independent primitives, which will further trigger the link dependent primitives.

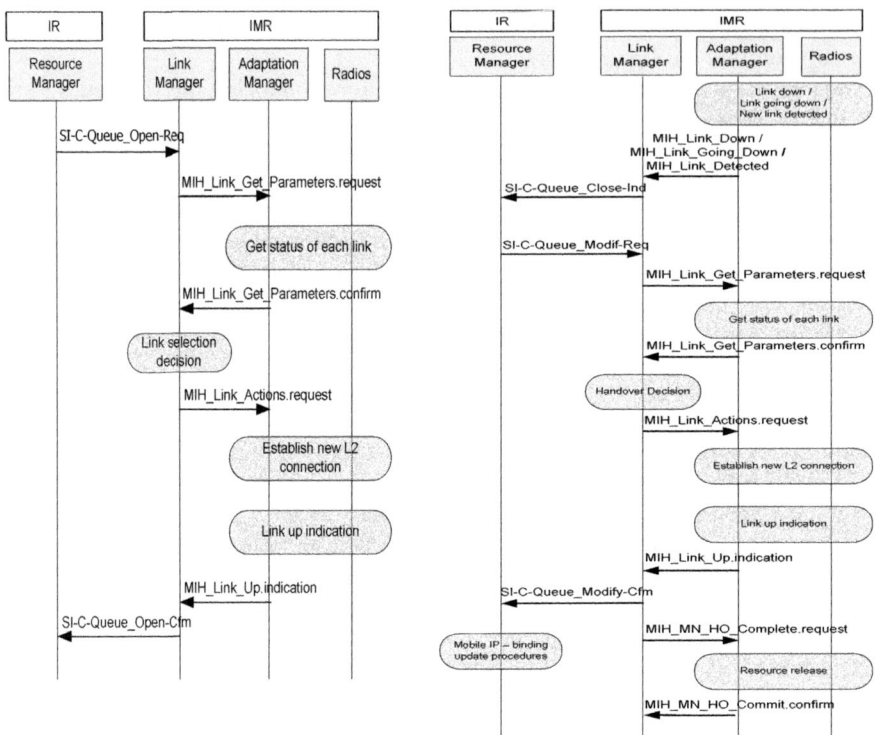

Fig. 5. General session establishment   **Fig. 6.** General Mobile controlled handover

From Fig. 5, the resource request in a new session establishment procedure is handled by the ETSI BSM SI-C-Queue_Open-Req primitive that demands specific QoS requirements to be fulfilled by the IMR link setting. Upon reception of this primitive, the IMR makes use of MIH primitives to check the link status of each available radio technology then perform the link selection function to establish L2 connection on the selected radio technology. Finally, the ESTI BSM

SI-C-Queue_Open-Cfm primitive is used by the IMR to confirm the establishment of L2 connection with the IR.

Fig. 6 presents the L2 connection establishment procedure for handover using the ETSI BSM SI-C-Queue_Modify-Req primitive that indicates a new queue modify request due to the unavailability of resources on a given link or the detection of a newly available link that triggers a handover event. Consequently, QoS re-negotiation is required on the new link. This phase is then accomplished by making use of both ETSI BSM and MIH primitives as can be seen from the first three signaling message exchanges between the IR and the IMR.

5 Performance Analysis of RRM Procedures

To perform the delay analysis the links and interfaces between different network components shown in Fig. 5 and Fig. 6 have been considered. All messages involved in the procedure like connection establishment and handover have been taken into account in calculating the different delay components. In general, the total time taken to transmit a single message over any given link, D_{Total} can be expressed as the sum of four delay components [11, 12]:

$$D_{Total} = D_{Prop} + D_{Proc} + D_{Trans} + D_{queue} \qquad (1)$$

Where, D_{Prop} is the propagation delay, D_{Proc} is the processing delay, D_{queue} is queuing delay and D_{Trans} is the transmission delay. The general queuing delay D_{queue} for any network entity, based on an M/M/1 queuing model can be expressed as $D_{queue} = \frac{1}{(\mu C - \lambda)}$ where μ is the service rate, C is communication channel capacity and λ is the arrival rate [13, 14].

In both the session establishment and handover procedures, signaling messages are exchanged between different entities of the IMR, between the IR and the IMR, and over the radio links. Signalling messages exchanged between entities within the IMR, although not shown in Fig. 5 and Fig. 6 will only account for the processing delay, $D_{proc.}$ Message exchanged between the IR and the IMR contributes to the total delay over the wired link as the IR and the IMR are connected via the Ethernet and that over the selected radio link also involved all four delay components as indicated in equation (1).

In evaluating the total delay for the two procedures, the values as shown in Table I apply. The channel capacity is assumed to be 45 Mbps for AeroMACS [15], 492Kbps for BGAN [5], 80 Mbps for DVB-S2 [16] and 100 Mbps for the wired ethernet [17]. Propagation delays of 250 ms for Geo satellites, 167 μs for AeroMACS, and 5 ns for wired Ethernet 1 meter long cable are assumed. The transmission delay has been derived by dividing the average packet size with data rate of particular technology. For example the size of an average packet being transmitted over wired link is 55 bytes and the data rate of an Ethernet link is 100 Mbps, therefore by converting packet length in to bits and dividing it by data rate gives a transmission delay of 0.0000044 sec or 4.4 μsec. The number of messsges, K_{sel}, exchanged over the selected link for session establishment is 6 for BGAN, 9 for AeroMACS and 2 for DVB-S2 [18]. In terms of session establishment, the total delay is expressed as follows:

$$D_{Total} = P(D_{Proc}) + 2\left\{D_{Prop} + D_{Trans} + D_{Proc} + \left(\frac{1}{\mu C_i - \lambda_{wired}}\right)\right\} +$$
$$(K_{sel})\left\{D_{Prop} + D_{Trans} + D_{Proc} + \left(\frac{1}{\mu C_j - \lambda_{wireless}}\right)\right\} \tag{2}$$

From [18], total number of messages represented by P, exchanged between entities within the IMR and their associated delay is represented by the first term on the right hand side of equation (2). Value of P is 26 for BGAN, 16 for AeroMACS and 6 for DVB-S2. The second term denotes the total delay between the IR and the IMR where there are two message exchanges as shown in Fig. 5. The third term denotes the total delay over the selected radio link, where K_{sel} denotes the number of messages required for session establishment over the selected wireless link. C_i and C_j denote the capacity of wired and wireless communication channel.

Similarly, the signaling delay for handover shown in Fig. 6 has been represented by equation (3). The symbol Q in the first term at the right hand side of equation (3) represents total number of messages exchanged within IMR entities. Value of Q is 34 for BGAN, 30 for AeroMACS and 22 for DVB-S2.

$$D_{Total} = Q(D_{Proc}) + 3\left\{D_{Prop} + D_{Proc} + D_{Trans} + \left(\frac{1}{\mu C_i - \lambda_{wired}}\right)\right\} +$$
$$\left[(K_{Sel})\left\{D_{Prop} + D_{Trans} + D_{Proc} + \left(\frac{1}{\mu C_j - \lambda_{wireless}}\right)\right\}\right] \tag{3}$$

Table 1. Parameter value chart

No.	Parameter Name	Symbol	Value
1	Arrival rate at wireless interface	$\lambda_{wireless}$	10 to 95 s^{-1}
2	Arrival rate at wired interface	λ_{wired}	10 to 95 s^{-1}
3	Service rate	μ	100 s^{-1}
4	Wireless channel capacity	C_j	45,0.492, 80 Mbps
5	Wired channel capacity	C_i	100 Mbps
6	Wireless link Propagation delay	$D_{Prop.}$	250ms, 167µs
7	Wired link propagation delay	$D_{Prop.}$	5ns
8	Wireless transmission delay	$D_{Trans.}$	293µs,2.4µs, 23µs
9	Wired transmission delay	$D_{Trans.}$	4.4 µs
11	Average packet processing delay	$D_{Proc.}$	5 ms
12	Number of messages exchange required between mobile node and network entity for session setup.	$K_{sel.}$	6(BGAN), 9(AeroMACS), Approx. 2(for DVB-S2 receiver synchronization)
13	Average packet length	BGAN	18 bytes
		DVB-S2	24 bytes
		AeroMACS	129 bytes
		Wired link	55 bytes

In the delay analysis the attributes of AeroMACS have been used for AeroMACS. Fig. 7 and Fig. 8 show the total signaling delay during session establishment and during seamless vertical handover respectively.

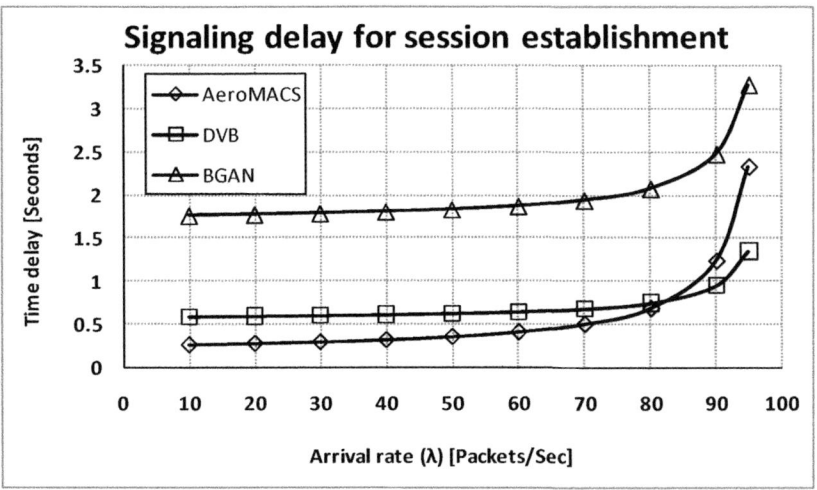

Fig. 7. Signaling delay for new session establishment on different technologies

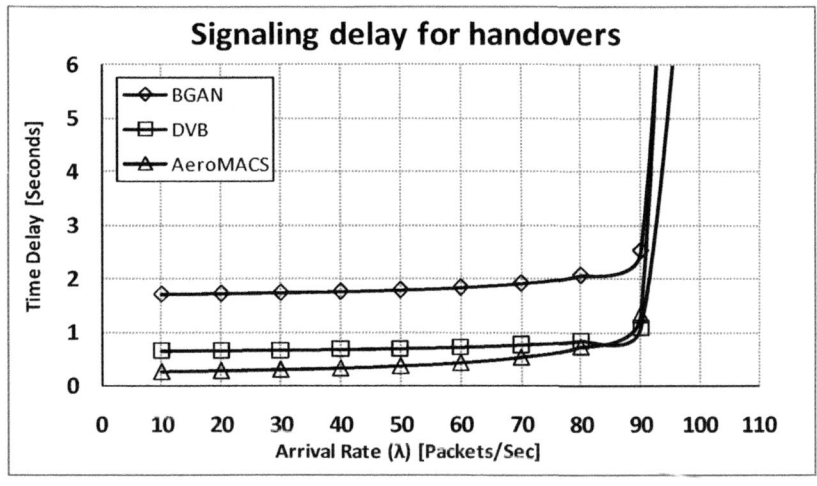

Fig. 8. Signaling delay to handover to different technologies

It can be seen from both figures that an increase in the arrival rate will cause an increase in the total signaling delay as a result of an increase in the queuing delay D_{queue}. Graphs in Fig. 7 illustrates AeroMACS exhibits the lowest delay for session establishment as its propagation delay is small and data rate is high. DVB-S2 has higher data rate than AeroMACS but incorporates high propagation delays. BGAN has the lowest data rate of 492 kbps and high propagation delays therefore it exhibits the highest total delay values in the graphs. The graphs also show that high data rate provides better results for high arrival rate. For example, the total delay for AeroMACS session establishment becomes more than that for DVB-S2 when the arrival rate goes beyond around 82 packets/sec. Similarly in case of handover DVB-S2 is proven better for higher arrival rate otherwise AeroMACS shows better results, having the lowest total delay values.

6 Conclusion

This paper presents the collaborative RRM mechanisms for an aeronautical communication network. It describes the sessions and QID concepts and shows how BSM and MIH standards are extended to enable signaling exchanges for the CRRM mechanism across different radio technologies. The paper also presents the RRM signaling procedures and the analytical model to measure the signaling delay for the different procedures. The results show that DVB-S2 offers more bandwidth and is more tolerant to an increase in arrival traffic. BGAN having lowest data rate and high propagation delay exhibits the highest total delays. AeroMACS, which will be used when an aircraft approaches the airport, having low propagation delay and high data rate, shows the lowest total delay. Since DVB-S2 has the same propagation delay as BGAN but with a higher data rate, its delay performance is better than AeroMACS under high arrival rate.

Acknowledgment. The research leading to these results has been partially funded by the European Community's Seventh Framework Programme (FP7/2007-2013) under Grant Agreement No. 233679. The SANDRA project is a Large Scale Integrating Project for the FP7 Topic AAT.2008.4.4.2 (Integrated approach to network centric aircraft communications for global aircraft operations).

References

1. http://www.sandra.aero
2. Technical Report, Reconfigurable Radio System (RRS); Software Defined Radio Reference Architecture for Mobile Device. ETSI TR 102 680 V1.1.1 (March 2009)
3. EUROCONTROL Long-Term Forecast, Flight Movements (2006 - 2025), EUROCONTROL STATFOR (December 2006)
4. Manual on VHF Digital Link (VDL) Mode 2, ICAO, Doc 9776 AN/970 (2001)
5. http://www.globalcoms.com/products_satellite_bgan.asp
6. ETSI EN 302 307 V1.2.1 (2009-08), Digital Video broadcasting (DVB) 2nd generation framing structure, channel coding & modulation systems for broadcasting, interactive services, news gathering and other broadbands satellite applications (DVB-S2) (2009)
7. IEEE Std. 802.16, Local and metropolitan area networks Part 16: Air Interface for Broadband Wireless Access Systems (Revision of IEEE Std 802.16-2004) (2009)
8. Network Mobility (NEMO) Basic Support Protocol. RFC3963 (January 2005)
9. IEEE Std. 802.21, Media Independent Handover Services (2009)

10. ETSI TS 102 357 V1.1.1 (2005-05), Technical Specification, Satellite Earth Stations and Systems (SES); Broadband Satellite Multimedia (BSM) Common air interface specification; Satellite Independent Service Access Point (SI-SAP) (2005)
11. Munasinghe, K.S., Jamalipour, A.: An analytical evaluation of mobility management in integrated WLAN-UMTS networks. Computers and Electrical Engineering 36, 735–751 (2010)
12. Pillai, P., Hu, Y.-F.: Performance analysis of EAP methods used as GDOI Phase 1 for IP multicast on Airplanes. In: International Conference on Advanced Information Networking and Applications Workshops, WAINA 2009 (June 2009)
13. Carrington, A., Harding, C., Yu, H.: Optimising Wireless Network Control System Traffic–Using Queuing Theory. In: Proceedings of the 14th International Conference on Automation & Computing, Brunel University, West London, UK (September 2008)
14. Griffiths, A.L., Carrasco, R.: IP Multiple Access between 3G/4G Mobile Radio and Fixed Packet Switched Networks. PhD Thesis, Staffordshire University (2004)
15. Philippe, L., Dietrich, B., Christophe, B., Laurence, F.: WiMAX, making ubiquitous high-speed data services a reality. Stratefy white paper from ALCATEL (June 2004)
16. Specification document for DM240-PIIC Digital Video Broadcast Modulator, http://www.comtechefdata.com/datasheets/legacy/radyne/dm240pic.pdf
17. IEEE std. 802.3u, Fast ethernet
18. Hu, Y.F., et al.: SANDRA WP4.2.2 D4.2.2.1 IMR Resource management interface definitions report, University of Bradford, UK (September 2010)

A Broadband CPW-Fed Printed Single-Patch Antenna for Galileo Applications

Constantinos T. Angelis and Spyridon K. Chronopoulos

Department of Informatics and Telecommunications
Technological Educational Institute of Epirus, Arta, Greece
kangelis@teiep.gr, schrono@cc.uoi.gr

Abstract. In this research work a novel and compact broadband U-shaped CPW-Fed single-patch antenna is proposed for Galileo applications. The enhanced performance of the single patch antenna relevant to Galileo signal reception and coverage is presented through simulation and analysis using Microwave Office for the purpose of establishing in depth the benefits of antennas in modern navigation apparatus and applications. The proposed antenna has a center frequency in 1.351 GHz and offers a return loss of better than -10dB from 1.1 to 1.6 GHz. Moreover, satisfactory radiation pattern is obtained through simulation. The proposed patch antenna is suitable for implementing low cost, high stable and well circular polarized Galileo antenna, as demonstrated by numerical simulations.

Keywords: Galileo, Printed antenna, CPW-feeding.

1 Introduction

In recent years, there has been tremendous growth in wireless communication technology, especially for the upcoming Galileo Satellite system. When Galileo will be finally operational, four carrier frequencies in the L band named Galileo E5a (1176.45 MHz), E5b (1207.14 MHz), E6 (1278.75 MHz) and E1 (1575.42 MHz) will be broadcasted. The E1 and E5a are designed with the same center frequency, and similar bandwidths compared to the GPS L1 C/A and L5 signals respectively [1], [2].

The upcoming Galileo system in order to provide more accurate navigation services will require modifications or even improvements on the navigation receiving systems. The previous receiving systems begin with the antenna or its terminal. For obtaining the best possible accuracy in navigation terms, along with the utilization of the fully offered capabilities of this new system, state-of-the-art antenna systems must be designed. Low profile antennas have therefore been a very hot research topic in antenna engineering. Many researchers work in designing antennas for Galileo and/or GPS [3-12] applications. In [3,4], dual-band and in [5] multiband printed antennas are designed for Galileo applications. Wideband antennas are designed in [6-8]. Smart Antennas are designed in [9,10]. Finally printed antennas for Safety-Critical Applications and Search and rescue applications are designed in [11,12].

G. Giambene and C. Sacchi (Eds.): PSATS 2011, LNICST 71, pp. 342–350, 2011.
© Institute for Computer Sciences, Social Informatics and Telecommunications Engineering 2011

A Galileo antenna is not selected in general, by choosing an antenna which satisfies all previous performance characteristics. This depends on the aimed application. So, the most suitable Galileo antenna is determined by considering the performance characteristics of the specified application, including its type, the frequency domain, the antenna cost, size and profile and whether single elements or an array of multiple elements are needed for satisfying the previous standards. Also, a passive or active antenna may be selected depending on different application types.

In this paper, a novel and compact broadband U-shaped coplanar waveguide CPW-Fed single-patch antenna is proposed for Galileo applications. The proposed antenna operates in the frequency band from 1176.45 to 1575.42 MHz and if broadband approach is considered, then the total bandwidth becomes 427 MHz centered at 1377.5 MHz, i.e. ≈ 31%. The CPW-fed antenna is intended for various applications due to its low-cost and the radiation losses. It is also light weighted and compatible with integrated circuits. The proposed antenna has the advantage of compact size, which makes it attractive for mobile devices. The broadband antenna design was chosen against the multiband design due to the fact that broadband antennas usually present better radiation pattern symmetry and higher polarization purity [10, 13].

2 Antenna Design

According to the Galileo Joint Undertaking, the Galileo Navigation Signals are located in the four frequency bands (in blue) in Figure 1. The frequency bands are the E5a, E5b, E6 and the L1 band, exhibiting a wide transmission bandwidth of the Galileo Signals.

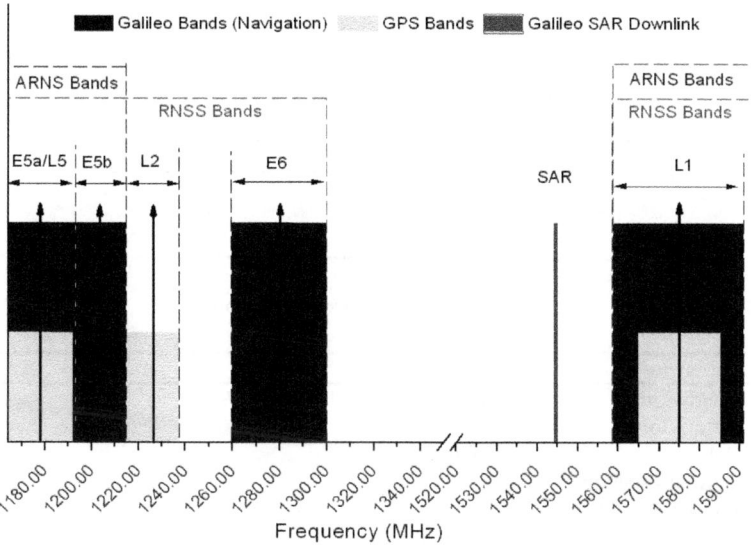

Fig. 1. Galileo frequency spectrum

RNSS (Radio Navigation Satellite Services) affect the selection of frequency bands in the relevant specified spectrum. Moreover, ARNS (Aeronautical Radio Navigation Services) include in their specified spectrum, E5a, E5b and L1 bands, as the previous services are employed by Civil-Aviation users, and allow dedicated safety-critical applications. The lower and upper frequency limits of each band are reported in Figure 1. Galileo carrier frequencies are shown in Table 1, while the Rx reference bandwidths and polarizations are specified in Table 2.

As it has been mentioned by many researchers, the development of an antenna which can operate in all Galileo frequency bands, shown by Fig. 1, and in the same time to fulfill all the design specifications given in Table 3 can prove to be a very difficult task, especially if other constrains like size, weight and cost are also to be taken into account. Another issue that has been faced by many researchers is whether the antenna should be broadband, covering the whole frequency range or multiband covering only the bands of interest separately from each other. Although the second approach seems to be more attractive for the first view due to a potentially better noise and interferer's suppression, it's known from the literature that multiband antennas usually present stronger radiation pattern asymmetry and lower polarization purity [10, 13]. Taking all the above into consideration, in this work we decided to design a broadband antenna to get the most of the performance.

Table 1. Carrier Frequency per Signal

Signal	Carrier Frequency
E5a	1176.450 MHz
E5b	1207.140 MHz
E5 (E5a+E5b)	1191.795 MHz
E6	1278.750 MHz
E1	1575.420 MHz

Table 2. Galileo signals Rx reference bandwidths

Signal	Rx Reference Bandwidth	Polarization
E5	51.150 MHz	RHCP
E6	40.920 MHz	RHCP
E1	24.552 MHz	RHCP

Table 3. Antenna specifications

Parameter	Specifications
Bandwidth	1164 MHz to 1591 MHz (31%)
Return loss	-10 dB min
Polarization	RHCP
Azimuth scanning	360°
Elevation scanning	from 30° to 90° (from 0° to 30° desired)
Gain	10 dBi min. over all scan angles (30° to 90° elevation)
Axial ratio	3 dB min. over all scan angles (30° to 90° elevation)
Cross-polarization	15 dB minimum, 25 dB or better is desirable

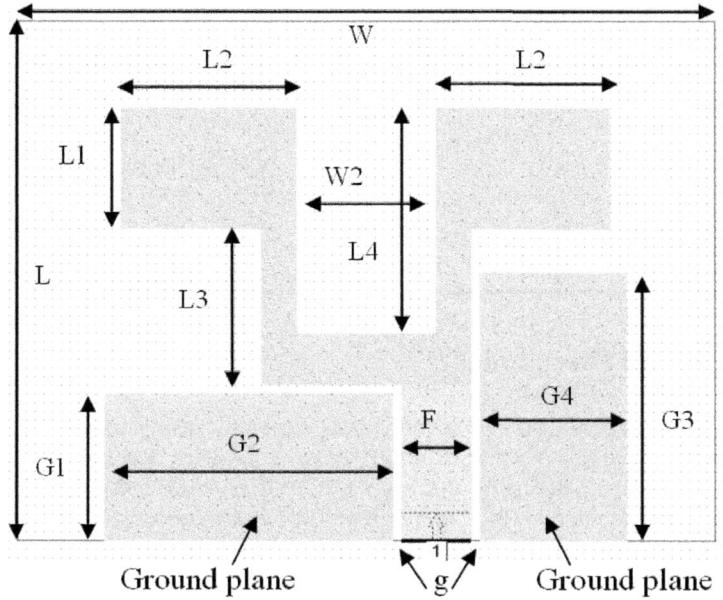

Fig. 2. Two dimensional structure of the proposed antenna

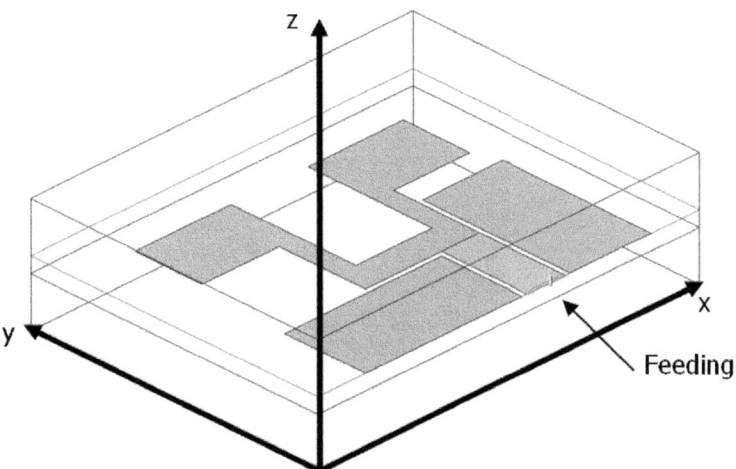

Fig. 3. Three dimensional structure of the proposed antenna

The proposed antenna is etched on a substrate of thickness 1.52 mm and dielectric constant of 3.38. The substrate has a compact dimension of 30 x 40mm^2. Fig. 2 shows the structure of the proposed antenna. The antenna consists of a single layer, which is the radiating layer as the ground plane of the antenna is located on the same side of the CPW-feed and radiating portion. The antenna is simulated using the parameters given in Table 4. The complete structure is presented in Fig. 3 where it is possible to observe the patch of the antenna and its feeding plate.

Table 4. Parameters of the proposed antenna

Parameter	Value	Parameter	Value
W	40mm	L4	14mm
L	30mm	W1	8mm
F	4mm	W2	2.5mm
g	0.5mm	G1	8.5mm
L1	7mm	G2	16.5mm
L2	10mm	G3	15.5mm
L3	9mm	G4	8.5mm

3   Results and Discussion

The reflection coefficient of the antenna was simulated using the AWR Microwave Office software. Using a VSWR≤2 (return loss ≤-9.5 dB) as benchmark, the measured result shows that the antenna covers from 1.160 GHz to 1.595 GHz frequency range; thus the impedance bandwidth is more than 30%. The frequency band that is covered by this antenna is suitable for Galileo as well as GPS bands. Full wave simulations for the antenna including its feeding system were performed showing good results. The reflection coefficient remains under 10 dB for the complete specified frequency range, as shown by Fig. 4.

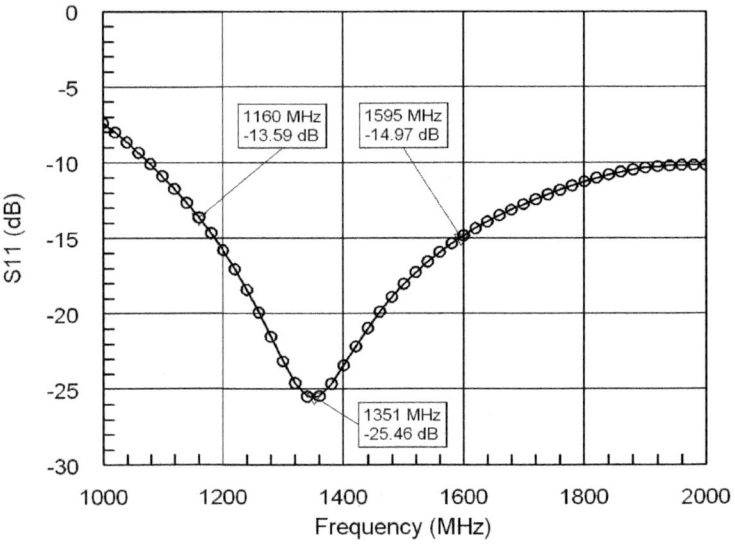

Fig. 4. Simulated reflection coefficient versus frequency

Figure 5 shows full 360 degree conic cuts at an elevation angle of 45 degrees. Both right-hand (RHCP) and left-hand (LHCP) circular polarizations are shown (a) for the E5a/b band, (b) for the E6 band and (c) for the L1 band.

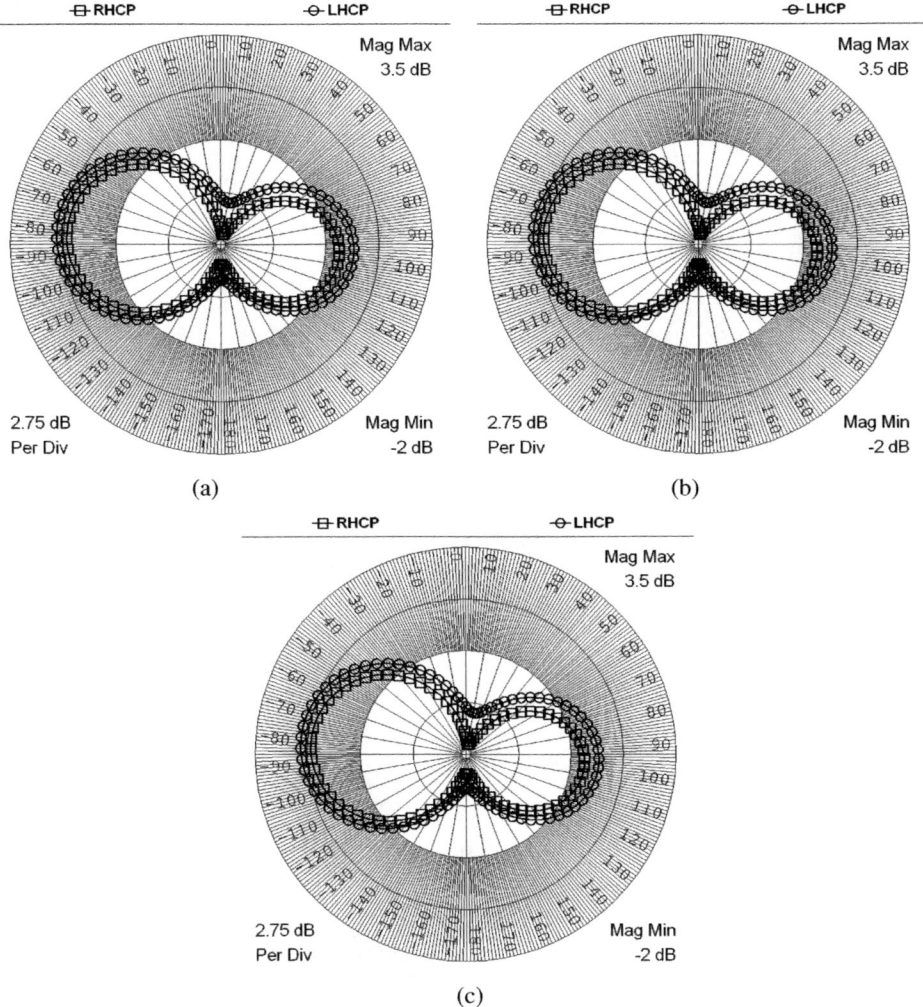

Fig. 5. RHCP and LHCP polarizations at an elevation angle of 45 degrees **(a)** for the E5a/b band, **(b)** for the E6 band and **(c)** for the L1 band

Figure 6 (a) shows Conic Cuts polarized along E_θ and F_φ at the center frequency of the complete Galileo band, $f_c = 1.380\,\text{GHz}$. The Theta (θ) and Phi (φ) components of the E-field are plotted.

Figure 6(b) shows Conic Cuts which capture the total power in all directions for fixed values of Frequency and Theta (θ) at the center frequency of the complete Galileo band, $f_c = 1.380\,\text{GHz}$. The total power is defined as the sum of the power contained in E_θ and E_φ.

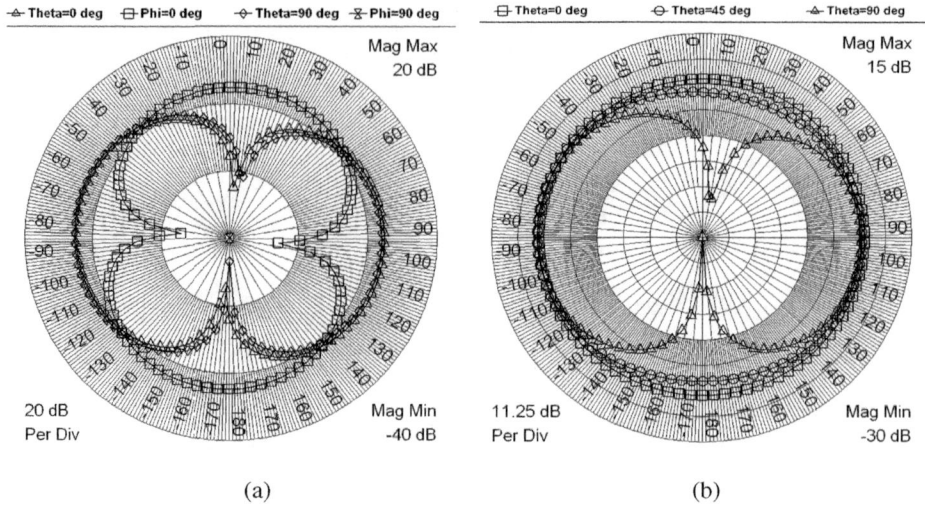

(a)   (b)

Fig. 6. (a) Conic Cuts polarized along E_θ and E_φ for the complete Galileo band. **(b)** Conic Cuts of the total power in all directions for the complete Galileo band.

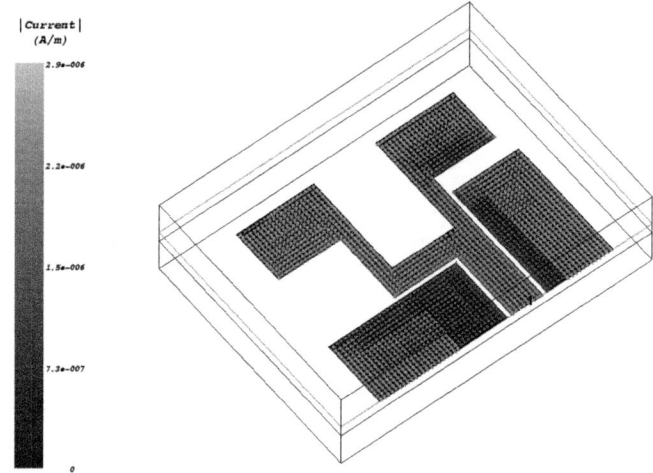

Fig. 7. Animation snapshot of the electric current in all directions for the complete Galileo band

Figures 7 and 8 display representations of the electric current and of the electric field respectively occurring on the specified 3D view of the proposed antenna. The current and the electric field are overlaid with the structure drawn in wire frame mode.

Fig. 8. Animation snapshot of the electric field in all directions for the complete Galileo band

4   Conclusion

In this paper, a compact U-shaped CPW-fed printed antenna is proposed for the Galileo bands. The proposed antenna has been successfully simulated in a small area of 30 x 40 mm². Acceptable return loss and satisfactory radiation patterns are obtained through simulation. Further investigations are undergoing. The next step is to finish the antenna construction and to measure it and afterwards to design an antenna array with the proposed antenna as the basic element.

References

1. European Space Agency/Galileo Joint Undertaking, Galileo Open Service Signal in Space Interface Control Document (OS SIS ICD) (1) (May 23, 2006)
2. GPS Joint Program Office, Navstar GPS Space Segment / Navigation User Interfaces, IS-GPS-200D, Revision D (December 7, 2004)
3. Ramírez, M., Parrón, J.: Design, Fabrication and measurement of a dual-band circularly-polarized stacked microstrip antenna for Galileo and GPS. In: Proceedings of the Fourth European Conference on Antennas and Propagation (EuCAP) (April 12-16, 2010)
4. Hebib, S., Aubert, H., Pascal, O., Fonseca, N., Ries, L., Lopez, J.M.: Pyramidal multi-band antennas for GPS/Galileo/MicroSat application. In: 2007 IEEE Antennas and Propagation International Symposium. United States, Honolulu (2007)
5. Ur Rehman, M., Gao, Y., Chen, X., Parini, C.: Dual-element Diversity Antenna for Galileo/GPS Receivers. In: 13th IAIN World Congress and Exhibition, Stockholm (October 27-30, 2009).
6. Robin, G., Peter, R., Steve, S.: The Development of a Professional Antenna for Galileo, Roke Manor Research Limited (2009-2010)

7. Sohyeun, Y., Dongpil, C., Inbok, Y.: Wideband Receive Antennas of Sensor Stations for GPS/Galileo Satellite. In: 4th Advanced Satellite Mobile Systems, ASMS 2008 (August 26-28, 2008)

8. Rens, B., Marta, M.-V., Jens, L., Sybille, H., Luca, S.D., de Peter, M.: Low Profile GALILEO Antenna Using EBG Technology. IEEE Transactions on Antennas and Propagation 56(3), 667 (2008)

9. Neves, E.S., Vita, P.D., Dreher, A.: Smart Antenna Array for GPS and Galileo Applications. In: 2nd ESA Workshop on Satellite Navigation User Equipment Technologies, Noordwijk, The Netherlands (December 2004)

10. Neves, E.S., Dreher, A.: An antenna array for the Galileo system with beamforming capabilities. Deutscher Luft- und Raumfahrtkongress (DLR), Germany (2006)

11. Cuntz, M., Konovaltsev, A., Hornbostel, A., Schittler Neves, E., Dreher, A.: GALANT – Galileo Antenna and Receiver Demonstrator for Safety-Critical Applications. In: Proceedings of the 10th European Conference on Wireless Technology, Munich Germany, p. 59 (October 2007)

12. Montero Jose, M., Esteban, C., Ana, T.: Search and rescue antenna for Galileo constellation. In: Antennas and Propagation Society International Symposium (APSURSI), Torondo, pp. 1–4 (2010)

13. Wong, K., Chiou, T.: Broad-band single-patch circularly polarized microstrip. IEEE Trans. Antennas Propagat 49, 41–44 (2001)

Design of MIMO Satellite System: Inter-antenna Spacing Determination and Possible Enhancement of Capacity

Boujnah Noureddine[1] and Issaoui Leyla[2]

[1] Politecnico di Torino, Electronic Department, Turin, Italy
boujnah_noureddine@yahoo.fr
[2] Ecole Superieure des Communications de Tunis, MEDIATRAN, Tunisie
leyla.issaoui@gmail.com

Abstract. MIMO(Multiple Input Multiple Output) technology was found to exploit the channel richness and spatial diversity, its possible to increase data throughput and reduce the error probability[1]. MIMO channel capacity was derived by Teletar in [2], MIMO find its application in many technology such as HSPA and WLAN. Recent research were devoted to applicability of MIMO to satellite communication and for high altitude platforms (HAPs)[3]. Previous works were concentrated on describing the MIMO model [4] by mean of ray tracing. This paper is organized as follow: first, we investigate MIMO-SAT model for clear sky case and presents new method of computation of the covariance matrix in order to determine the inter-antenna spacing for the MIMO-SAT system at ground and in the satellite side by mean of capacity maximization. The second part will be devoted to MIMO-SAT model with atmospheric effect and impairment mitigation using precoding and dual polarization scheme.

Keywords: MIMO satellite, antenna inter-spacing, Covariance matrix, Capacity maximization.

1 Introduction

MIMO satellite system consist of one or more satellite in the same orbit communicating with one or more fixed ground station. MIMO enables system capacity to be increased in proportion to the number of transmitting and receiving antennas, a result of which is improved spectrum efficiency. application of MIMO to satellite system, with appropriate model definition, raise capacity of links and reduce error probability. In satellite communication, frequency is always above 10 GHz where wave attenuation in free space is much higher than for terrestrial wireless system, distance between transmitter and receiver is at least 500Km for low orbit satellite(LEO) and reach 36000 for Geostationary orbits(GEO), the radio link depend also on the earth to satellite observation angle which is variable from LEO satellite. In this paper we assume that satellites and ground stations are in the same orbital plane containing the center of the earth and the

G. Giambene and C. Sacchi (Eds.): PSATS 2011, LNICST 71, pp. 351–364, 2011.

satellite orbits is circular. Waves for frequency above 10 GHz are subject of deep mitigation such as attenuation and scattering from rain and snow, exploiting spatial diversity can reduce this effect.

2   MIMO Channel with No Atmospheric Effect

2.1   MIMO Model

In MIMO system the discrete time model is commonly used:

$$Y = HX + B \tag{1}$$

Where, Y is $r \times 1$ received signal, X is $t \times 1$ transmitted signal. H is the $r \times t$ channel matrix. B is $r \times 1$ is the additive noise vector.

Using results from Information theory, MIMO system achieve the maximum capacity when the elements of the transmitted vector X are zero mean independent identically distributed Complex Gaussian variables. The covariance matrix of the transmitted signal is given by:

$$\Gamma_{XX} = E(XX^*) \tag{2}$$

The total transmitted power is given by:

$$P = tr(\Gamma_{XX}) \tag{3}$$

When the channel is unknown, components of X are independent and have equal power distribution, thus the covariance is:

$$\Gamma_{XX} = \frac{P}{t}I_t \tag{4}$$

In general we assume that components of B are zero mean Gaussian variable with independent and equal variance real and imaginary part. The covariance matrix of the noise is:

$$\Gamma_{BB} = E(BB^*) = N_0 I_r \tag{5}$$

Where, N_0 is the noise level at receiver side.

Let define the signal to noise ratio by: $\gamma = \frac{E(||X||^2)}{tE(||B||^2)}$

2.2   Covariance Matrix: First Approach

The block diagram of MIMO system is presented in Fig.1. To describe the channel by its covariance matrix let first transform H into a colon vector $\tilde{H}$ with length $rt \times 1$ as follow:

$$\tilde{H}_{t(j-1)+i} = H_{i,j} \tag{6}$$

The covariance matrix of the channel is given by:

$$\Gamma_{\tilde{H}\tilde{H}} = E(\tilde{H}\tilde{H}^*) \tag{7}$$

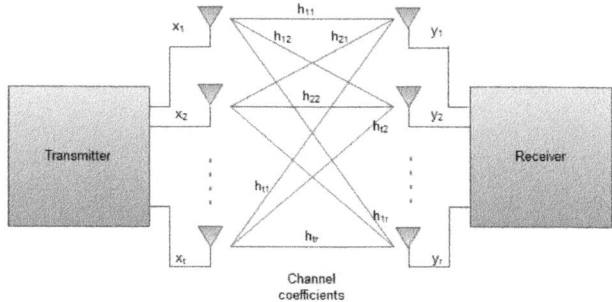

Fig. 1. Block diagram of MIMO system

and,

$$\Gamma_{\tilde{H}\tilde{H}}(a, b) = E(\tilde{H}_a \tilde{H}_b^*) = E(H_{kl} H_{mn}{}^*) \tag{8}$$

with $a = t(l-1)+k$ and $b = r(n-1)+m$. Size of $\Gamma_{\tilde{H}\tilde{H}}$ is $rt \times rt$. The computation of $\Gamma_{\tilde{H}\tilde{H}}$ depend on path difference of MIMO channel.

$$\Gamma_{\tilde{H}\tilde{H}}(a, b) = E(\{\frac{c_0}{2\pi f_c R}\}^2 \exp\{-j\frac{2\pi f_c}{c_0}(R_{lk} - R_{lk})\}) \tag{9}$$

Consider a MIMO channel with t transmit and r receive antennas, with antenna separation of d_t and d_r, let assume that transmitter and receiver are separated by a distance h.

Let ϵ_{kl}^t the algebraic angle of the direction of departure (AOD) and ϵ_{lk}^r be the algebraic angle of the direction of arrival (AOA), and denote by R_{kl} the distance between antenna k from one TX/RX side and antenna l from the other side.

Using the model with only free space attenuation $H_{kl} = \frac{c_0}{2\pi f_c R_{kl}} \exp\{-j\frac{2\pi f_c}{c_0} R_{kl}\}$, using the fact that $h \gg d_t$ and $h \gg d_r$, each entry of the channel matrix, taking into account previous conditions, can be approximated by:

$$H_{kl} = K \exp\{-j\frac{2\pi f_c}{c_0} R_{kl}\} \tag{10}$$

Were, $K = \frac{c_0}{2\pi f_c h}$. The entries of the covariance matrix $R_{\tilde{H}\tilde{H}}$ are given by:

$$\Gamma_{\tilde{H}\tilde{H}}(a, b) = K^2 E(\exp\{-j\frac{2\pi f_c}{c_0}(R_{kl} - R_{mn})\}) \tag{11}$$

the computation of $R_{kl} - R_{mn}$ could be performed geometrically and using some approximation.

$$R_{kl} - R_{mn} = (R_{kl} - R_{ml}) + (R_{ml} - R_{mn}) \tag{12}$$

Let compute first $R_{kl}^2 - R_{ml}^2$ and $R_{ml}^2 - R_{mn}^2$:

$$R_{kl}^2 - R_{ml}^2 = (k - m)^2 d_t^2 + 2(m - k)d_t R_{ml} \sin(\epsilon_{ml}^t) \tag{13}$$

And,

$$R_{ml}^2 - R_{mn}^2 = (l-n)^2 d_r^2 + 2(l-n)d_r R_{mn} \sin(\epsilon_{mn}^r) \tag{14}$$

Eq.13 can be simplified as:

$$R_{kl} - R_{ml} = \frac{(k-m)^2 d_t^2}{R_{kl} + R_{ml}} + 2(m-k)d_t \frac{R_{ml}}{R_{kl} + R_{ml}} \sin(\epsilon_{ml}^t) \tag{15}$$

using previous approximation, Eq.15 become:

$$R_{kl} - R_{ml} = (m-k)d_t \sin(\epsilon_{ml}^t) \tag{16}$$

Using the same approach to compute $(R_{ml} - R_{mn})$, we obtain:

$$R_{ml} - R_{mn} = (n-l)d_r \sin(\epsilon_{mn}^r) \tag{17}$$

and finally,

$$R_{kl} - R_{mn} = (m-k)d_t \sin(\epsilon_{ml}^t) + (n-l)d_r \sin(\epsilon_{mn}^r) \tag{18}$$

Then Eq.11 become:

$$\Gamma_{HH}(a,b) = K^2 E(\exp\{-j\frac{2\pi f_c}{c_0}((m-k)d_t \sin(\epsilon_{ml}^t))\} \tag{19}$$

$$\exp\{-j\frac{2\pi f_c}{c_0}((n-l)d_r \sin(\epsilon_{mn}^r))\}) \tag{20}$$

Equation Eq.19 point out that the covariance matrix of the channel depend on distances between antennas in transmitter and receiver side and arrival and departure angles.

Expectation operator concern only angles, distance between antenna are fixed. In practical MIMO communication system d_r is approaching to Zeros, the covariance matrix will depend only on the distances between antennas in the transmitter side and the AOD statistics in presence of scatters.

When the channel is random and AOD and AOA are assumed to be statistically independent, the covariance matrix expressed in Eq.19 is often assumed to have a simpler separable Kronecker structure[8]. In rich scattering environment the Non-Line of Sight(NLOS) dominate, such as in indoor and in some outdoor scenarios [9] [10].

2.3   Covariance Matrix: Second Approach

Let first study the case where only the free space attenuation is considered.

The MIMO system consist of t ground stations and r satellites in the same orbits, the system model is given by:

$$Y = HX + B \tag{21}$$

Y, H, X and B have the same dimension as considered previously. Fig.2 show the geometrical presentation of 2×2 MIMO satellite system.

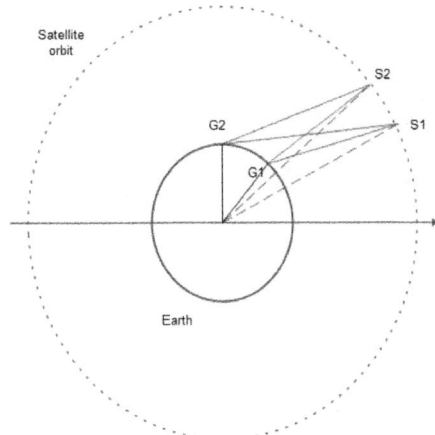

Fig. 2. Block diagram of 2×2 MIMO satellite system

Each entry of H, H_{kl} with $k \in 1 \ldots r$ and $l \in 1 \ldots t$, is given by:

$$H_{lk} = \frac{c_0}{2\pi f_c R_{lk}} \exp\left\{-j\frac{2\pi f_c}{c_0} R_{lk}\right\} \tag{22}$$

The distance R_{lk} between the lth ground station to the kth satellite antenna is given by:

$$R_{lk} = \sqrt{(h + r_e)^2 + r_e^2 - 2(h + r_e)r_e \cos\left(\theta_{G_l} - \theta_{S_k}\right)} \tag{23}$$

Where G stand for ground station and S antenna satellite, one satellite can contain more than one antenna, h is the satellite height and r_e the effective earth radius(8500Km).

Distance between earth grounds and satellites is higher than distance between antenna in each side of MIMO system, H_{lk} is then approximated by:

$$H_{lk} = \frac{c_0}{2\pi f_c R} \exp\left\{-j\frac{2\pi f_c}{c_0} R_{lk}\right\} \tag{24}$$

Where $R_{lk} \approx R, \forall l, k$ For sake of simplicity we choose $R = h$. For the MIMO satellite communication distance R_{lk} depend on two independents angle θ_{G_l} and θ_{S_k}, for circular orbit, θ_{S_k} is a linear function of time t and the angle of the ground station is fixed.

Ground stations are uniformly separated by the same distance d_G, and satellite antenna are separated by the same distance d_S. Using a geometrical approach, we compute the covariance matrix of the channel Γ_{HH} given by Eq.8. Using the same approach as in paragraph 2.2, we need first to calculate $R_{lk} - R_{nm}$. Denote by $\psi_{nm} = \cos\left(\theta_{G_n} - \theta_{S_m}\right)$

$$R_{lk}^2 - R_{nm}^2 = 2(h + r_e)r_e[\psi_{nm} - \psi_{lk}] = 2(h + r_e)r_e[\psi_{nm} - \psi_{lm} + \psi_{lm} - \psi_{lk}] \tag{25}$$

Applying the identity:

$$\cos{(p)} - \cos{(q)} = 2\sin{(\frac{p+q}{2})}\sin{(\frac{q-p}{2})}$$

We get:

$$\psi_{nm} - \psi_{lm} = 2\sin{(\frac{\theta_{G_l} + \theta_{G_n}}{2} - \theta_{S_m})}\sin{(\frac{\theta_{G_l} - \theta_{G_n}}{2})} \tag{26}$$

and

$$\psi_{lm} - \psi_{lk} = 2\sin{(\theta_{G_l} - \frac{\theta_{S_m} + \theta_{S_k}}{2})}\sin{(\frac{\theta_{S_m} - \theta_{S_k}}{2})} \tag{27}$$

Let d_{ln} the distance between G_l and G_n and D_{mk} distance between antenna satellite S_m and S_k. We have:

$$d_{ln} = (n - l)d_G = 2r_e\sin{(\frac{\theta_{G_l} - \theta_{G_n}}{2})} \tag{28}$$

And

$$D_{mk} = (m - k)d_S = 2(r_e + h)\sin{(\frac{\theta_{S_m} - \theta_{S_k}}{2})} \tag{29}$$

Combining Eq.27 and Eq.28 we get:

$$\psi_{nm} - \psi_{lm} = \frac{d_{ln}}{r_e}\sin{(\frac{\theta_{G_l} + \theta_{G_n}}{2} - \theta_{S_m})} \tag{30}$$

And combining Eq.26 and Eq.29 we get:

$$\psi_{lm} - \psi_{lk} = \frac{D_{mk}}{r_e + h}\sin{(\theta_{G_l} - \frac{\theta_{S_m} + \theta_{S_k}}{2})} \tag{31}$$

Let θ_1 and θ_2 the angular separation between two consecutives ground stations and two satellites antenna respectively, we have:

$$\begin{cases} \theta_{G_l} = (l - 1)\theta_1 + \theta_{G_1} \\ \theta_{G_n} = (n - 1)\theta_1 + \theta_{G_1} \\ \theta_{S_m} = (m - 1)\theta_2 + \theta_{S_1} \\ \theta_{G_k} = (k - 1)\theta_2 + \theta_{S_1} \end{cases} \tag{32}$$

Eq.30 become:

$$\psi_{nm} - \psi_{lm} = \frac{(n - l)d_G}{r_e}\sin{\left[\theta_{G_1} - \theta_{S_1} + \frac{l + n - 2}{2}\theta_1 + (m - 1)\theta_2\right]} \tag{33}$$

and, Eq.31 become:

$$\psi_{lm} - \psi_{lk} = \frac{(m - k)d_S}{r_e + h}\sin{\left[\theta_{G_1} - \theta_{S_1} + (l - 1)\theta_1 + \frac{m + k - 2}{2}\theta_2\right]} \tag{34}$$

Assume that $d_G << r_e$ and $d_S << (r_e + h)$, Eq.33 and Eq.34 can be writen as:

$$\psi_{nm} - \psi_{lm} =$$

$$\frac{(n-l)d_G}{r_e} \sin \left[\theta_{G_1} - \theta_{S_1} + \frac{(l+n-2)d_G}{2r_e} + \frac{(m-1)d_S}{h+r_e} \right] \tag{35}$$

$$\psi_{lm} - \psi_{lk} =$$

$$\frac{(m-k)d_S}{h+r_e} \sin \left[\theta_{G_1} - \theta_{S_1} + \frac{(l-1)d_G}{r_e} + \frac{(m+k-2)d_S}{2(h+r_e)} \right] \tag{36}$$

$$R_{lk}^2 - R_{nm}^2 =$$

$$2(n-l)d_G(h+r_e) \sin \left[\theta_{G_1} - \theta_{S_1} + \frac{(l+n-2)d_G}{2r_e} + \frac{(m-1)d_S}{h+r_e} \right] \tag{37}$$

$$+2(m-k)d_S r_e \sin \left[\theta_{G_1} - \theta_{S_1} + \frac{(l-1)d_G}{r_e} + \frac{(m+k-2)d_S}{2(h+r_e)} \right]$$

Let assume that $R_{kl} + R_{mn} \approx 2h$ [23] and for sake of simplicity let assume that $\theta_{G_1} = 0$, Eq.37 become:

$$R_{lk} - R_{nm} =$$

$$(n-l)d_G(1 + \frac{r_e}{h}) \sin \left[\frac{(l+n-2)d_G}{2r_e} + \frac{(m-1)d_S}{h+r_e} - \theta_{S_1} \right] \tag{38}$$

$$+(m-k)d_S \frac{r_e}{h} \sin \left[\frac{(l-1)d_G}{r_e} + \frac{(m+k-2)d_S}{2(h+r_e)} - \theta_{S_1} \right]$$

The first approach is convenient in presence of scatter in atmosphere such as rain, in this case, angle of arrival and angle of departure are random variable, the second approach is adequate for clear sky scenarios in the MIMO-SaT links.

3 Capacity Optimization

3.1 MIMO-SAT Capacity for Single User Communication

MIMO Capacity is the maximum of mutual information between the transmitted vector X and received vector Y, when component of the transmitted vector X have equal power distribution the explicit form of capacity, derived by Teletar in [2], and given by: [17]

$$C(H) = \log(\det(I_{r \times r} + \gamma H H^*)) \tag{39}$$

Using the property: $\det(I + AB) = \det(I + BA)$ we can write:

$$C(H) = \log(\det(I_{t \times t} + \gamma H^* H)) \tag{40}$$

Let denote by:

$$\begin{cases} W_1 = HH^* \\ W_2 = H^*H \end{cases} \tag{41}$$

entries of W_1 and W_2 are given by:

$$W_1(k, l) = \sum_{m=1}^{t} H_{k,m} H_{l,m}^* \tag{42}$$

And

$$W_2(k, l) = \sum_{m=1}^{r} H_{m,k} H_{m,l}^* \tag{43}$$

Matrix W_1 and W_2 are both definite positive.

Link between ground and satellite is fixed for GEO system and variable for NGEO system with respect to elevation angle in clear sky condition.

3.2 Inter-spacing Computation for Clear Sky Conditions

Inter antenna spacing is a convex optimization problem, where the objective function is the system capacity. Using results from Hadamrad inequality for definite positive matrix, when $\min(r, t) \geq 2$ capacity reach its maximum when W_1 is diagonal. For the particular case of 2×2 MIMO system, the optimization problem can expressed by the following equations:

$$\begin{cases} \{\frac{2\pi f_c}{c_0}(R_{11} - R_{12})\} - \{\frac{2\pi f_c}{c_0}(R_{21} - R_{22})\} = (2\mu + 1)\pi \\ \{\frac{2\pi f_c}{c_0}(R_{11} - R_{21})\} - \{\frac{2\pi f_c}{c_0}(R_{12} - R_{22})\} = (2\nu + 1)\pi \end{cases} \tag{44}$$

where, $\nu, \mu \in \mathbf{Z}$. For GEO system $\theta_{S_1} = \frac{\pi}{36}$, solution to Eq.51 and Eq.49 d_G and d_S can be performed numerically.

The equation system of Eq.44 can be reduced to the following equation:

$$R_{11} - R_{12} - R_{21} + R_{22} = (\mu + \frac{1}{2})\frac{c_0}{f_0} \tag{45}$$

GEO System. AOD and AOA for GEO system are fixed, hence, capacity is maximized when $W_1 = tI_{t \times t}$, for 2×2 MIMO-SAT system, inter-spacing distance d_G and d_S are solution of the following equations:

$$W_1(1, 2) = 0 \tag{46}$$

and

$$W_2(1, 2) = 0 \tag{47}$$

this is equivalent to the equations:

$$\frac{2\pi f_c}{c_0} d_S \frac{r_e}{h} \left[\sin(\theta_{S_1}) + \sin(\frac{d_G}{r_e} - \theta_{S_1}) \right] = (2\nu + 1)\pi \tag{48}$$

And

$$\frac{2\pi f_c}{c_0} d_G (1 + \frac{r_e}{h}) \left[\sin(\theta_{S_1}) + \sin(\frac{d_S}{h + r_e} - \theta_{S_1}) \right] = (2\mu + 1)\pi \qquad (49)$$

Using first order Taylor's expansion of $\sin(\frac{d_S}{h+r_e} - \theta_{S_1})$ and $\sin(\frac{d_G}{r_e} - \theta_{S_1})$ we get:

$$\frac{f_c}{c_0} \frac{d_S d_G \cos(\theta_{S_1})}{h} = (\nu + \frac{1}{2}) \qquad (50)$$

Finally,

$$d_S d_G = \frac{h c_0}{\cos(\theta_{S_1}) f_c} (\nu + \frac{1}{2}) \qquad (51)$$

NGEO System. For NGEO system satellite describe an elliptical orbit(only circular case will be considered), and the orbit period is less than 24 hours, the satellite is in orbit lower than for GEO system. Using the concavity property of $C(H)$, capacity is upper bounded by its Jensen bound:

$$\mathbf{E}(C(H)) < C(\hat{W}_1) \qquad (52)$$

Where, $\hat{W}_1$ is the mean value over the angle θ_{S_1} which can variate from two angular positions: $\theta_{min} = \theta_0 - \Delta\theta$ and $\theta_{max} = \theta_0 + \Delta\theta$, $\Delta\theta$ is positive value that depend on the height of the satellite.

For the particular case of 2×2 MIMO, let compute first $\hat{W}_1(2,1)$. Using results from paragraph 2.3, one can write:

$$\begin{aligned} \hat{W}_1(2,1) = \\ \frac{1}{2\Delta\theta} \left(\frac{c_0}{2\pi f_c R} \right)^2 \int_{\theta_{min}}^{\theta_{max}} \left[\exp \left(j \frac{2\pi f_c}{c_0} \frac{d_S r_e}{h} \sin\theta \right) \right. \\ \left. + \exp \left(-j \frac{2\pi f_c}{c_0} \frac{d_S r_e}{h} \sin(\frac{d_G}{r_e} - \theta) \right) \right] d\theta \end{aligned} \qquad (53)$$

Let $\Psi(d_S, d_G, \theta) = \exp \left(j \frac{2\pi f_c}{c_0} \frac{d_S r_e}{h} \sin\theta \right) + \exp \left(-j \frac{2\pi f_c}{c_0} \frac{d_S r_e}{h} \sin(\frac{d_G}{r_e} - \theta) \right)$ Applying the triangular inequality to Eq.53, one get:

$$|\hat{W}_1(2,1)| \leq \frac{1}{2\Delta\theta} \left(\frac{c_0}{2\pi f_c R} \right)^2 \int_{\theta_{min}}^{\theta_{max}} |\Psi(d_S, d_G, \theta)| d\theta \qquad (54)$$

The expression inside the integral of Eq.54 can be simplified as:

$$|\Psi(d_S, d_G, \theta)| = 2 \left| \cos \left(\frac{c}{c_0} \frac{d_S d_G \cos(\theta)}{h} \right) \right|$$

Eq.54 became:

$$|\hat{W}_1(2,1)| \leq \frac{1}{\Delta\theta} \left(\frac{c_0}{2\pi f_c R} \right)^2 \int_{\theta_0 - \Delta\theta}^{\theta_0 + \Delta\theta} \left| \cos \left(\frac{\pi f_c}{c_0} \frac{d_S d_G \cos(\theta)}{h} \right) \right| d\theta \qquad (55)$$

4   MIMO SAT Channel with Atmospheric Effect

By contrast to MIMO for terrestrial communication, MIMO satellite depend on the geographic position of the ground station on the earth, climatic condition variate from one region to another, recent research are performed in some countries to draw their rain or snow map, in order to determine the average rain fall rate and rainy period [22].

Using inter-distances results from the previous section, its possible to choose, by mean of similation, the best design that mitigate the best the atmospheric effect.

4.1   Impact of Rain

Impact of rain links was studied for the scenario where one satellite with 2 transmitting antenna (inter antenna-spacing is very small) and it is was shown [23] that capacity depend only on modulus of the rain attenuation. The rain attenuation for a given path p_R is given by [15]:

$$L^r(dB) = aR^b sp_R \qquad (56)$$

And the attenuation coefficient T^r:

$$T^r = 10^{\frac{-L^r}{20}} \qquad (57)$$

Where, a and b two constant depending on frequency, polarization and elevation angle, p_R the path length and s the path correcting factor.

4.2   Depolarization

By passing through raindrops, the electric field tend to have a new component along the orthogonal axis. Depolarization is strongly correlated with rain attenuation, and standard models of depolarization use this fact to predict L^d directly from the attenuation L^r . One such model takes the form:

$$L^d(dB) = \alpha - \beta \log(L^r) \qquad (58)$$

α and β two constants depending on frequency. For frequencies above 10 GHz, $\alpha = 35,8$ and $\beta = 13,4$ [7].

5   Mitigation to Atmospheric Effect

5.1   Impact on System Capacity

In presence of atmospheric impairment, antenna inter-spacing to get optimal capacity is not sufficient, because matrices W_1 and W_2 are not diagonals. Two technique can be proposed to reduce this effect and enhance the capacity.

5.2 Dual Polarization

Transmission of dual independent orthogonal polarized channels in the same frequency band increase channel capacity. The transmitted electromagnetic wave can be impaired by atmospheric medium by transferring an amount of energy from one polarization to another, resulting in interference between the two channels.

The 2×2 channel matrix $H^a(k, l)$ when using dual polarization for each link (k, l) in MIMO-SAT system in presence of atmospheric impairment is expressed as function of the polarization matrix $T(k, l)$ by:

$$H^a = H(k, l)T \tag{59}$$

Where,

$$T = \begin{pmatrix} \tilde{T}_r^{(1)} & \frac{\tilde{T}_r^{(2)}}{\tilde{T}_d^{(2)}} \\ \frac{\tilde{T}_r^{(1)}}{\tilde{T}_d^{(1)}} & \tilde{T}_r^{(2)} \end{pmatrix} \tag{60}$$

Where,

- (1): horizontal polarization
- (2): vertical polarization
- $\tilde{T}_r$:complexe rain attenuation coefficient in the link (k, l)
- $\tilde{T}_d$: complexe depolarization loss in the link (k, l)

Using climatic parameters such as, rain rate, rain region dimension it is possible to determine the modulus of rain attenuation and the modulus of polarization loss, we assume that phase shift caused by atmospheric impairment is uniformly distributed in $[0, 2\pi]$ [21] and hence can be added to the AOA or AOD for each polarization. The resulting channel has dimension of $2t \times 2r$.

5.3 Power Allocation: Linear Precoding

Precoding design for MIMO wireless has been an active research area in recent year and is now finding applications in emerging wireless standard, the linear precoder functions as an input shaper and a beam-former with one or multiple beams with per-beam power allocation at the transmitting side. Precoder must match signal from Space-Time encoder and transmit antenna, a precoder which is matrix L is a subject of constraint optimization of a certain function, for example channel capacity, pairwise error probability. Generally we assume that the precoder satisfy the following power constraint:

$$tr(LL^*) = 1 \tag{61}$$

Assuming a power normalized code word C with covariance matrix Γ_{CC}, the transmitted signal at the transmitter is decomposed as : $\mathbf{X} = \sqrt{\frac{P}{t}}\mathbf{LC}$, capacity is expressed as:

$$C(H) = \log(\det(I_{t \times t} + \gamma HLT_{CC}L^*H^*)) \tag{62}$$

Where P is the total transmitted signal power.

6   Simulation Results

6.1   GEO System

Simulation parameters for 2×2 MIMO for GEO system are fixed as follow:

- $f_c = 12GhZ$
- GEO satellite $\theta_{S_1} = \frac{\pi}{36}$ (in earth position near equatorial region)
- $d_S = 6m$ antenna are embedded in the same satellite

The distance between antenna at ground is then given by: $d_G(Km) = 90.3438 \times (\mu + \frac{1}{2})$ and $\mu = 0, 1, 2, ...$ The antenna separation depend mainly on the region and taking into account installation cost of the ground station, its possible to determine the best choice of μ.

6.2   NGEO System

Simulation parameters for 2×2 MIMO for NGEO system are fixed as follow:

- $f_c = 12GhZ$
- NGEO satellite: The angular position of the satellite toward earth center is variating between $\theta_0 - \Delta\theta$ and $\theta_0 + \Delta\theta$
- $d_S = 6m$ antenna are embedded the same satellite

The determination of distance between the earth station d_G can be performed numerically by minimization of the upper bound of $|\hat{W}_1(2,1)|$, in figure 3 are presented the variation of the upper bound of $|\hat{W}_1(2,1)|$ as function of $K = \frac{\pi f_c d_S d_G}{c_0 h}$.

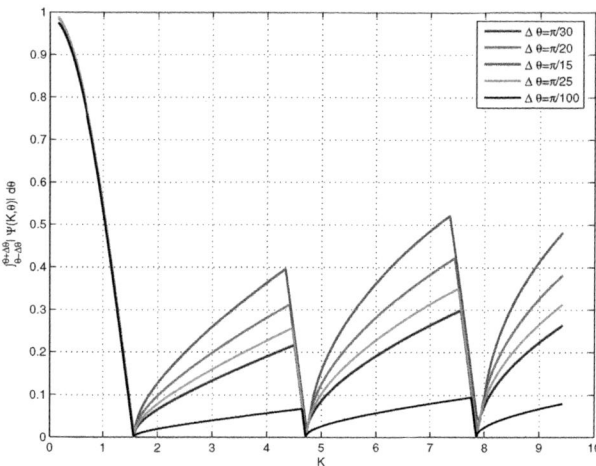

Fig. 3. Variation of $|\hat{W}_1(2,1)|$ as function of K for deferents values of $\Delta\theta$

7 Conclusion

We present in this paper a theoretical study of the design of MIMO-SAT system by computing the inter-spacing distance between antenna maximize the system capacity for both GEO an NGEO system in the case of clear sky conditions, the solution of the optimization problem is not unique and depend on geographical position. It is possible to perform a second stage optimization of capacity in order to reduce effects of the atmosphere on communication link, two techniques are proposed in this case: linear precoding and dual polarization techniques.

References

1. Marzetta, T.L., Hochwald, B.M.: Capacity of a mobile multiple-antenna communication link in Rayleigh flat fading. IEEE Trans. Inform. Theory 45, 139–157 (1999)
2. Telatar, E.: Capacity of multi-antenna Gaussian channels, Bell Lab. Technical Memorandum (October 1995), http://mars.bell-labs.com/papers/proof/, (Europe Trans. Telecomm. ETT 10(6), 585–596 (November 1999)
3. Perez-Neira, A.I., Zorba, N., Realp, M.: TR2-MIMO Applicability to Satellite Systems, internal report of MIMO Applicability to Satellite Networks with reference: AO/1-5146/06/NL/JD (July 2007)
4. Michailidis, E.T., Kanatas, A.G.: Capacity Optimized Line-of-Sight HAP-MIMO Channels for Fixed Wireless Access. In: International Workshop on Satellite and Space Communications IWSSC 2009, Siena-Tuscany, Italy (September 9-11, 2009)
5. Cover, T., Thomas, J.: Elements of Information Theory. John Wiley & Sons, Chichester (1991)
6. Sampath, H., Paulraj, A.: Linear precoding for space-time coded systems with known fading correlations. IEEE Communications Letters 6(6), 239–241 (2002)
7. Flock, W.L.: Propagation effects on satellite systems at frequencies below 10 GHz, NASA Reference Publication 1108 (1983)
8. Shiu, D., Foschini, G., Gans, M., Kahn, J.: Fading correlation and its effect on the capacity of multielement antenna systems. IEEE Trans. Comm. 48(3), 502–513 (2000)
9. Yu, K., Bengtsson, M., Ottersten, B., McNamara, D., Karlsson, P., Beach, M.: Second order statistics of NLOS indoor MIMO channels based on 5.2 GHz measurements. In: Proc. IEEE Global Telecomm. Conf., vol. 1, pp. 25–29 (November 2001)
10. Kermoal, J., Schumacher, L., Pedersen, K., Mogensen, P., Frederiksen, F.: A stochastic MIMO radio channel model with experimental validation. IEEE J. Selected Areas in Comm. 20(6), 1211–1226 (2002)
11. Tarokh, V., Seshadri, N., Calderbank, A.R.: Space-time codes for high data rate wireless communication: Performance criterion and code construction. IEEE Trans. Inform. Theory 44, 744–765 (1998)
12. Alamouti, S.M.: A simple transmit diversity technique for wireless communications. IEEE J. Select. Areas Commun. 16, 1451–1458 (1998)
13. Alamouti, S.M.: A simple transmit diversity technique for wireless communications. IEEE J. Sel. Areas Comm. 16(8), 1451–1458 (1998)
14. Tarokh, V., Jafarkhani, H., Calderbank, A.: Spacetime block codes from orthogonal designs. IEEE Trans. Inform. Theory 45, 1456–1467 (1999)

15. Saunders, S.R., Zavala, A.A.: Antenna and propagation for wireless communication systems, 2nd edn. JohnWiley & Sons Ltd., The Atrium (2007)
16. Tarokh, V., Jafarkhani, H., Calderbank, A.: Spacetime block codes from orthogonal designs. IEEE Trans. Inform. Theory 45, 1456–1467 (1999)
17. Cover, T., Thomas, J.: Elements of Information Theory. John Wiley and Sons, Chichester (1991)
18. Jngren, G., Skoglund, M., Ottersten, B.: Combining beamforming and orthogonal space-time block coding. IEEE Trans. on Info. Theory 48(3), 611–627 (2002)
19. Liu, L., Jafarkhani, H.: Application of quasi-orthogonal space-time block codes in beamforming. IEEE Trans. on Signal Processing 53(1), 54–63 (2005)
20. Gallager, R.: Information Theory and Reliable Communication. Wiley and Sons, Chichester (1968)
21. Conrat, J.M., Pajusco, P.: A Versatile Propagation Channel Simulator for MIMO Link Level Simulation. EURASIP Journal on Wireless Communications and Networking 2007, Article ID 80194, 13 (2007)
22. Moupfouma, F.: Improvement of rain attenuation prediction method for terrestrial microwave links. IEEE Trans. Antennas Propag. 32(12), 1368–1372 (1984)
23. Schwarz, R.T., Knopp, A., Lankl, B.: The Channel Capacity of MIMO Satellite Links in a Fading Environment: A Probabilistic Analysis. In: International Workshop on Satellite and Space Communications, IWSSC 2009, Siena-Tuscany, Italy (September 9-11, 2009)

Unified Multibeam Satellite System Model for Payload Performance Analysis

Ricard Alegre-Godoy, Maria-Angeles Vázquez-Castro, and Lei Jiang

Department of Systems and Telecommunications Engineering
Universitat Autònoma de Barcelona, Spain
{ricard.alegre,angeles.vazquez,jiang.lei}@uab.es

Abstract. This paper presents a novel unified multibeam satellite system model for the performance analysis of different satellite payloads. The model allows the analysis in terms of Signal to Interference plus Noise Ratio (SINR) and Co-Channel Interference (CCI). Specifically we formulate the SINR as a function of the multibeam geometry for a given user location granularity. Furthermore, we apply our model to analyze the performance of two novel satellite payloads with respect to current conventional (CONV) ones using fixed frequency reuse and per-beam frequency/time assignment: the so-called "flexible" (FLEX) payload and the "beam-hopping" (BH) which allow a flexible per-beam frequency assignment and a flexible per-beam time assignment respectively. Our results show that CONV payloads achieve higher SINR values than BH and FLEX payloads at the expense of lower bandwidth assignment to the beams. Leading, therefore, to a trade-off, between received signal quality and resource management flexibility.

Keywords: Co-channel, interference, payload, multibeam and model.

1 Introduction

Current trends of multibeam satellite systems focus on the design of more efficient systems in order to achieve not only larger throughputs but also flexible resource management. There already exists an amount of work corresponding to this topic, such as the implementation of new techniques, e.g. power control [1], Forward Error Correction (FEC) codes at physical or link layer [2] and Adaptive Coding and Modulation (ACM) techniques [2]. In addition, another way to achieve larger throughputs is by increasing the number of beams. However, this leads to an increment of the CCI since the same frequency is reused by a subset of beams.

This effect was noticed in reference [3] where pre-coding schemes were used in order to overcome the CCI. Also references [1] and [4] focus on algorithms for satisfying user requirements and performing the multiple access respectively taking into account the minimization of the CCI. Therefore, studying the CCI is of relevant importance in satellite systems in order to validate new multibeam satellite payload models.

G. Giambene and C. Sacchi (Eds.): PSATS 2011, LNICST 71, pp. 365–377, 2011.

The aim of this paper is to formulate the unified expressions for multibeam satellite systems in order to compare the performances of any payload models. Based on the general expressions we compare the performance of three different satellite payloads, a conventional payload model (CONV) and two novel payload models named flexible (FLEX) and beam-hopping (BH). All three models are designed for the multimedia broadband satellite services. This comparison is carried out in terms of SINR and CCI.

The remainder of the paper is organized as follows: Section 2 introduces the derivation of the general system model. In Section 3 we introduce three payload models which are designed for the broadband multimedia and IP services. Finally in Section 4 we evaluate the performance of the payload models. Section 5 draws the conclusions.

2 Derivation of a Unified System Model

In this section, we first depict preliminary issues for the general system model derivation, i.e. the multibeam geometry and chosen antenna models. Subsequently, we express the steps to model the system in a general and unified way.

2.1 Multibeam Satellite Geometry

The multibeam satellite geometry interested in this paper is shown in Fig. 1, without loss of generality we focus on an example with two beams. Assuming that the position of a specific user (e.g. P) in a beam i, a beam center BC_j and the satellite SL is known, we can compute the distances (P,BC_j), (P,SL) and (SL,BC_j). Therefore, we can derive the angle θ_{ij} between the link (P,SL) and (SL,BC_j) by applying the cosine law.

$$\theta_{ij} = \arccos\left(\frac{(P,BC_j)^2 - (SL,BC_j)^2 - (SL,P)^2}{-(SL,BC_j)(SL,P)} \right). \tag{1}$$

where θ_{ij} is the angle between the user location inside the beam i and the corresponding beam center BC_j (e.g. θ_{12} shown in Fig. 1). Note that we can also obtain θ_{ii} (e.g. θ_{11} in Fig. 1) in the same way.

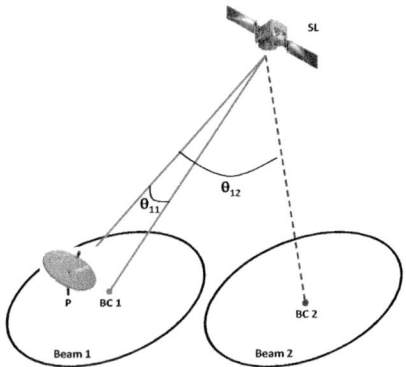

Fig. 1. Considered geometry

2.2 Analytical Antenna Models

We now present the two analytical antenna models that generate the multibeam coverage.

The first one is the Single Feed per Beam Network (SFBN) antenna model with combined transmission and reception antennas, i.e. each of the beams has a dedicated feed element to generate the beam and the same antennas are used for signal transmission and reception. In the analyzed payload models, both CONV and BH structures implement SFBN antenna model which is analytically expressed as in [3]:

$$G(\theta) = G_{max} \left(\frac{J_1(u)}{2u} + 36 \frac{J_3(u)}{u^3} \right)^2 .$$ (2)

where $u = 2.07123 \frac{\sin \theta}{\sin \theta_{-3dB}}$, being θ_{-3dB} the half angle power bandwidth, G_{max} the antenna boresight gain and J_1 and J_3 are the Bessel functions of first and third kind respectively.

The other one is the Array Fed Reflector (AFR) antenna model using separated transmission and reception antennas, i.e. we have fewer antenna elements than beams, and the beams are generated through a Digital Beam Forming Network (DBFN). Different antennas are used for transmission and reception of the signal. FLEX payload structure implement the AFR model which can be modeled as in [5]:

$$G(\theta, \phi) = \sum_{i=1}^{N} c_i g_i(\theta, \phi) \quad for \quad \sum_{i=1}^{N} |c_i|^2 .$$ (3)

where N are the number of elements in the AFR antenna, c_i is a complex excitation coefficient and $g_i(\theta, \phi)$ is the secondary component beam directivity.

2.3 SINR Derivation

In this subsection we introduce the formulation of the general multibeam satellite system model. We first define the overall channel matrix $\mathbf{H} \in C^{kxk}$ which is composed of two terms: (1) the satellite antenna gains matrix $\mathbf{G} \in C^{kxk}$ which depends on the angle θ, (2) the link budget matrix $\mathbf{A} \in C^{kxk}$. Subsequently, the received signal model and SINR can be formulated to study the CCI.

We adopt following notations:

- Vectors are set in bold lowercase letters.
- Matrixes are set in bold uppercase letters.
- Superscript $(.)^T$ denotes the transpose of a vector or matrix in (.).
- diag($\mathbf{x}$) stands for a diagonal matrix with the elements of $\mathbf{x}$ on its main diagonal.

The scenario is shown in Fig. 2, where a user in the interested beam i (e.g. beam 1 in the figure) is being interfered by any number of beams, e.g. k. The desired signal power level depends on the angle, θ_{ij}, where $i = j$ (θ_{11} in the figure), of the user with its beam center. The interference signal power level depends on the θ_{ij}'s where $i \neq j$ (θ_{12} to θ_{1K} in the figure).

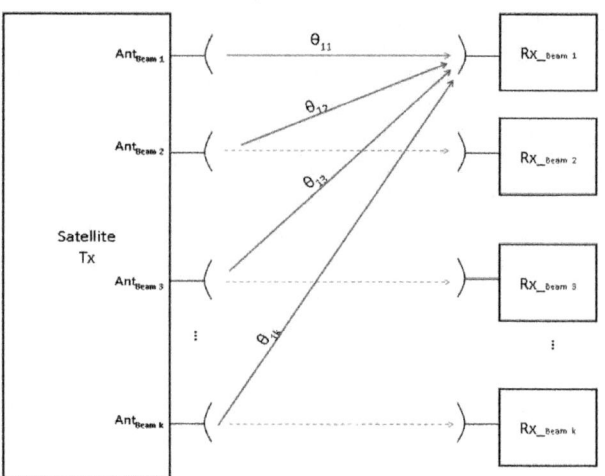

Fig. 2. Considered scenario

Let the symbols transmitted to user i inside the coverage of beam i be defined as $\mathbf{x}_i = [x_{i1}, x_{i2}, \dots, x_{iM}]$.

Let also define the link budget matrix $\mathbf{A} \in C^{k \times k}$ and the channel matrix gain $\mathbf{G} \in C^{k \times k}$ which includes the satellite antennas gains as:

$$\mathbf{A} = diag(\sqrt{\beta_1}, \sqrt{\beta_2}, \dots, \sqrt{\beta_k}).$$ (4)

$$\mathbf{G} = \begin{pmatrix} g_{11} & g_{12} & \cdots & g_{1k} \\ g_{21} & g_{22} & & g_{2k} \\ \vdots & \vdots & \ddots & \vdots \\ g_{k1} & g_{k2} & \cdots & g_{kk} \end{pmatrix}.$$ (5)

where:

- $\beta_i = OBO_{hpa} L_{sat} L_{down} G_{gt}$ where the parameters show the gain and losses which is the gain and losses terms that do not depend on the angle θ, OBO_{hpa} is the Output Back-Off of the High Power Amplifier (HPA), L_{sat} is the satellite repeater output losses, L_{down} is the free space losses and the additional rain, polarization, atmospheric and scintillation losses of the FWD downlink, and G_{gt} is the ground terminal antenna gain .

- $g_{ij} := \sqrt{g(\theta_{ij})}$ is the square root of the antenna gain between the satellite transmitter antenna for beam j and beam i, being θ_{ij} the angle that forms the receiver in beam i towards the spot beam center j as seen from the satellite.

Hence, we can formulate the overall channel matrix H as:

$$\mathbf{H=AG=}\begin{pmatrix} h_{11} & h_{12} & \cdots & h_{1k} \\ h_{21} & h_{22} & & h_{2k} \\ \vdots & \vdots & \ddots & \vdots \\ h_{k1} & h_{k2} & \cdots & h_{kk} \end{pmatrix}. \tag{6}$$

where the element of $\mathbf{H}$, e.g. h_{ij} is defined as $h_{ij} = \sqrt{\beta_i g(\theta_{ij})}$. Note that the definition of matrix $\mathbf{A}$ and matrix $\mathbf{G}$ can help us to separate the overall channel matrix into two terms, one does not depend on the θ (i.e. β_i) whilst the other one depends on θ (i.e. $g(\theta_{ij})$). The reason being, we can evaluate the CCI at beam level as a function of the angle θ.

Subsequently we can express the received symbols $\mathbf{y}_i(\theta) \in C^{Mxl}$ for a user in a beam i by separating the received signal from non-desired signal as in equation (7).

$$\mathbf{y}_i(\theta) = \sqrt{P_{sat}}\, h_{ii}\, \mathbf{x}_i + \sum_{j=1, j\neq i}^{j=k} \sqrt{P_{sat}}\, h_{ij}\, \mathbf{x}_j + \mathbf{n}_i. \tag{7}$$

where the term $\sqrt{P_{sat}}\, h_{ii}\, \mathbf{x}_i$ is our desired signal, the term $\sum_{j=1, j\neq i}^{j=k} \sqrt{P_{sat}}\, h_{ij}\, \mathbf{x}_j$ is the co-channel interference and the term $\mathbf{n}_i$ is a column vector of zero mean and complex circular noise with variance N.

By replacing h_{ij} with $\sqrt{P_{sat}\beta_i g(\theta_{ij})}$ we can obtain the following expression:

$$\mathbf{y}_i(\theta) = \sqrt{P_{sat}\beta_i g(\theta_{ii})}\, \mathbf{x}_i + \sum_{j=1, j\neq i}^{j=k} \sqrt{P_{sat}\beta_i g(\theta_{ij})}\, \mathbf{x}_j + \mathbf{n}_i. \tag{8}$$

The SINR can be derived from equation (7) for a specific user in beam i by assuming that the power of the transmitted symbols is normalized, $E[|\mathbf{x}_i|^2]=1$.

$$SINR_i(\theta) = \frac{P_{sat}\,|h_{ii}(\theta)|^2}{\sum_{j=1, j\neq i}^{j=k}(P_{sat}\,|h_{ij}(\theta)|^2) + N_i} \tag{9}$$

By replacing h_{ij} with $h_{ij} = \sqrt{\beta_i g(\theta_{ij})}$, equation (9) can be reformulated as:

$$SINR_i(\theta) = \frac{P_{sat}\beta_i g(\theta_{ii})}{\sum_{j=1, j\neq i}^{j=k}(P_{sat}\beta_i g(\theta_{ij})) + N_i} \tag{10}$$

Regarding the obtained expressions (9) and (10) we have to note that:

- The expression of the received signal, i.e. $\mathbf{y}_i(\theta)$, and the signal to interference plus noise ratio, i.e. $SINR_i(\theta)$, depend not only on the θ_{ii} where $i = j$ (i.e. the angle between user i and its beam center i) but also on the θ_{ij} where $i \neq j$ (i.e. the angle between user i and each of the interferers j). Thus, we have expressed the received signal and the SINR as a function of θ, which is the objective of this subsection.

- $\beta_i = OBO_{hpa}L_{sat}L_{down}G_{gt}$ depends on the system payload design. We can extract specific SINR expressions for each of the payloads, $SINR_i^{CONV}(\theta)$, $SINR_i^{FLEX}(\theta)$ and $SINR_i^{BH}(\theta)$, by replacing β_i with β_i^{CONV}, β_i^{FLEX} and β_i^{BH} respectively where the superscripts $CONV$, $FLEX$ and BH stand for the acronym of each of the payloads we will present in Section 3.

3   Payload Models

The aim of this section is to describe three different payloads models which are designed for the multimedia and IP broadcasting services in a multi-star access network, using Digital Video Broadcasting over Satellite second generation (DVB-S2) in the FWD link and Digital Video Broadcasting Return Channel over Satellite (DVB-RCS) in the return (RTN) link.

We first study the current operating payloads in multibeam satellite systems (Conventional payload or CONV in the equations) in order to have reference for the comparison with the other two payloads. Subsequently, we study the flexible payload model where the carrier allocation is fully flexible for each beam (Flexible Payload or FLEX in the equations). Finally we introduce the beam-hopping payload model, in which a subset of beams can be illuminated simultaneously during each timeslot (Beam-hopping payload or BH in the equations).

Regarding to the satellite payloads configuration and performance evaluation, more results can be found in [6].

3.1   Conventional Payload

Conventional payload, abbreviated CONV, is used for the classical MF-TDM transmission schemes where the total bandwidth is divided into a fixed number of portions. Each beam can be assigned one of the portions. Portions of the bandwidth (carrier slots) can be reused or not. The elements forming part of the conventional FWD link payload can be seen in Fig. 3.

After the uplink signal filtering of each polarization output, the antenna elements are connected to a 2 for 1 redundant Low Noise Amplifier (LNA). Depending on the frequency plan, more than one type of Down Converter (DOCON) could be needed, so the splitter performs the action of sending the signal to the correct DOCON. Then, the DOCONs down-convert each of the frequency segments. Depending on the number of gateways and the number of polarizations, the number of inputs and outputs of the DOCONs could change.

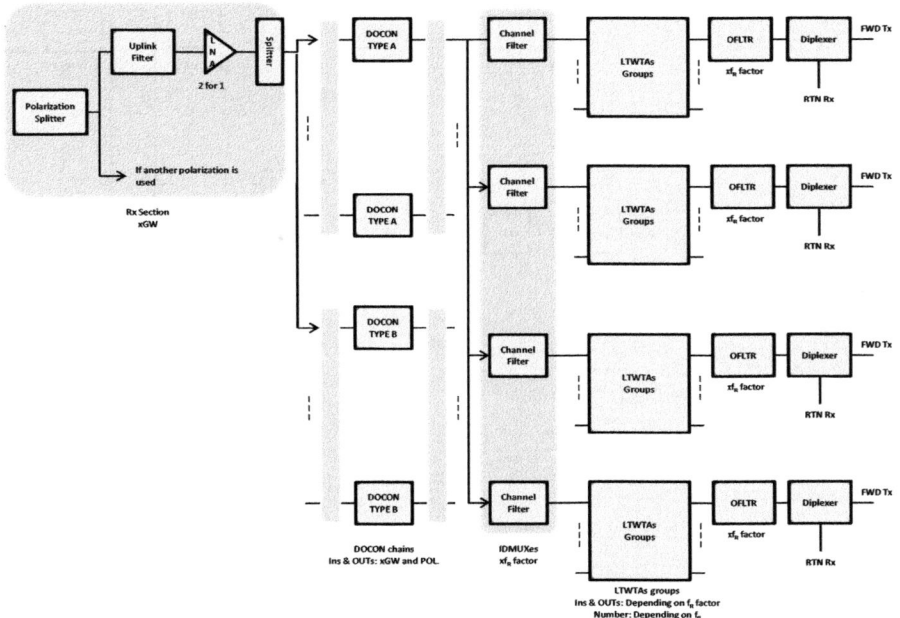

Fig. 3. Conventional payload

Then Input Demultiplexers (IDMUXes) separate the channels assigned to each user link beam, the needed number of IDMUXes is at least the same as the frequency reuse factor. A group of Linear Traveling Wave Tube Amplifiers (LTWTAs) are used to provide the final amplification of the channels and Output Filters (OFLTRs) are used to limit the inter-modulation and harmonics high amplification effects.

3.2 Flexible Payload

Flexible payload, or just FLEX, is used in Non Orthogonal Frequency Reuse (NOFR) air interfaces where a ground cell can allocate a variable number of carriers depending on the traffic requirement. The elements constituting the flexible FWD link payload can be seen in Fig. 4.

In the FWD link, firstly each polarization output signal is amplified by a LNA, then the DOCONS down-convert the received signals to the C-band frequency, consequently the On Board Processor (OBP) can process the converted signals. The Intermediate Filters (IFLTRs) are applied to limit the out of band spurious emissions.

The OBP performs the following actions:

- Spectral isolation of the individual modulated user channels that compose each Frequency Division Multiplexing (FDM) multiplexed, multicarrier, gateway signal.
- Routing and steering of the complex samples that compose the uplink carriers signals received on FWD uplink to the destined FWD downlink Digital Beam Forming Network (DBFN) in order to generate the subsequent FWD downlink signals.

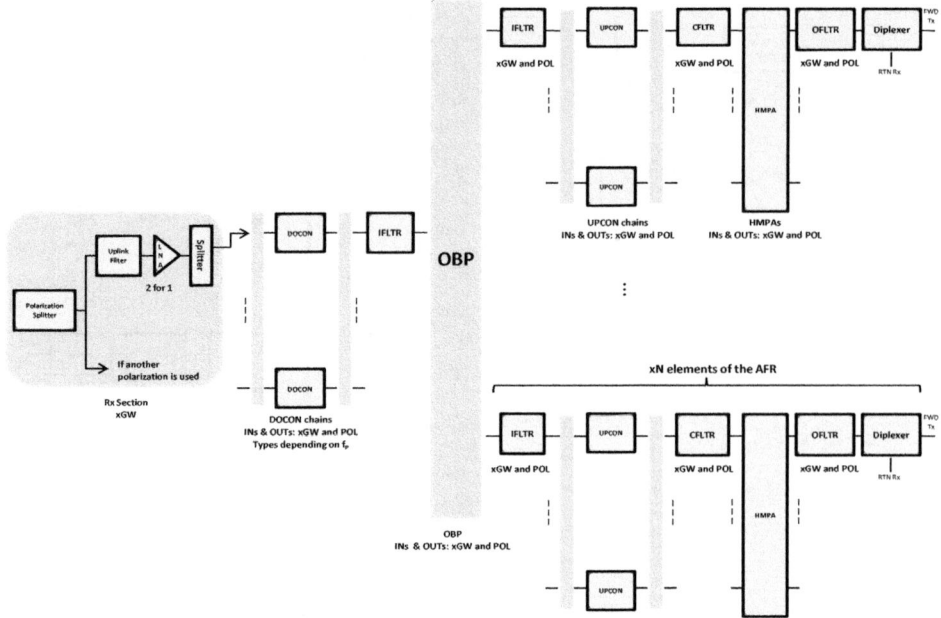

Fig. 4. Flexible payload

- Spatial filtering of the complex samples that compose the uplink carriers signals to generate the subsequent constituent beam signals to be applied to the antenna elements.
- Frequency synthesis of the spatially filtered element beam signals to generate the FDM multiplexed, multicarrier element signal to be applied to each of the antenna elements that compose the transmission antenna array.

The signals from the output of the OBP are then up-converted to the downlink frequencies by the Up Converters (UPCONs) and filtered by the Chanel Filters (CFLTRs) to limit the out of band spurious emissions. Hybrid Matrix Power Amplifiers (HMPAs) composed of LTWTAs are used to amplify the signals that feed the antenna elements. Signals are filtered with OFLTRs before transmitted to limit the noise in the receive frequency band and to limit the spurious emissions.

3.3 Beam-Hopping Payload

Beam-hopping payload, abbreviated BH, is used in air interfaces where the total bandwidth is used in some specific beams during a timeslot. The elements in the beam-hopping FWD link payload can be seen in Fig. 5.

In the FWD link the signals go through the 2 for 1 LNAs, then are down-converted to the OBP C-band and processed by the IFLTRs to limit the out of band spurious emissions.

The OBP performs the following actions:

- Spectral isolation of the individual, phase modulated carriers signals that constitute each FDM multiplexed, multicarrier gateway signal.
- Grouping the carriers received on the FWD uplink into FWD downlink sets.
- Frequency synthesis of the FWD downlink carrier sets to generate the subsequent FDM multiplexed, multicarrier signals. These synthesized multicarrier signals are identified as beam-hopping signals.
- Application of the beam-hopping signals to the antenna elements.

The signal at the output of the OBP is up-converted from the OBP C-band to the FWD downlink frequency by the UPCON, filtered and amplified by HMPAs. The signal is filtered with OFLTRs before sending to the antenna feed elements to limit noise and harmonic distortion.

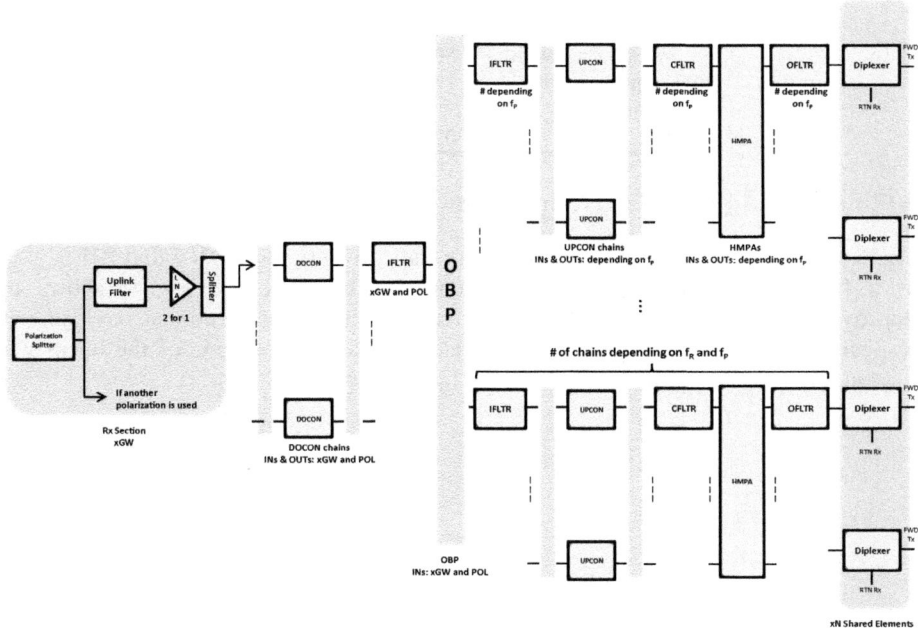

Fig. 5. Beam-hopping payload

4 Numerical Results

In order to evaluate and compare performance of the payload models presented above, in a realistic multibeam scenario, we will study the CCI and the $SINR(\theta)$. Note that, given space constraints, herein we only discuss the comparative performance evaluation of CONV and BH payloads. However, we note that the authors in [7] have shown that the flexible payload and beam-hopping payload are dual of each other.

We assume a 70-beam multi-star access system scenario. For each of the beams we analyze:

- The effect of the interference in the received SINR with respect to the number of adjacent interfering beams.
- The effect of the interference in the SINR with respect to all non-adjacent beams.

The system parameters are shown in Table 1 and the payload parameters of the CONV and BH payload parameters are extracted from [6].

Table 1. System parameters for the simulations

Parameter	Value
Orbit	GEO
Satellite Position	$0°$ Long, $0°$ Lat
Frequency Band	19.50GHz
Modulation	8PSK
System Bandwidth	500MHz
Frequency reuse factor	17.5
θ_{-3dB}	0.249°

4.1 Effect of Adjacent Interfering Beams

Fig. 6 shows the average received SINR for the conventional CONV and BH payload as a function of the number of adjacent interfering beams. Fig. 7 shows the improvement of the average SINR in percentage in the conventional payload with respect to the BH payload for a user located inside the coverage of the beam with coordinates $14.25°$ Latitude and $50.75°$ Longitude.

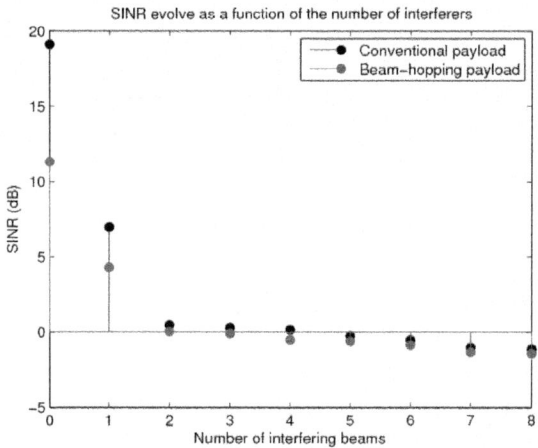

Fig. 6. Average SINR as a function of the number of interferers

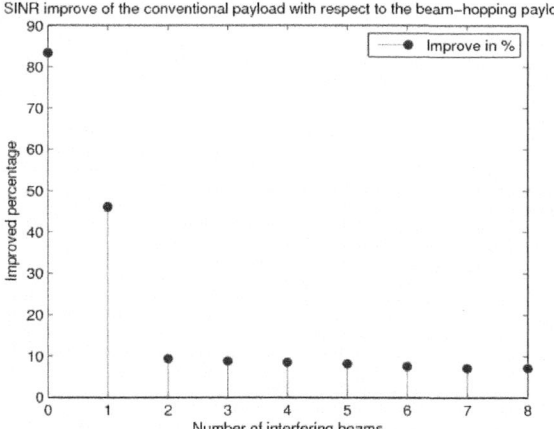

Fig. 7. SINR increase of the conventional payload with respect to the beam-hopping payload

From the figures above it can be observed:

- Case with 0 interferers is equivalent to the received SNR in the beam of interest.
- Adjacent interfering beams cause a fast decrease of the received SINR for both payloads; however, differences are lower as the number of adjacent beams increase.
- For any number of adjacent interfering beams CONV payload is less affected by CCI than BH payload and hence by results obtained in [7] than FLEX payload since the bandwidth assigned to beams is higher in BH and FLEX schemes.

Therefore, it should be avoided to assign the same frequency band to adjacent beams in the CONV payload, and to illuminate at the same time adjacent beams in the BH payload. Besides we can note that CONV payload achieves higher SINR's basically because the amount of bandwidth assigned to each beam is lower than in the BH payload where we assign all the bandwidth to each beam. This bandwidth assignment is done in order to satisfy user requirements in a more efficient and flexible way rather the fixed conventional way used in CONV model. Hence there is a clear trade-off between bandwidth assignment and signal strength and by extension between throughput and signal strength. This means, if we want to achieve larger throughputs, we have to assign more bandwidth to each beam, in order to deal with broadband traffic, but received signal power will be lower because of the noise bandwidth. It is worth mentioning that this trade-off is not a bad feature for the novel payloads, as a uniform quality throughout the coverage might not be necessary.

4.2 Effect of Non-adjacent Interfering Beams

In this subsection we show the effect on the SINR in the beam of interest when we set non-adjacent beams following a typical 4 colored frequency reuse scheme in the conventional payload (for 70 beam frequency reuse factor 17.5). In order to obtain comparison we will illuminate the same beams in the beam-hopping payload and compare the obtained results.

Obtained SINR for CONV payload can be seen in Fig. 8.

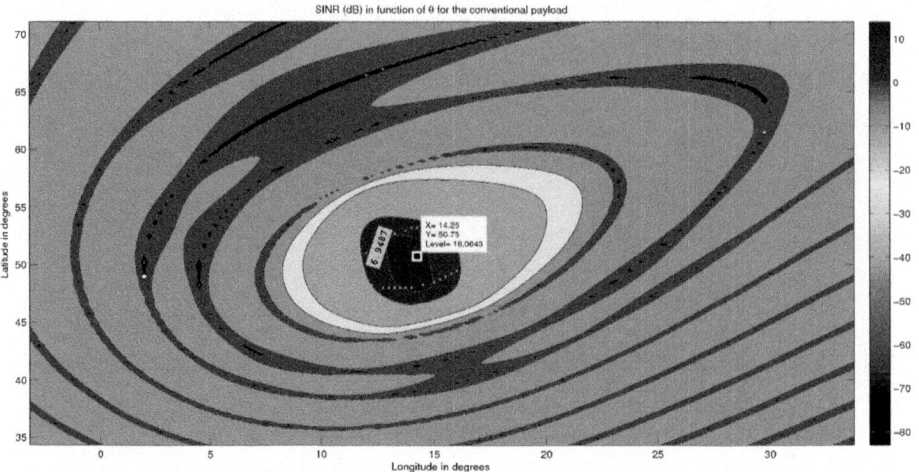

Fig. 8. SINR in the beam of interest for a set of non-adjacent interfering beams using a four colored frequency reuse scheme

In the Fig. 8, the grey line indicates the original contour of the beam when there are no interferers (3dB loss). Note that within the original contour of the beam, now the SINR values can differ in 10dB as show the contours. However the SINR levels received in zones close to the center of the beam are big enough to ensure the correct reception of the signal.

Simulations results (not shown here for matter of lack of space) let us draw the same conclusions for the BH payload, a similar interference pattern is obtained. Nevertheless, when illuminating the same set of beams assigning all the bandwidth to each of them, the SINR value obtained in the center of the beam is 6dB under the value of the conventional payload, i.e. as in the adjacent interfering beams case, BH payload is more affected by the CCI than CONV payload. As explained before, this is produced because BH and FLEX schemes assign larger bandwidth to each beam, hence the noise bandwidth is bigger. So, under this scenario we also find the bandwidth signal strength throughput trade-off only that the SINR decrease is not produced in such a drastic way as the interfering beams are now further.

5   Conclusions

In this paper we have presented a unified system model for multibeam satellite systems. The model allows the performance analysis of different payloads in terms of received signal strength and co-channel interference. The model is easy to use as it identifies the key parameters of the payloads to be analyzed and how they should be included in the model.

We have applied our model for the performance analysis of two novel satellite payloads with respect to current conventional (CONV) ones: the so-called "flexible" (FLEX) payload and the "beam-hopping" (BH), both described in the paper. The first one allows a flexible per-beam frequency assignment while the second one allows a flexible per-beam time assignment. This flexibility is lacking in current CONV payloads, with fixed frequency reuse and per-beam frequency/time assignment.

The numerical results we have obtained with our developed unified model indicate that the CONV payload achieves better received signal strength and co-channel interference management throughout the coverage than the BH novel payload. This means that a trade-off exists between received signal quality and resource management flexibility. The reason for this is that the new payloads can accommodate larger bandwidths per beam, which is an advantageous feature for handling broadband traffic. This trade-off is actually not a bad feature for the novel payloads as a uniform quality throughout the coverage might not be necessary. For lack of space we have not included the numerical results for the FLEX payload, which show the same trend as the BH and it can be further justified by the duality between both payloads ([7]).

Our results are fully in line with the results in the related papers [1] and [7], which focus on the resource management algorithms of the proposed payloads.

References

1. Lei, J., Castro, M.A.V.: Joint Power and Carrier Allocation for the Multibeam Satellite Downlink with Individual SINR constraints. In: Proc. IEEE Int. Conf. on Commun., Cape Town, South Africa, pp. 1–5 (May 2010)
2. Maral, G., Bousquet, M.: Satellite Communications Systems: Systems, Techniques and Tehcnology, 4th edn. Wiley, Chichester (April 2002)
3. Díaz, M.A., Courville, N., Mosquera, C., Liva, G., Corazza, G.E.: Non-Linear Interference Mitigation for Broadband Multimedia Satellite Systems. In: International Workshop on IWSSC 2007, pp. 61–67 (September 2007)
4. Liu, Q., Li, J.: Multiple Access in Broadband Satellite Networks. In: IEEE International Conference on ICC, vol. 1, pp. 417–421 (May 2003)
5. Ueno, K.: Multibeam Antenna using a Phased Array Reflector. In: Antennas and Propagation Society International Symposium, vol. 2, pp. 840–843 (July 2007)
6. Alberti, X., Cebrian, J.M., Del Bianco, A., Katona, Z., Lei, J., Vázquez- Castro, M.A., Zanus, A., Gilbert, L., Alagha, N.: System Capacity Optimization in Time and Frequency for Multibeam Multi-media Satellite Systems. In: Proc. 5th Advanced Satellite Multimedia Systems Conference and the 11th Signal Processing for Space Communications Workshop (ASMS/SPSC) 2010, Cagliari, Italy (September 2010)
7. Lei, J., Castro, M.A.V.: Frequency and Time-Space Duality Study for Multibeam Satellite Communications. In: Proc. IEEE Int. Conf. on Commun., Cape Town, South Africa, pp. 1–5 (May 2010)
8. Cover, T.M., Thomas, J.A.: Elements of Information Theory. John Wiley & Sons Ltd., Chichester

Galileo E1 and E5a Link-Level Performances in Single and Multipath Channels

Jie Zhang and Elena-Simona Lohan

Tampere University of Technology,
Korkeakoulunkatu 1, 33101 Tampere, Finland
www.cs.tut.fi/tlt/pos
{jie.zhang,elena-simona.lohan}@tut.fi

Abstract. The emerging global satellites system Galileo has gained much public interest regarding location and positioning services. Two new modulations, Composite Binary Offset Carrier (CBOC) and Alternate Binary Offset Carrier (AltBOC) will be used in the E1 and E5 band in the Galileo Open service (OS), respectively. The AltBOC modulation has the advantage that the E5a and E5b band can be processed independently as traditional BPSK signal or together, leading to a better tracking performance in terms of noise and multipath mitigation at the cost of a large front-end bandwidth and increased complexity. The theoretical study of the signal tracking in each band, separately, has been addressed before, but a comparison between the E1 and E5 signals and validation through the simulation with the realistic channel are still lacking in the current literature. In this paper, the tracking performance between the Galileo E5a signal and Galileo E1 signal with different noise level and multipath profiles are compared by using the Simulink-based simulators built within our department at Tampere University of Technology. The simulation results are shown in terms of Root Mean Square Error (RMSE). The probability distribution of code tracking error is also investigated.

Keywords: Galileo, E5a/E5 signal, E1 signal, Multiplexed Binary Offset Carrier (MBOC), AltBOC, error distribution, multipath channel, open source, Simulink Galileo simulator.

1   Introduction

During the second half of the last century, Global Navigation Satellite Systems (GNSS) have been widely used in personal devices, public transportation and industries. A GNSS device can point out the exact location of any user on the surface of the earth anytime and anywhere, provided that it is placed in a direct Line Of Sight (LOS) with at least four satellites. As one of the emerging GNSS, Galileo is going to provide more services, higher availability and higher accuracy than the only fully operational GNSS nowadays, Global Position System (GPS). Galileo will provide worldwide services depending on user needs. One of them is Open Service (OS), which is designed for mass-market and will be

G. Giambene and C. Sacchi (Eds.): PSATS 2011, LNICST 71, pp. 378–390, 2011.

free of user charge. Two frequency bands, E5, consisting of two sub-bands E5a and E5b with carrier frequency at 1176.45 MHz and 1207.14 MHz and E1 with carrier frequency 1575.42 MHz, will be used for transmitting OS signals. Multiplexed Binary Offset Carrier or MBOC are defined to be the common modernized GPS and Galileo modulations for civilian use. MBOC introduces more power on higher frequencies compared with BOC(1,1) case, by multiplexing with a high frequency BOC(6,1) component, which improves the performance in tracking [1]. The MBOC implementation for Galileo is adding simultaneously a BOC(1,1) and BOC(6,1), defined as Composite-BOC (CBOC). The AltBOC modulation is designed to be used in Galileo OS E5 band. AltBOC(15,10) modulated E5 signal is by far the most sophisticated signal among all the signals used for GNSS. Four signal component are modulated into a wideband signal by AltBOC modulation [2]. Two of them will carry navigation messages and the remaining two are data-free pilot channels. The AltBOC modulation provides such advantage that E5a and E5b can be processed independently, as traditional BPSK(10) signal, or together, leading to a better tracking performance in terms of noise and multipath mitigation at the cost of a large front-end bandwidth and increased complexity [3]. In addition, E5 signal has chip rate of 10.23 MHz, which is ten times higher than the E1 signal's chip rate f_c=1.023 MHz. The higher chip rate may provide better tracking performance. Recently only E5a band has attracted attention in the context of dual/multi frequency Galileo receives. E5a can be acquired independently and the requested front-end bandwidth is less than half of the bandwidth for the whole E1 signal. It has also been proved that combining E1/E5a is the best choice for dual frequency receiver and has the additional property that it overlaps with GPS frequency L1/L5.[8] This property also provides the advantage of an easier integrability of a joint Galileo/GPS receiver. Many publications have addresses E5 acquisition strategies[7], [10], and code tracking noise based on mathematic formula [3], [11], [12]. However, very few studies have been published about the comparative performance of E1 with E5a in terms of signal tracking accuracy and the validation of potential performance of E5 signal in realistic channel at link level. In this paper, the authors evaluate and compare the signal tracking performance of E1 and E5a in link-level Simulink simulators.

This paper is organized as follows: first, the E1 and E5a signal simulators used in the paper are described. Then, the performance of code tracking with E1 and E5a is presented in terms of Root Mean Square Error (RMSE). Finally, the code tracking error distribution is analyzed.

2 Simulink Model Overview

2.1 Generic Structure

Simulation is a powerful method in the analysis and design of communication devices. The performance of new signals, new algorithms can be assessed before it is implemented on a real model. The E1 signal simulators and E5a signal simulators used in this paper for evaluating the tracking performance with E1 and E5a signal were created at Tampere University of Technology (TUT).

The generic structure of the simulators is shown in Fig. 1, which consists of five blocks: transmitter, propagation channel, front-end, acquisition and tracking block. More detail of E1 and E5a signal Simulink simulators will be described in the following sections.

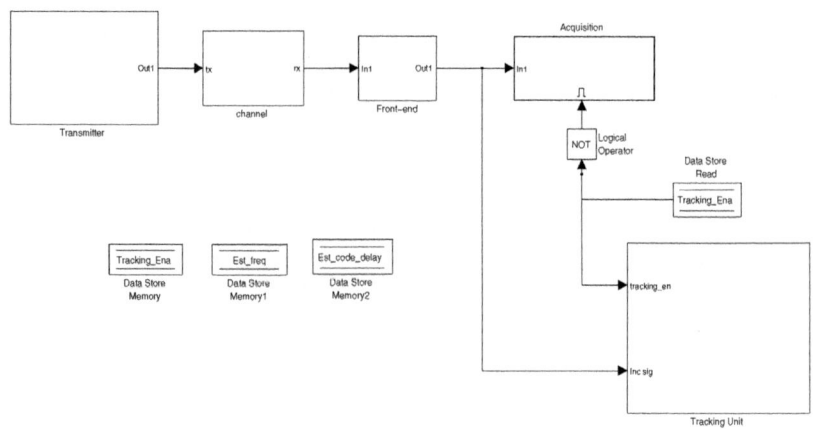

Fig. 1. Generic Simulink block of Galileo simulator at TUT

2.2   Galileo E1 Simulink Model

Transmitter. The E1 signal transmitter block is implemented based on CBOC modulation, including primary code and secondary code, in the accordance with the latest Galileo OS SIS ICD [2]. The snapshot of E1 signal transmitter block is shown in Fig. 2. In the transmitter block, E1B is CBOC(+) modulated signal with navigation data and E1C is CBOC(-) modulated signal with a pre-defined bit sequence of CS25 (i.e.,pilot channel). The E1 signal is formed as the difference between those two signals. The signal at the output of the transmitter is at Intermediate Frequency (IF).

Channel. The channel block generates the multipath signals and complex noise for a user-defined C/N_0. The interference from GPS or other sources, excepting noise and multipath are not considered here. Fig. 3 shows the snapshot of the channel block. The multipath delay and power are other two input parameters for channel block. Two channel configurations can be used: static and time variant. The input parameters for static channel are user defined, and for time variant channel, the path delay and power are defined through a Land and Mobile Multipath Channel Model from DLR [9].

Front-end The front-end block in E1 signal simulator is used for receiver front-end filtering. Several front-end bandwidths can be used, i.e., infinite bandwidth for the ideal case, 4 MHz which covers the main lobe of E1 signal.

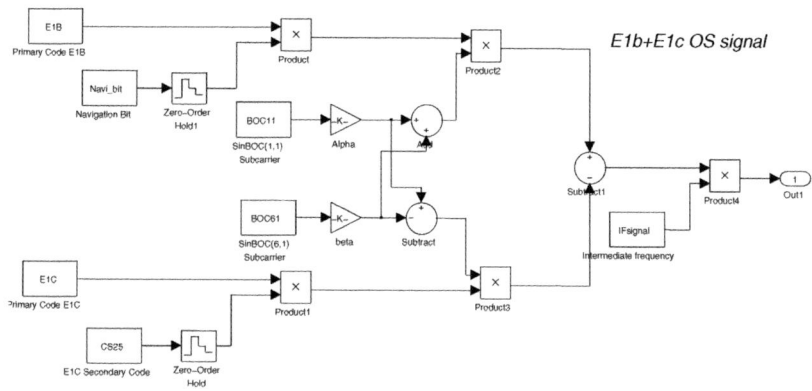

Fig. 2. The transmitter model in Galileo E1 signal simulator at TUT

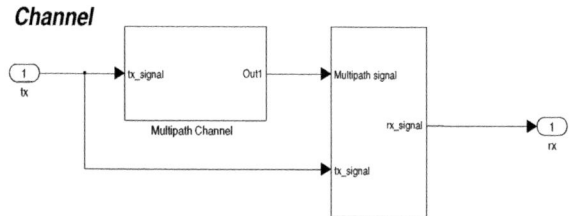

Fig. 3. The channel model in Galileo E1 simulator at TUT

Tracking. When a signal is detected in the 'Acquisition Block', a control signal 'Tracking Ena' will activate the 'Tracking Unit'. The tracking unit consists of three main blocks: carrier wipe-off block, code Numerically controlled Oscillator (NCO) block and dual channel correlation and discriminator block as in Fig. 5.

The task of the carrier wipe-off block is to down convert the incoming signal with the estimated frequency and phase from PLL and FLL in the tracking loop. After the carrier wipe-off, the real part and the imaginary part of the complex signal are separated as the in-phase (i.e., I channel in Fig. 5) and the quad-phase (i.e., Q channel in Fig. 5) channels in baseband. The 'code NCO' block is used to generate the local PRN reference code, which is shifted by the estimated code phase from DLL. According to the correlator offset and the status of phase holding shifter, the primary code and the sub-carrier offset can be determined. The reference code sequences are generated separately for E1B and E1C channel. Since the CBOC modulation combines two sub-carrier wave components, the tracking can be done either with CBOC modulated reference codes (i.e., CBOC(+) for E1-B data channel and CBOC(-) for E1-C pilot channel), or with SinBOC(1,1) modulated reference code for both E1-B and E1-C channels. The simulations in this paper are using SinBOC(1,1) modulated E1 reference code. In the dual channel correlation and discriminator block, the E1B

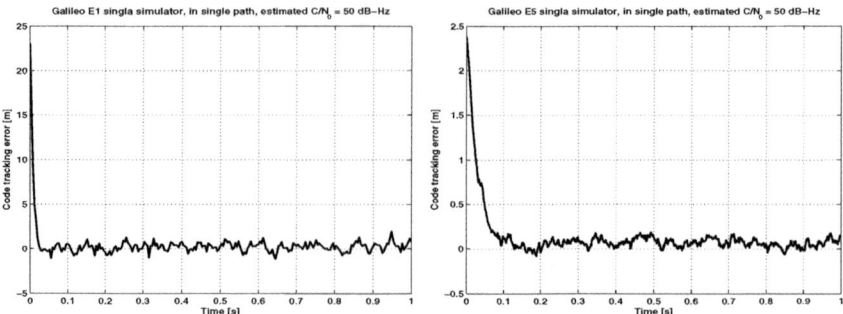

Fig. 4. Left: Code tracking error versus simulation time of E1 signal at estimated $C/N_0 = 50dB - Hz$ with non-coherent integration over 4 ms; Right: Code tracking error versus simulation time of E5a signal at estimated $C/N_0 = 50dB - Hz$ with non-coherent integration over 1 ms

and E1C channels are implemented separately. In both channels, FLL, PLL and DLL are included. In the DLL discriminator block, various conventional DLL discriminator functions are implemented, such as Narrow Correlator [5] and High Resolution Correlator (HRC) [6]. The C/N_0 estimator is also implemented. The C/N_0 estimation is performed based on the ratio of the signal's wideband power to its narrowband power as described in Fig. [14].

The code tracking error is calculated after the simulation is finished. An example of the code tracking error versus simulation time is shown as left figure in Fig. 4. The main parameters used in the E1 signal Simulink simulator are summarized in Table 1.

Table 1. E1 signal simulator parameters

Parameters	Typical value
Sampling frequency f_s in whole simulator	13 MHz, 26 MHz
Intermediate Frequency IF	3.42 MHz, 6.7 MHz
Early-Late correlator spacing	bandwidth dependent, 0.1 chips (infinite BW), 0.1 chips (13 MHz)
Reference code	E1 code with BOC(1,1) modulation

2.3   Galileo E5a Simulink Model

In the Galileo E5a Simulink simulator, the whole E5 signal is generated in the transmitter. At the receiver side, only the E5a band in processed.

Transmitter. The E5a signal transmitter generates E5 signal by using the AltBOC(15,10) 8-PSK modulation, as described in [2]. The snapshot of the E5 transmitter block is in Fig. 6. The transmitted signal at the output of transmitter block is shifted to Intermediate Frequency (IF), as shown in Fig. 7(a).

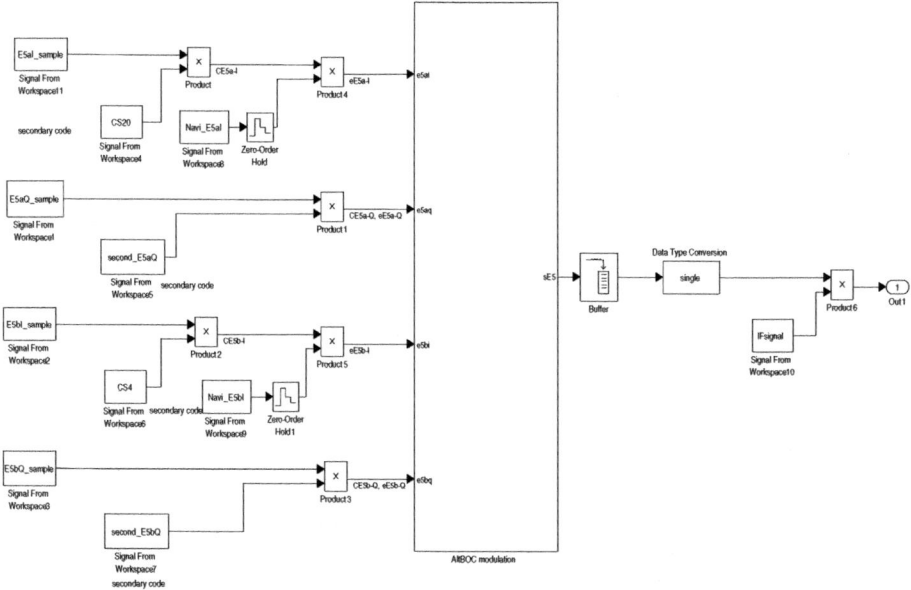

Fig. 5. The tracking block in Galileo E1 simulator at TUT

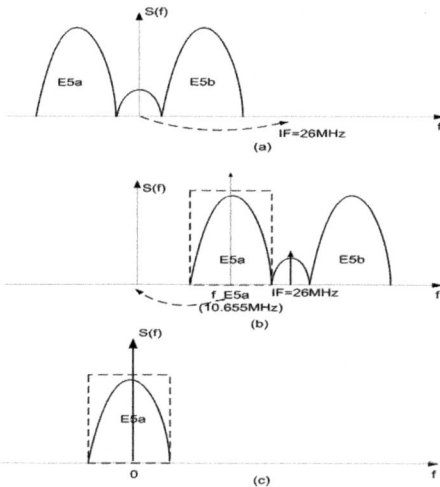

Fig. 6. The transmitter model in Galileo E5 simulator at TUT

Channel. The channel model used in E5a Simulink model has the same structure as that used in E1 simulator channel block. The complex noise and multipath are generated. The static and time variant channel can also be used.

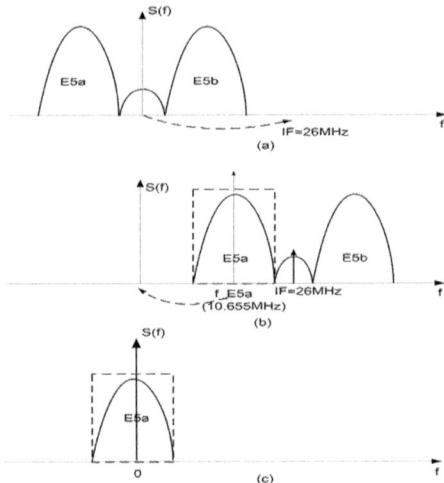

Fig. 7. (a)E5 signal spectra shifted to IF at transmitter; (b) Front-end filtering; (c) E5a signal down convert to baseband

Front-end. The front-end block in E5a simulator is used for filtering and down-sampling. The front-end filter has various bandwidth, i.e., 20.46 MHz bandwidth to cover the main lobe of E5a signal. The reason for using down-sample here is that the interested signal is only the main lobe of E5a signal, which has a much narrower band than the whole E5 band. Therefore, a lower sampling frequency used in the blocks after the filter without losing useful information can be realized.

Acquisition. The acquisition unit is also using FFT technique. Since the receiver only acquires the E5a signal, the E5aI without the sub-carrier is used as the reference code to estimate the frequency of E5a main lobe. The same as in E1 simulator, this estimated frequency will be used in the tracking unit to shift the filtered E5a signal.

Tracking. In the tracking unit, The structure and the functionality is the same as in E1 signal simulator. The carrier wipe-off down converts the E5a signal component to baseband, as shown in Fig. 7 (b) and (c). In the 'Code NCO' block, only the E5aI signals are generated in the code NCO. There is one channel in the 'Channel Correlation and discriminators' block to track E5a signal. PLL, FLL and DLL are implemented. Currently, Narrow Correlator [5] and HRC [6] are used in DLL as discriminator functions. The C/N_0 estimator is also implemented based on [14]. An example of code tracking error versus simulation time, which is calculated after the simulation is shown as right figure in Fig. 4 . The main parameter used in E5 signal simulator are summarized in Table 2.

Table 2. E5a simulator parameters

Parameters	Typical value
Sampling frequency f_s in transmitter and channel	126 MHz
Sampling frequency f_s in acquisition and tracking	42 MHz
Intermediate Frequency IF	26 MHz
E5a frequency f_{E5a}	10.655 MHz
Early-Late correlator spacing	bandwidth dependent, 0.1 chips (infinite BW), 0.8 chips (13 MHz)
Reference code	E5aI code

3   Simulation Results and Analysis

The tracking performances are evaluated with the E1 and E5a signal simulator, which have been described in the previous section. Due to the sensitivity of the receiver, the tracking units in both E1 and E5a signal simulators are enabled all the time in order to test the performance below the sensitivity. Different double-sided front-end bandwidths are considered in the simulations: 1. ideal infinite bandwidth; 2. 4 MHz for E1 band and 20.46 MHz for E5a band, which cover the main lobe the signals, respectively; 3. 13 MHz for both signals, which is a bandwidth chosen between 4 MHz and 20.46 MHz in order to have a fair comparison between two signals. The $E - L$ correlator spacing Δ are defined by the rule of $\Delta \geq f_c/BW$ [15], where f_c is the chip rate and BW is double-sided front-end bandwidth. RMSEs between the estimated delay and the true Line-Of-Sight (LOS) delay are calculated. The parameters used in the simulations are summarized in Table. 3.

Table 3. Simulation parameters for multipath scenario

Signal	E1			E5a		
Channel	Static channel					
f_s (MHz)	13	13	26.3	126	126	126
Bandwidth (MHz)	inf	4	13	inf	13	20.46
E-L spacing (chips)	0.1	0.1	0.26	0.1	0.5	0.8
Multipath distance (chips)	0.1	0.1	0.1	0.1	0.1	0.1

3.1   Delay Tracking in Single Path-Simulation Results

The tracking performances with E1 and E5a signal are first evaluated in a single path static channel profile. The simulation parameters used in the simulation can be found in the Table 3.

The tracking errors in meter versus estimated C/N_0 in a single path scenario are shown in Fig. 8. As can be seen from the figures, tracking E5a signal has better performance than tracking E1 signal with infinite bandwidth and the 20.46 MHz bandwidth, which covers the main lobe of the E5a band. With 13 MHz, the performance of tracking E5a signal outperforms than that of E1 signal most of the times. When the estimated C/N_0 drops to around 33 dB-Hz, the tracking performance with E5a signal degrades and is worse than the performance of E1 signal due to the signal energy loss on E5a band.

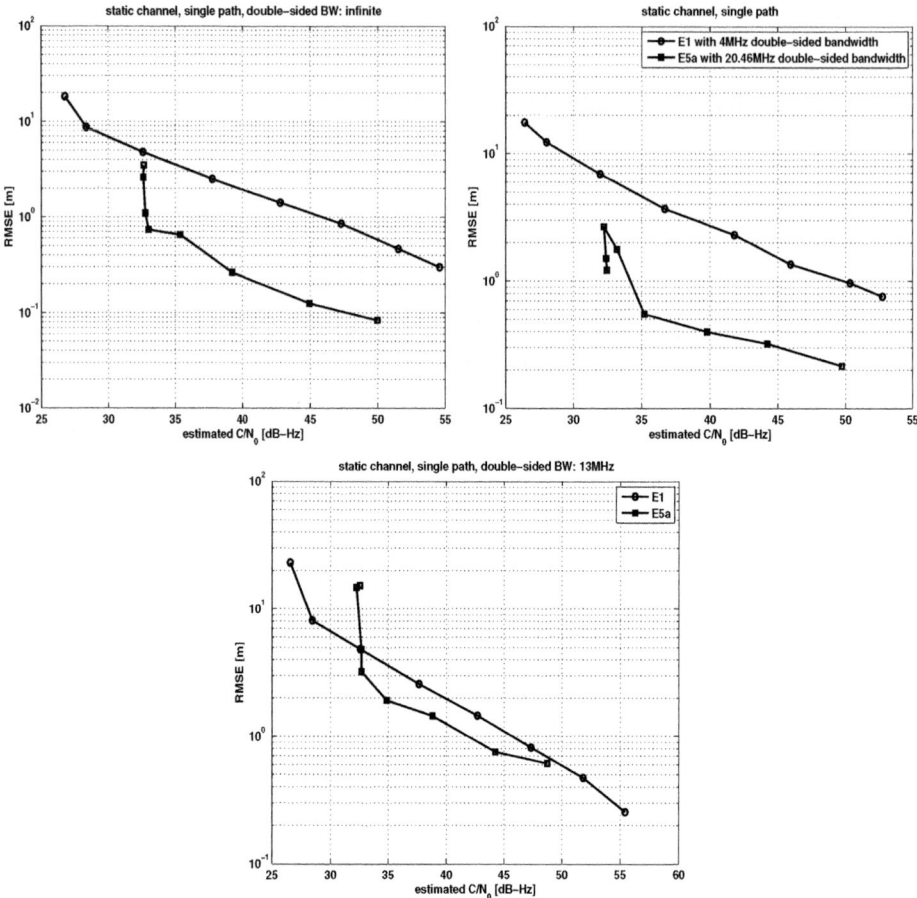

Fig. 8. RMSE simulation results with single path in static channel; Left: infinite bandwidth for both E1 and E5a signals; Right: 4 MHz double-sided bandwidth for E1 signal, 20.46 MHz double-sided bandwidth for E5a signal; Lower: 13 MHz double-sided bandwidth for both E1 and E5a signal

3.2   Delay Tracking in Multipath-Simulation Results

The performance of tracking E1 and E5a signal in a multipath scenario is shown in Fig. 9. The parameters used in the simulation are shown in Table. 3. As expected, E5a signal has better tracking performance than E1 signal if the front-end bandwidth is wide enough to cover the E5a band. With narrower band 13 MHz, E5a signal losses the benefit at the lower C/N_0.

A special case is also considered here, which is assumed that E1 signal is transmitted through a good channel (the LOS signal has much higher power than NLOS signal) and E5 signal is transmitted through a bad channel (the LOS signal has very weak power and LOS and NLOS has similar power). The channel profiles are generated with DLR channel model. The result is as given in the lower right figure in the Fig. 9. It can be observed that the tracking performance of E5a signal is much worse than E1 signal most of the time. Although the transmitted signal in both simulator have the same nominal C/N_0, the performance becomes worse because of the channel condition.

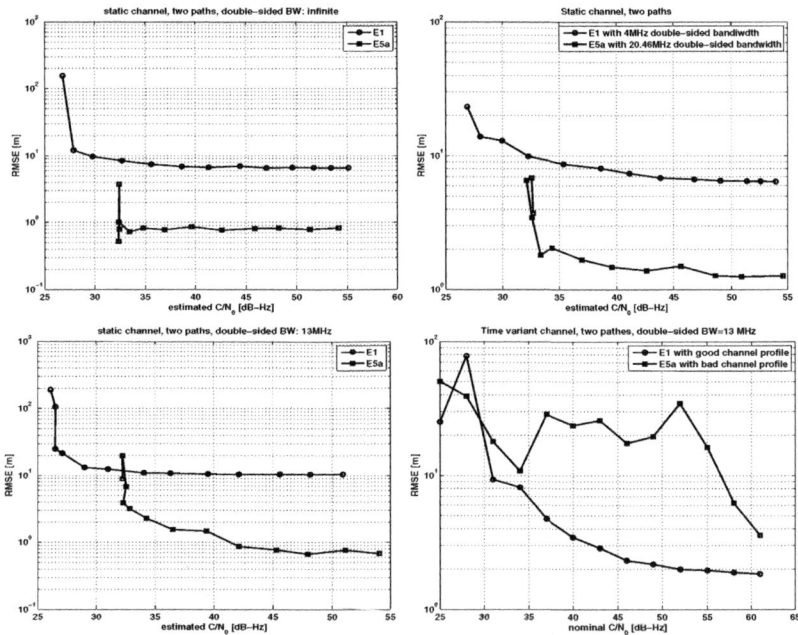

Fig. 9. RMSE simulation results with two paths in static channel; Upper left: infinite bandwidth for both E1 and E5a signals; Upper right: 4 MHz double-sided bandwidth for E1 signal, 20.46 MHz double-sided bandwidth for E5a signal; Lower left: 13 MHz double-sided bandwidth for both E1 and E5a signals; Lower right: 13 MHz double-sided bandwidth for E1 and E5a with DLR channel model

3.3   Delay Errors Distribution

The histograms of tracking error obtained from above simulations are presented
in this section. Since the noise added in the channel block is Gaussian white
noise, the histograms are compared with the Gaussian distribution, of which the
mean and variance are calculated from the corresponding code tracking error.
In order to ignore the effect from filtering, the histogram of the tracking error
under the infinite bandwidth are considered here, as in Fig.10 and Fig.11. As it
can be seen from the figures, the Gaussian distribution is more fit to E1 signal
no matter if the signal is transmitted in a single path or multipath scenario. The
tracking error of E5a signal in single path scenario has Gaussian-like distribution,
however, it is not like Gaussian distribution any more in multipath scenario,
which could be the effect from down-sampling.

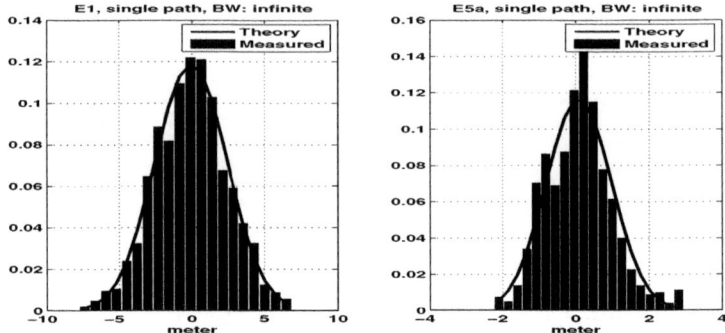

Fig. 10. Histogram of code tracking error of E1 and E5a signal at nominal $C/N_0 = 40$
dB-Hz

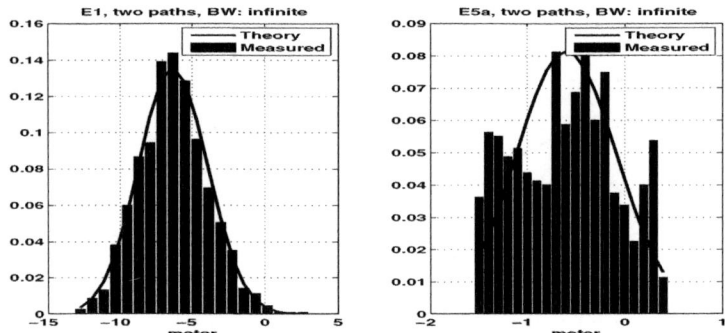

Fig. 11. Histogram of code tracking error of E1 and E5 signal at nominal $C/N_0 = 40$
dB-Hz

4   Conclusions

In this paper, the tracking performance with E1 and E5a signals has been evaluated in Simulink-based simulator built at TUT. The E1 and E5a signal simulators are described in the context of the paper. The tracking performances were evaluated in different channel profiles and receiver front-end configurations. They were shown that the tracking performances with E5a signal are better than those with E1 signal most of the time, especially at high C/N_0. In certain cases, when the E5a signal was transmitted through a much worse channel than that for E1 signal, tracking E5a lost its benefit .

The histogram of the code tracking error showed that the Gaussian distribution is more fit to E1 signal than E5 signal. It also indicated that in the E5a chain, there was not only the noise and multipath as error source, but also other aspects in the chain, such as the down-sampling.

For future work, it remains to be investigated how to combine E1 and E5a results for better accuracy results in a dual-frequency receiver mode . In addition, the E1 signal simulator is an open source, which is available at $www.cs.tut.fi/tlt/pos$.

Acknowledgments. The research leading to these results has received funding from the European Union's Seventh Framework Programme (FP7/2007-2013) under grant agreement n227890 (GRAMMAR project) and from Academy of Finland, which are gratefully acknowledged. The authors would also like to thank Nokia Foundation for their support.

References

1. Hein, G.W., Godet, J., Issler, J.L., Martin, J.C., Erhard, P., Rodridus, R.L., Pratt, T.: Status of Galileo Frequency and Dignal Design. ION GPS, pp. 266–277 (2002)
2. Galileo Open Service Signal In Space Interface control Document. OS SIS ICD (1) (2010)
3. Sleewaegen, J.M., De Wilde, W., Hollreiser, M.: Galileo AltBOC Receiver. In: ENC-GNSS 2004, Rotterdam (May 17, 2004).
4. Hein, G.W., Avila-Rodriguez, J.A., Wallner, S., Pratt, A.R., Owen, J., Issler, J.L., Betz, J.W., Hegarty, C.J., Lenahan, L.S., Rushanan, J.J., Kraay, A.L., Stansell, T.: MBOC: The New Optimized Spreading Modulation Recommendation for GALILEO L1 OS and GPS L1C. In: IEEE/ION PLANS, San Diego, California, USA (2006)
5. Dierendonck, A.V., Fenton, P., Ford, T.: Theory and performance of narrow correlator spacing in a GPS receiver. Journal of the Institute of navigation 39, 265–283 (1992)
6. McGraw, G.A., Collins, R., Braasch, M.S.: GNSS Multipath Mitigation Using Gated and High Resolution Correlator Concepts, pp. 333–342. ION NTM (1999)
7. Dovis, F., Mulassano, P., Margaria, D.: Multiresolution Acquisition Engine Tailored to the Galileo AltBOC Signals. In: Proceedings of ION GNSS 2007 (2007)
8. Hurskainen, H., Lohan, E.S., Nurmi, J., Sand, S., Mensing, C., Detratti, M.: Optimal Dual Frequency Combination for Galileo Mass Market Receiver Baseband. In: CDROM Proc. of SIPS 2009, Tampere, Finland, pp. 261–266 (October 2009)

 9. Lehner, A., Steingaß, A.: A novel channel model for land mobile satellite navigation. In: Institute of Navigation Conference ION GNSS, Long Beach, USA (2005)
10. Shivaramaiah, N.C., Dempster, A.G.: Galileo E5 Signal Acquisition Strategies. In: ENC-GNSS, Toulous, France (2008)
11. Margaria, D.: M.Sc thesis: Galileo AltBOC Receivers, analysis of Receiver Architectures, Acquisition Strategies and Multipath Mitigation Techniques for the E5 AltBOC signal (2007), http://mdavide.interfree.it/ Thesis_Galileo_AltBOC_Receivers_MARGARIA_2007.pdf.
12. Shivaramaiah, N.C., Dempster, A.G., Rizos, C.: A Hybrid Tracking Loop Architecture for Galileo E5 Signal. In: European Navigation Conference ENC-GNSS, Naples, Italy (2009)
13. Kaplan, E.D., Hegarty, C.J.: Understanding GPS: Principles and Applications, 2nd edn. Artech House, Boston (2006)
14. Parkinson, B.W., Spilker Jr., J.J.: Global Positioning System: Theory and Applications, vol. 1, pp. 390–392. American Institute of Aeronautics, 370 L.Enfant Promenade, SW, Washington, DC (1996)
15. Betz, J.W., Kolodziejski, K.R.: Extended Theory of Early-Late Code Tracking for a Bandlimited GPS Receiver, to be Published in Navigation. Journal of The Institute of Navigation (Fall 2000)

Galileo Dual-Channel CBOC Receiver Processing under Limited Hardware Assumption*

Elena Simona Lohan and Heikki Hurskainen

Tampere University of Technology,
Korkeakoulunkatu 1, 33101 Tampere, Finland
`www.cs.tut.fi/tlt/pos`
{elena-simona.lohan,heikki.hurskainen}@tut.fi

Abstract. Composite Binary Offset Carrier (CBOC) modulation is currently proposed for future Open Service (OS) Galileo signals in E1 frequency band. CBOC consists of a weighted sum or difference of two sine Binary Offset Carrier (BOC) waveforms: a sine BOC(1,1) and a sine BOC(6,1) component. The transmitted OS signal has both data and pilot channels. Data and pilot channels use slightly different modulation, namely CBOC(+) (i.e., weighted sum of BOC(1,1) and BOC(6,1)) and CBOC(-) (i.e., weighted difference of those). At the receiver side, depending on the number of channels available, several approaches are possible: processing either data or pilot, or processing both channels with any of the BOC(1,1) and BOC(6,1) components, or with a weighted or time-multiplexed combination of those. Therefore, a significant number of receiver processing variants is possible. The focus here is on the architectures having a limited hardware available, when we assume that only two channels per satellite and per E1 Open Service signal are used at the receiver and when we have one-bit processing only. This allows us to either process both data and pilot channels with a single sine-BOC(1,1) reference, or to process only the data channel with both BOC(1,1) and BOC(6,1) components, and then combine them with appropriate weights. The question we address here is which of these two variants is better in terms of performance. The novelty of our solution comes from an analytical approach of this problematic and from the comparison of the two architectures in terms of tracking performance at various bandwidths. Our analysis focuses both on narrowband receiver cases (i.e., low front-end receiver bandwidts, of interest in mass-market applications), and on wideband receiver cases (more suitable for professonal receivers). The tracking results are analyzed in terms of tracking error variances and multipath error envelopes.

Keywords: Binary Offset Carrier (BOC), Composite Binary Offset Carrier (CBOC), Galileo, Global Navigation Satellite Systems (GNSS), Non-coherent Early Late Power discriminator (NELP) discriminator, narrowband GNSS receiver.

* The research leading to these results has received funding from the European Union's Seventh Framework Programme (FP7/2007-2013) under grant agreement number 227890 (GRAMMAR project). This research work has also been supported by the Academy of Finland.

G. Giambene and C. Sacchi (Eds.): PSATS 2011, LNICST 71, pp. 391–401, 2011.

1   Background and Motivation

For receiver manufacturers and vendors the upcoming changes in signal domain opens new market potential since new products are needed to get benefit of the increased accuracy and availability of GNSS signals. On the other hand, additional challenges are created by increased complexity needs of receiver. It may become impossible to implement optimal structures to receivers due short time-to-market requirements or dependency of 3rd party receiver Intellectual Property. Composite Binary Offset Carrier (CBOC) modulation has now been selected for the Galileo Open Service signal in E1 band [1]. CBOC modulation is a weighted superposition of two sine BOC-modulated signals: a BOC(1,1) and a BOC(6,1) component. The higher modulation BOC(6,1) requires more bandwidth than BOC(1,1), but has the ability to enhance the tracking performance. CBOC signal is a four-level signal, while BOC(1,1) and BOC(6,1) components are two-level signals, able to be implemented via 1-bit receiver processing.

Traditionally, CBOC signals have been processed either with a CBOC receiver (if a large bandwidth, e.g., 24.552 MHz, is available) [2,3,4], or with a sine BOC(1,1) receiver (for narrowband GNSS receivers) [5,6]. In the first case (CBOC-based processing), implementation is based on at least 2 bits. If we assume some hardware restrictions, such as 1-bit receiver and limited number of channels, processing the incoming CBOC signal with BOC(1,1) and/or BOC(6,1) componets separately makes more sense because it reduces the receiver complexity. Therefore, this is the problem we address in this paper: how to choose the dual-channel processing of Galileo CBOC signal from the limited hardware point-of-view. One-bit processing architectures for CBOC signal have been previously proposed in [4,7].

The novelty of these paper comes from analytical model of the two receiver architectures (data plus pilot processing versus data-only processing) and from the architecture we propose for processing the data-only channel in a dual-channel receiver (which is slightly different from the ones proposed in [4,7], as described in Section 2). To the best of the authors' knowledge such investigation of the CBOC receiver architectures with limited number of channels has not been made yet. We believe that the results reported here are important from the designer point of view, because it allows him or her to choose the best architecture according to the available receiver bandwidth.

This paper is organized as follows: the dual-channel architectures for GNSS signal tracking are discussed in Section 2, the analytical model used in our studies is presented in Section 3, the tracking error performance of discussed algorithms is studied in Section 4, and finally the multipath behavior of them is illustrated in Section 5.

2   Dual-Channel Architectures

The collaborative (or composite) tracking of both data and pilot components is discussed in detail in [8]; both non-coherent combining and coherent combining with relative sign recovery are discussed. In our work we assume a non-coherent combining channel architecture, where the code generators are capable of creating only binary outputs. The non-coherent architecture for the traditional architecture, namely the data-pilot tracking, is illustrated in Fig. 1. There, the upper channel tracks data signal and the lower is tracking pilot signal, both using only sine-BOC(1,1)-modulated replicas. The code generators have multiple, differently delayed outputs, each feeding its own correlator. The number of correlators per channel is typically three (e.g. narrow early-minus-late correlator (NELP) [9]) to five (e.g. high resolution correlator (HRC) [10]). The discriminator function outputs are combined, filtered and fed back to code generation. Since our paper focuses on the code tracking, the carrier tracking with its in-phase and quadrature-phase branches of correlators is omitted from figures to gain clarity.

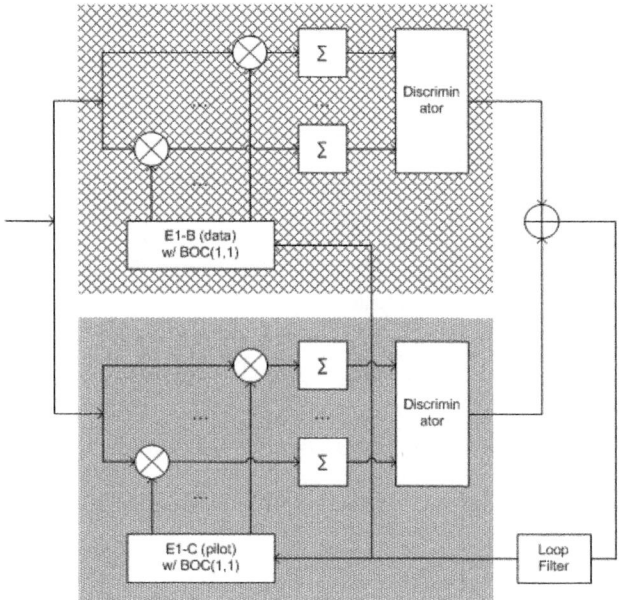

Fig. 1. Dual channel tracking for data and pilot signals (architecture 1)

Similarly, the tracking of four-level CBOC signal with a CBOC-modulated replica code becomes impossible due to binary generator outputs. In [4], a dual-channel approach for CBOC signal tracking has been presented as an alternative to the 4-level CBOC receiver tracking, where two channels are used: the first one for sine-BOC(1,1) and the second for sine-BOC(6,1) tracking. The discriminator results are weighted and combined to reach CBOC signal performance while still using binary subcarrier generation with some hardware overhead. The architecture of [4] has been the starting point of our proposed architecture from Fig. 2, with the main difference that, in our case, no time-multiplexing is used (we use a code-multiplexed approach, where the correlator outputs are weighted and combined as shown in Fig. 2).

The architecture for dual channel tracking for CBOC signal is illustrated in Fig. 2. Here, the differences comppared with the architecture 1 of Fig. 1 are: the code generator on the lower channel produces the spreading code for data signal with BOC(6,1) component and amplitude weighting factors ($\sqrt{1 - \alpha^2}$ and α) are applied to discriminator outputs from BOC(1,1) and BOC(6,1) channels, respectively. The α^2 factor refers to the power percentage of BOC(6,1) component.

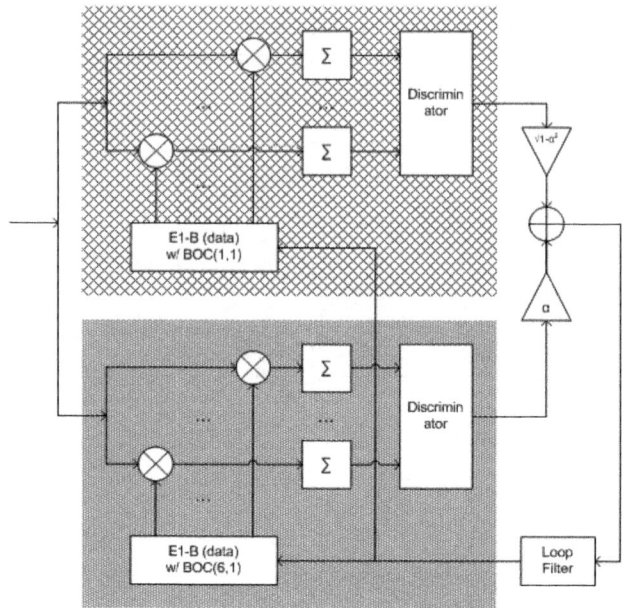

Fig. 2. Dual channel tracking of CBOC signal (architecture 2)

3   Analytical Model

We will adopt here the model introduced in [11] and detailed in [5]. The block diagram of the transmitter-receiver chain for Galileo data or pilot channel is illustrated in Fig. 3. The modulation at the receiver side is dependent on the channel type (CBOC(+) if data, and CBOC(-) if pilot channel) and it is characterized by the $H_{tx}(f)$ transfer function. The modulation at the receiver side, characterized by $H_{rx}(f)$ transfer function can be either a sine BOC(1,1)-modulation (in case we process both data and pilot channels), or a weighted combination of sine BOC(1,1) and sine BOC(6,1) components. The weighting factor is to be decided separately, according to the receiver tracking variance, as explained further on. The correlation block contains a multiplier and an Integrate ad Dump (I&D) block.

The overall effects of the channel $H_c(f)$ and the bandwidth-limiting filter $H_f(f)$ can be lumped in a single term $H(f) = H_c(f)H_f(f)$. This paper focuses on single-path static channel case, therefore assuming that $|H_c(f)| = 1$, in order to find out the maximum achievable performance. This analysis can be straightforwardly extended to multipath fading channels, under a variety of scenarios.

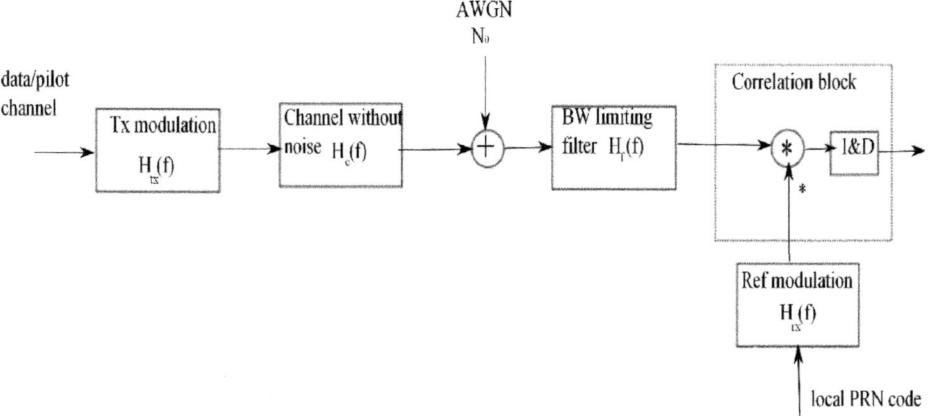

Fig. 3. Transmitter-receiver block diagram in terms of transfer functions

The normalized Power Spectral Density (PSD) of the received signal $\overline{G}_s(f)$ and, respectively, of the noise $\overline{G}_n(f)$ after the correlation can be therefore written as:

$$
\begin{cases}
\overline{G}_s(f) = \dfrac{|H_{tx}(f)H_{rx}(f)||H(f)|^2}{\displaystyle\int_{-\infty}^{\infty} |H_{tx}(f)H_{rx}(f)||H(f)|^2 df} \\[4mm]
\overline{G}_n(f) = \dfrac{|H_{rx}(f)|^2|H(f)|^2}{\displaystyle\int_{-\infty}^{\infty} |H_{rx}(f)|^2|H(f)|^2 df}
\end{cases}
\tag{1}
$$

The impulse response of a sine-BOC(m,n) modulation is given by [11]:

$$
h_{tx}(t) = p_{T_B}(t) \circledast \sum_{i=0}^{N_B-1} (-1)^i \delta\left(t - i\frac{T_c}{N_B}\right)
\tag{2}
$$

where $\circledast$ is the convolution operator, $c(t)$ is the pseudorandom code sequence, $N_B = \frac{2m}{n}$ is the BOC modulation order, $T_c = 1/f_c$ is the chip interval, f_c is the chip rate, and $p_{T_B}(t)$ is the convolution between a rectangular pulse of support $T_B = \frac{T_c}{N_B}$ and the front-end filter used to limit the signal bandwidth.

The transfer functions, $H_{tx}(f)$ and $H_{rx}(f)$ can be computed according to the modulation type, using eq. (2) and following the derivations similar to [5] (the full derivations are not included here due to lack of space):

1. Data and pilot channel processing with sine BOC(1,1):

$$
\begin{cases}
H_{tx,data} = e^{-j\pi fT_c}\dfrac{sin(\pi fT_c)}{\pi f}\left(\sqrt{\dfrac{10}{11}}e^{-j\pi f\frac{T_c}{2}}tan(\dfrac{\pi fT_c}{2}) + \sqrt{\dfrac{1}{11}}tan(\dfrac{\pi fT_c}{12})\right) \\[2mm]
H_{tx,pilot} = e^{-j\pi fT_c}\dfrac{sin(\pi fT_c)}{\pi f}\left(\sqrt{\dfrac{10}{11}}e^{-j\pi f\frac{T_c}{2}}tan(\dfrac{\pi fT_c}{2}) - \sqrt{\dfrac{1}{11}}tan(\dfrac{\pi fT_c}{12})\right) \\[2mm]
H_{rx} \quad = e^{-\frac{3j\pi fT_c}{2}}\dfrac{sin(\pi fT_c)}{\pi f}tan(\pi f\frac{T_c}{2})
\end{cases}\tag{3}
$$

2. Data-only channel processing with a weighted combination of sine BOC(1,1) and sine BOC(6,1):

$$
\begin{cases}
H_{tx,data} = e^{-j\pi fT_c}\dfrac{sin(\pi fT_c)}{\pi f}\left(\sqrt{\dfrac{10}{11}}e^{-j\pi f\frac{T_c}{2}}tan(\dfrac{\pi fT_c}{2}) + \sqrt{\dfrac{1}{11}}tan(\dfrac{\pi fT_c}{12})\right) \\[2mm]
H_{rx} \quad = e^{-j\pi fT_c}\dfrac{sin(\pi fT_c)}{\pi f}\left(\sqrt{1-\alpha^2}e^{-j\pi f\frac{T_c}{2}}tan(\dfrac{\pi fT_c}{2}) + \alpha tan(\dfrac{\pi fT_c}{12})\right)
\end{cases}\tag{4}
$$

Above, α^2 is the power percentage of sine BOC(6,1)-component in the reference signal (and $1-\alpha^2$ is the power percentage of sine BOC(1,1)-component).

4   Tracking Error Variances and Optimal Weighting Factor

The code tracking error variance, given in s^2 (squared seconds) for a signal with normalized PSD $\overline{G}_s$ used in a non-coherent early-minus late power delay tracker or correlator (NELP), in the presence of a colored Gaussian noise with normalized PSD $\overline{G}_n$ and operating at a carrier-to-noise density ratio C/N_0 is, according to [12]:

$$
\sigma^2_{NELP} = \frac{B_L(1 - 0.5B_LT)}{(2\pi)^2C/N_0I_2^2}\left(I_1 + \frac{I_3 - I_4}{4C/N_0T(I_5)^2}\right)\tag{5}
$$

where $I_i, i = 1,\ldots,5$ are the following integrals:

$$
I_1 = \int_{-\frac{B_W}{2}}^{\frac{B_W}{2}} \overline{G}_{y_s}(f)\overline{G}_{y_n}(f)sin^2(\pi f\Delta_{EL}T_c)df
$$

$$
I_2 = \int_{-\frac{B_W}{2}}^{\frac{B_W}{2}} f\overline{G}_{y_s}(f)sin(\pi f\Delta_{EL}T_c)df
$$

$$
I_3 = \int_{-\frac{B_W}{2}}^{\frac{B_W}{2}} \overline{G}_{y_s}(f)\overline{G}_{y_n}(f)df
$$

$$
I_4 = \int_{-\frac{B_W}{2}}^{\frac{B_W}{2}} \overline{G}_{y_s}(f)\overline{G}_{y_n}(f)e^{j2\pi f\Delta_{EL}T_c}df
$$

$$
I_5 = \int_{-\frac{B_W}{2}}^{\frac{B_W}{2}} \overline{G}_{y_s}(f)cos(\pi f\Delta_{EL}T_c)df
$$

$$\tag{6}$$

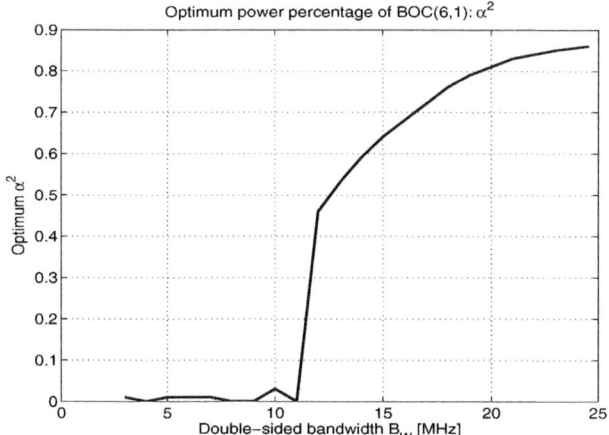

Fig. 4. Optimum power percentage of component sine BOC(6,1) in data channel processing with combined BOC(1,1)/BOC(6,1)

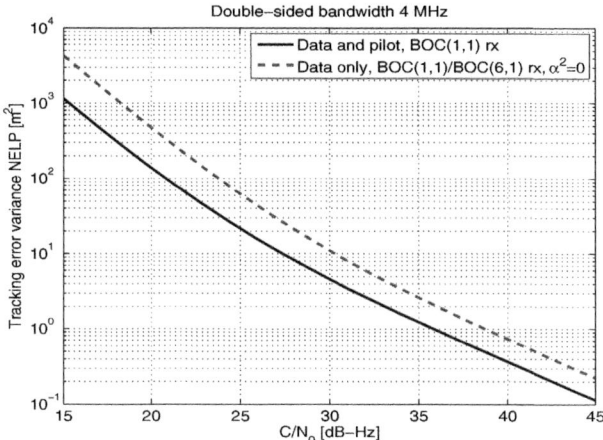

Fig. 5. Tracking error variance of the two dual-channel architectures at $B_W = 4$ MHz

Above, B_L is the NELP loop bandwidth (in Hz), Δ_{EL} is the early-late spacing (in chips), and $T = Nc*10^{-3}$ is the coherent integration in seconds, N_c being the number of codewords used in the coherent integration. For example, under the ideal rectangular filter assumption, the margins of the integrals in 5 can simply be replaced by $-B_W/2$ and $+B_W/2$, where B_W is the double-sided bandwidth at the receiver.

By replacing eqns. 1, 3, and 4 into eqn. 5, we obtain a traking error variance as a function of the receiver weighting factor α. A numerical optimization has been performed, as shown in Fig. 4, and the optimum α factor (for the case when data-only channel is processed with a weighted combination of BOC(1,1)/BOC(6,1))

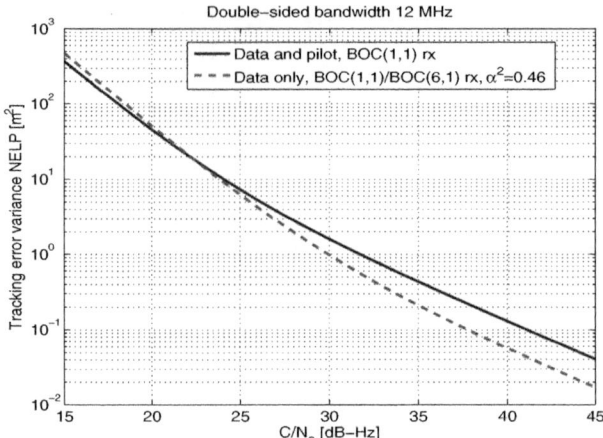

Fig. 6. Tracking error variance of the two dual-channel architectures at $B_W = 12$ MHz

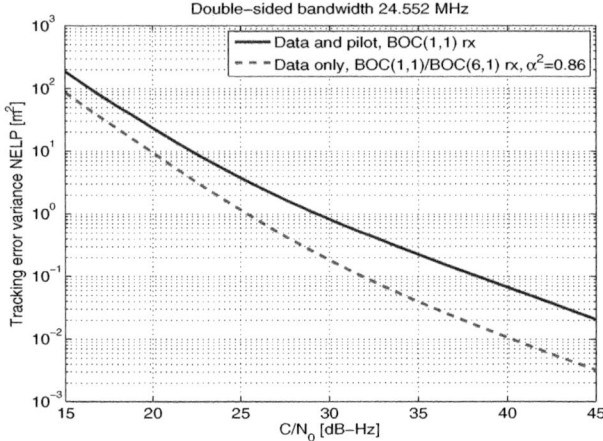

Fig. 7. Tracking error variance of the two dual-channel architectures at $B_W = 24.552$ MHz

was found to be bandwidth dependent. For double-sided bandwidths up to 12 MHz, the optimum processing is with a very weak or no sine BOC(6,1) component ($\alpha^2 = 0 - 0.02$). For sufficiently high bandwidths (i.e., larger than 12 MHz, a strong sine BOC(6,1) component was found beneficial (α^2 ranging from 0.46 at 12 MHz till 0.86 at 24.552 MHz).

The next question addressed here is whether processing the data and pilot together, both with a sine BOC(1,1) signal is better (in terms of tracking error variance) than processing the data channel alone with a weighted combination of sine BOC(1,1)/BOC(6,1). The optimal power weight α^2 found above will be considered in this comparison. Illustrative plots at two extreme double-sided bandwidths and a middle-case bandwidth are shown in Figs. 5, 6, and 7.

Clearly, from the tracking error variance point of view, architecture 1 is better that architecture 2 at low bandwidths (up to about 12 MHz double-sided bandwidth), while architecture 2 (which makes use also of the BOC(6,1) component) is better at higher bandwidths.

5   Multipath Behaviour

The cross-correlation shape at the receiver side can be computed via the inverse Fourier transform of the PSD $\overline{G}_s(f)$ given in eq. (1). The Multipath Error Envelopes (MEE) for a second path 3 dB weaker than the first one are shown

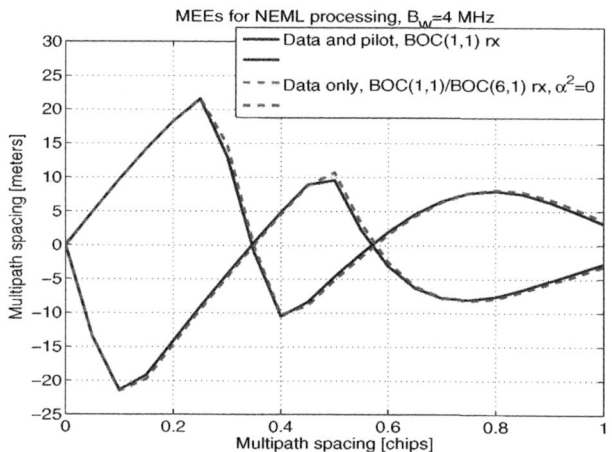

Fig. 8. Multipath Error Envelope (MEE) at $B_W = 4$ MHz

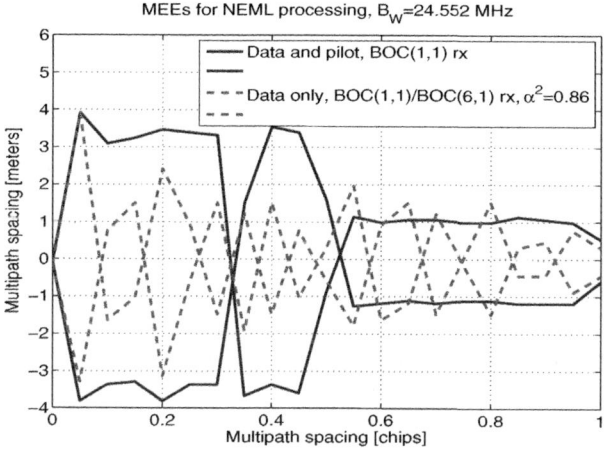

Fig. 9. Multipath Error Envelope (MEE) at $B_W = 24.552$ MHz

in Figs. 8 and 9, for double-sided bandwidths of 4 and 24.552 MHz. A MEE shows the robustness in the presence of multipaths [13], when the channel has two paths, spaced at the distance showed on x-axis of Figs. 8 and 9, and with the second path with 3 dB smaller than the first one. The double lines in the egend of the plots in Figs. 8 and 9 are due to the upper and lower multipath error envelopes. As seen in Fig. 8, at low bandwidths (i.e., narrowband receiver), there is no difference in the multipath performance between the two architectures. When the double-sided bandwidth increases (as seen in Fig. 9), the second architecture when data-only channel is processed via both BOC(1,1) and BOC(6,1) receivers, has a clear advantage also in terms of MEE, due to the higher resolution capability of BOC(6,1) component. Indeed, we can get few meters better multipath resistance with the joint BOC(1,1)/BOC(6,1) architecture of Fig. 2.

6   Conclusions and Design Issues

The processing of CBOC signal with 1-bit receiver and limited number of channels (e.g., two channels per satellite for E1 signals) can be done in two main architectures: either by processing both the data and and the pilot channels with a sine BOC(1,1) receiver (and ignoring completely the BOC(6,1) component) as shown in Fig. 1, or by processing only the data channel (needed to extract navigation data) and using both BOC(1,1) and BOC(6,1) correlations, as shown in Fig. 2. In the second approach, the power division between BOC(1,1) and BOC(6,1) components at the receiver should be optimized separately. Our studies show that, when the receiver double-sided bandwidth is less than about 12 MHz, the optimum processing is given by the first architecture: data and pilot are processed via BOC(1,1) component, and the BOC(6,1) component is completely ignored. The gain in terms of tracking error variance is about 3 dB (over the second architecture), as seen in Fig. 5. This is a completely intuitive result, since data and pilot powers are split in half at the transmitter, and we expect to lose 3 dB when processing data-only or pilot-only channels.

The interesting and novel result comes when higher bandwidths are available: in here, there is a tracking error variance performance gain when BOC(6,1) processing is also used (i.e., architecture of Fig. 2) and this performance gain can compensate for the 3 dB loss when only the data channel is processed. Indeed, according to the available receiver bandwidth, and by choosing properly the weighting factor between BOC(1,1) and BOC(6,1) components, as illustrated in Fig. 4, we can get up to 5 dB performance gain in the tracking error variance with the second architecture and also an enhacement in the multipath error bias (i.e., smaller multipath error envelopes).

Therefore, the design recommendations we adopt are bandwidth dependent: for a narrowband receiver as those employed in mass-market solutions, the classical architecture of processing data and pilot channels with a sine BOC(1,1) receiver is the best choice, while in a wideband receiver (here, meaning with the receiver double-sided bandwidths higher than 12 MHz), a processing of the CBOC signal with both BOC(1,1) and BOC(6,1) channels is the best option.

References

1. European GNSS (Galileo) Open Service, Signal in space interface control document (OS SIS ICD) (February 2010)
2. Rodriguez, J., Wallner, S., Hein, G., Rebeyrol, E., Julien, O., Macabiau, C., Ries, L., DeLatour, A., Lestarquit, L., Issler, J.: CBOC - An Implementation of MBOC. In: First CNES Workshop on Galileo Signals and Signal Processing, Toulouse, France (October 2006)
3. Hein, G., Avila-Rodriguez, J., Wallner, S., Betz, J.W., Hegarty, C., Rushanan, J., Kraay, A., Pratt, A., Lenahan, S., Owen, J., Issler, J., Stansell, T.: MBOC: The New Optimized Spreading Modulation Recommended for GALILEO L1 OS and GPS L1C. In: Inside GNSS - Working Papers, vol. 1(4), pp. 57–65 (2006)
4. Julien, O., Macabiau, C., Avila Rodriguez, J.-A., Wallner, S., Paonni, M., Hein, G.W., Issler, J.-L., Ries, L.: On Potential CBOC/TMBOC Common Receiver Architectures. In: Proc of. ION GNSS 20th International Technical Meeting of the Satellite Division, Fort Worth, TX, pp. 1530–1542 (September 2007)
5. Lohan, E.: Analytical performance of CBOC-modulated Galileo E1 signal using sine BOC(1,1) receiver for mass-market applications. In: Proc. of IEEE Position Location and Navigation Symposium, PLANS (May 2010)
6. Artaud, G., Ries, L., Dantepal, J., Issler, J., Grelier, T., Delatour, A.: CBOC performances using software receiver. In: 2nd Workshop on GNSS Signals & Signal Processing - GNSS SIGNALS 2007, Nordwijk, Netherlands (October 2007)
7. Macabiau, C., Issler, J.-L., Ries, L., Julien, O.: Two for one: tracking Galileo CBOC signal with TMBOC. Inside GNSS journal (Spring 2007)
8. Borio, D., Mongredien, C., Lachapelle, G.: Collaborative code tracking of composite gnss signals. IEEE Journal of Selected Topics in Signal Processing 3, 613–626 (2009)
9. Dierendonck, A.V., Fenton, P., Ford, T.: Theory and performance of narrow correlator spacing in a GPS receiver. Journal of the Institute of navigation 39, 265–283 (1992)
10. McGraw, G., Braasch, M.: GNSS multipath mitigation using high resolution correlator concepts. In: Proc. of ION National Technical Meeting, San Diego, CA, pp. 333–342 (January 1999)
11. Lohan, E.S., Lakhzouri, A., Renfors, M.: Binary-Offset-Carrier modulation techniques with applications in satellite navigation systems. Wiley Journal of Wireless Communications and Mobile Computing (July 2006), doi:10.1002/ wcm.407
12. Betz, J., Kolodziejski, K.: Extended theory of early-late code tracking for a bandlimited GPS receiver. Journal of Navigation 47(3), 211–226 (2000)
13. Avila-Rodriguez, J., Wallner, S., Hein, G.: How to Optimize GNSS Signals and Codes for Indoor Positioning. In: ION GNSS 19th International Technical Meeting of the Satellite Division (September 2006)

Comparison of Single and Dual Frequency GNSS Receivers in the Presence of Ionospheric and Multipath Errors

Danai Skournetou and Elena-Simona Lohan

Tampere University of Technology, P.O. Box 553 Tampere, Finland
{danai.skournetou,elena-simona.lohan}@tut.fi
http://www.cs.tut.fi/tlt/pos/

Abstract. In this paper, we test the performance of single and dual frequency GPS and Galileo GNSS receivers in terms of satellite-receiver range estimation. In particular, we focus on the effects caused by ionosphere and multipath propagation. Therefore, the available pseudoranges are assumed to be contaminated with first-order ionospheric delay and measurement errors produced in the code tracking stage. We used three dual-frequency methods, two ionospheric models (Klobuchar and NeQuick for GPS and Galileo receivers, respectively) and compared their ionosphere-corrected ranges on a Root Mean Square Error basis. The simulation results showed that a dual-frequency receiver is superior to a single-frequency one, only when the standard deviation of the measurement error is small and when the correlation factor between the two available pseudoranges is higher than −0.4.

Keywords: Global Navigation Satellite System, ionosphere, multipath, dual-frequency receivers.

1   Introduction

Almost three decades have passed since the first Global Navigation Satellite System (GNSS), commonly known as Global Positioning System (GPS), became available to the public. Since then, the amount of devices which are equipped with a GPS receiver has been continuously increasing. The growing interest on the position information has initiated the creation of new GNSSs with improved performance characteristics; among those, Galileo represents Europe's initiative to build it's "own" GNSS, expected to be ready by 2014 [1].

The main principle of satellite-based positioning lies on the trilateration method which requires the computation of at least three satellite-receiver ranges (a minimum of four ranges is needed in order to estimate the receiver clock bias). For an accurate range estimation, it is not enough to measure the difference between the received and transmitted times. Instead, the various error sources affecting the transmitted signal shall be accounted for and mitigated. Among those, the ionosphere is responsible for the signal's biggest delay [2]. More precisely, when the satellite signal travels through the ionospheric layer (located 50-1000 km above the Earth's surface) it is delayed due to

G. Giambene and C. Sacchi (Eds.): PSATS 2011, LNICST 71, pp. 402–410, 2011.
© Institute for Computer Sciences, Social Informatics and Telecommunications Engineering 2011

the presence of charged particles (ions and electrons). The amount of the delay depends on two parameters: the frequency of the signal the Total Electron Content (TEC) [3].

In single-frequency receivers, TEC is estimated with the help of mathematical models whose accuracy is typically counterbalanced by their complexity. Among the various model reported in the literature, Klobuchar model is the one employed in most GPS receivers and which makes use of eight broadcast coefficients [4]. The NeQuick model is adopted by ITU-R and proposed for the future Galileo receivers [5]. Unlike Klobuchar model, NeQuick makes use of only three broadcast coefficients and it is claimed to be more accurate [6].

In dual-frequency receivers no modelling of the ionosphere is required because the availability of two signals which have undergone the same ionospheric effects is exploited. In the absence of measurement errors, first-order ionospheric delay can be estimated and fully mitigated via linear combination of the available pseudorange measurements [2]. The afore-mentioned advantage in combination with the advent of new GNSS signals (e.g., future Galileo and modernised GPS signals) and the decreasing cost of GNSS receivers can meet the growing demand of higher accuracy in mass-market receivers.

However, in practise the presence of measurement errors (i.e., due to multipath propagation effects) degrades the accuracy of ionospheric delay estimation. Moreover, the existing studies on the impact of multipath errors in the estimation accuracy of the ionospheric delay and consequently of the range estimation in the case of dual frequency methods have been rather limited [7]. In this paper, we attempt to shed some light on the above-mentioned problem. More precisely, we examine the effect of multipath errors in the estimation of ionoshere-corrected ranges by theoretically modelling the performance of single- and dual- frequency receivers and comparing the ionosphere-corrected ranges of various methods in terms of Root Mean Square Error (RMSE).

The remainder of this manuscript is organised in the following manner: Section 2 includes the description of the model used to analyse the effects of ionosphere and multipath propagation. Section 3 describes the simulation setup and presents the results. Finally, Section 4 summarises the most important findings of our study.

2   Model Description

In this section we describe the mathematical model used to represent dual-frequency receivers and the methods employed by such receivers to estimate and correct the delay caused by ionosphere. In what follows, we consider only the first-order ionospheric effects since they account for 99% of the total delay [3] and because the effect of the higher order terms can be considered negligible for the accuracy requirements of mass-market GNSS receivers.

The first-order ionospheric delay is defined as [2]

$$I_i = \frac{40.3}{f_i^2} TEC \tag{1}$$

where TEC is the total electron content measured in TEC Units (TECUs) with 1 TECU= 10^{16} electrons/m^2 and f_i is the i−th frequency for $i = 1, 2$. The first order ionospheric delay versus TEC for different carrier frequencies can be seen in Fig. 1.

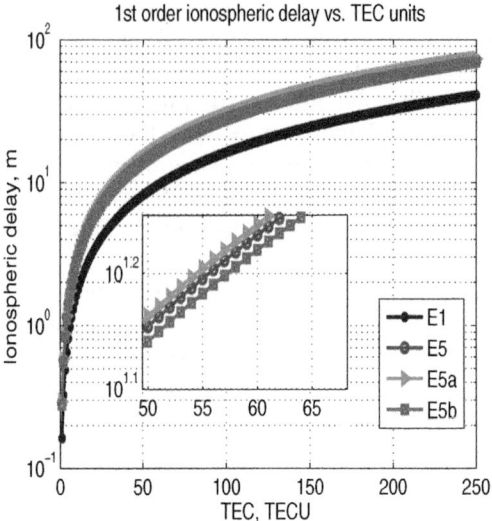

Fig. 1. First order ionospheric delay vs. TEC for E1 (1575.42 MHz), E5 (1189 MHz), E5a (1176.45 MHz) and E5b (1207.14 MHz) carrier frequencies

Considering a dual-frequency GNSS receiver, it is possible to model the available pseudoranges into matricial format as [7]

$$\begin{bmatrix} \rho_1 \\ \rho_2 \end{bmatrix} = \begin{bmatrix} 1 & \frac{40.3}{f_1^2} \\ 1 & \frac{40.3}{f_2^2} \end{bmatrix} \begin{bmatrix} \rho \\ TEC \end{bmatrix} + \begin{bmatrix} e_1 \\ e_2 \end{bmatrix} \tag{2}$$

For simplicity, we neglect other error sources in Eq. (2) because we want to focus on the impact of multipath errors in the estimation of the satellite-receiver range which is characterised by ionospheric delay. Moreover, we notice that most of the errors sources affecting the transmitted signals from a single satellite are the same since they travel through the same medium. So, common errors (e.g., ephemeris, tropospheric and clock errors) can be easily removed by subtracting one of the two pseudoranges from the other [2].

Equivalently, Eq. (2) can be represented in a compact manner as

$$\mathbf{r} = \mathbf{Ax} + \mathbf{e} \tag{3}$$

where $\mathbf{r}$ is the observation vector that contains the pseudorange measurements, $\mathbf{A}$ is a 2×2 matrix, $\mathbf{x}$ is the unknown parameter vector to be estimated and $\mathbf{e}$ is the measurement error vector. The measurement error represents the residue of the processing done in the code tracking stage. We notice that the code tracking error is different for different signals because it depends on signal-specific characteristics such as type (i.e. data or pilot), modulation, frequency, etc. and it represents mostly the effects of multipath propagation [7].

Unlike in the case of single-frequency receivers where the total electron content has to be modelled, dual-frequency receivers exploit the availability of two pseudoranges and discard the need for TEC modelling. In the absence of errors, TEC can be accurately estimated by a proper linear combination of the pseudoranges [8]. Then, the ionospheric-free ranges can be computed in a straightforward manner.

In the presence of errors (i.e., due to multipath delay tracking errors), the linear Least Square (LS) solution can be used. In this case, the unknown TEC and true range parameters can be estimated as [9]

$$\hat{\mathbf{x}}_{LS} = (\mathbf{A}^T \mathbf{A})^{-1} \mathbf{A}^T \mathbf{r} \tag{4}$$

where T denotes the operation of transposition.

One limitation of LS method is that it does not account for physically invalid solutions. For example, it was found that the estimated TEC parameter can be a negative value which is against its physical meaning (i.e., the electron content can be only equal or greater to zero) [7]. In order to avoid the above-mentioned scenario, we can impose certain constraints for the estimated vector. This leads to the method commonly known as Constrained Least Square (CLS). More precisely, the idea is to minimise the squared difference between the observed data and $\mathbf{Ax}$, subject to the linear inequality constraint $\mathbf{A}\hat{\mathbf{x}}_{CLS} \geq \mathbf{b}$ (see Section 3 for the constraint chosen in our study).

In order to avoid the computationally heavy approach of CLS, a new method, called Brute Force Constraint (BFC), was proposed in [7]. The main idea of BFC is that within only two iterations we are able to estimate TEC and true range subject to the constraint of non-negative TEC. So, the complexity of BFC is reduced compared to the one of CLS method the BFC-based TEC estimates do not violate any physical rule.

3   Simulation Profile and Results

In this section, we compare the range estimation performance of single and dual-frequency receiver methods in terms of Root Mean Square Error (RMSE). The satellite systems of interest are the existing Navstar GPS and future Galileo. In the case of single frequency receivers, we assume that a GPS receiver employs the Klobuchar model [10, 4] for the estimation of TEC and a Galileo receiver utilises the Ne-Quick model [6, 11]. Unlike Klobuchar model, NeQuick does not make use of the thin-shell assumption; instead, it is a three-dimensional model that exhibits higher degree of realism.

For the testing of single-frequency methods, real data are required as the inputs to a certain ionospheric model. However, because we want to compare the theoretical performance of range estimation methods, we model the ionospheric delay estimation error of the single frequency methods as $e_{I,i} = \alpha I_i$, where α represents the percentage of the ionospheric delay which is not corrected (i.e., the estimation error of ionospheric delay). While the reported percentage values vary in the literature, we have here the most representative ones. More precisely, in the case of Klobuchar model, it is claimed that 50% of the ionospheric delay is corrected [12] (so, $\alpha = 50\%$). When NeQuick model is employed, up to 70% of the ionospheric delay can be corrected [13], leading to an ionospheric delay estimation error of $\alpha = 30\%$.

For the case of dual-frequency receivers, we focus on mass-market Galileo receivers. More precisely, we have chosen the E1-E5a frequency combination which appears to be

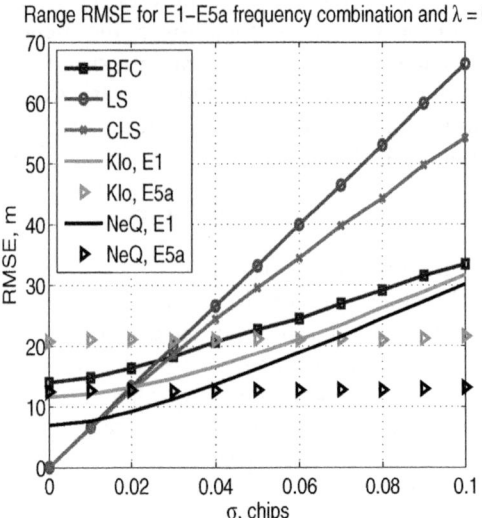

Fig. 2. RMSE vs. standard deviation of error for $i = 1, 2)$ and $\lambda = 0$

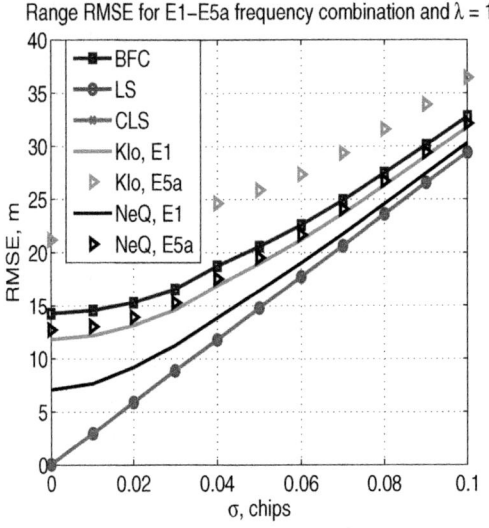

Fig. 3. RMSE vs. standard deviation of error for $\lambda = 1$

the most suitable [7]. In addition, this choice of frequencies is advantageous because it is overlapping with the existing L1 and L5 bands, thus facilitating the design of a joint GPS/Galileo receiver [14]. For the mitigation of ionospheric delay we used the three methods described in Section 2: LS, CLS with constraint vector $\mathbf{b} = [0\ 0]^T$ and BFC (more detailed description of these algorithms and the reasoning for choosing the above constraint vector are included in [7]).

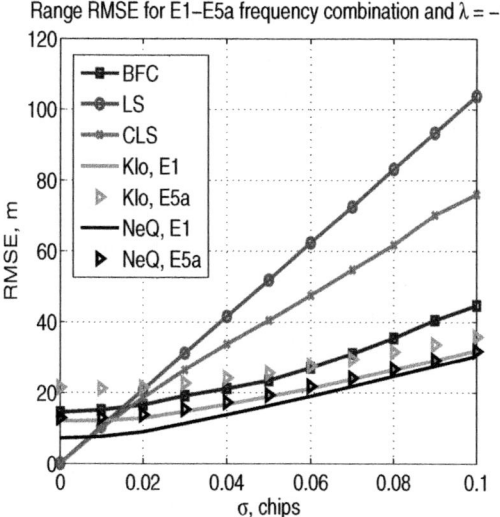

Fig. 4. RMSE vs. standard deviation of error for $\lambda = -1$

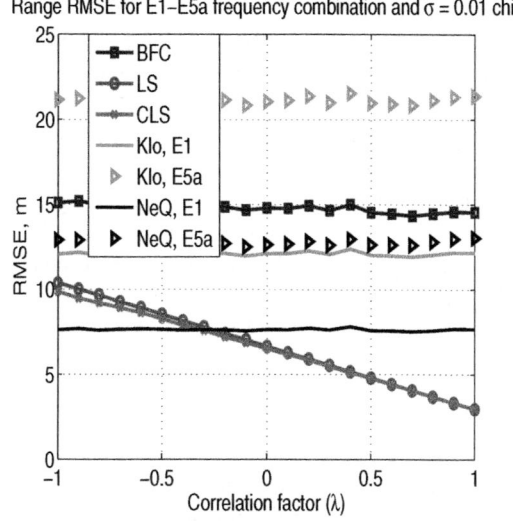

Fig. 5. RMSE vs. correlation factor for $\sigma = 0.01$ chip

In order to compute the RMSE performance values, we generate 2000 random real-isations of the signal and of the measurement errors. More precisely, the true range ρ is uniformly distributed between 18000 and 25000 km and the TEC is uniformly distributed between 1 and 250 TECU. The limits of the TEC parameter have been chosen in such a way that typical values encountered in various latitudes are included [15, 16, 17]. In our simulations, we don't employ a specific multipath channel profile. Instead, we model the measurement error as the tracking error attributed to the multipath propagation effects. More precisely, the error e is modelled as a random variable that follows the normal distribution (the assumption of normal distribution has been commonly encountered in the literature [18, 19, 20, 21] and is used here for simplicity; study of different error distributions belongs to our future plans). More precisely, the errors are distributed according to $e_i \sim \mathcal{N}(\mu_{e_i}, \sigma_{e_i})$, where $\mu_{e_i} = 0$ and σ_{e_i} takes values from 0 to 0.1 chips with a step of 0.01 chip. We remark that for the sake of simplicity, the measurement errors in E1 and E5a are assumed to follow the same distributions (however, it was recently found in [22] that the tracking error distributions of L1 and E5a signals may differ in different channel profiles and the effect of such difference is another interesting research topic). In addition, we notice that because E1 signal has smaller chip rate than the other three, a standard deviation error of 0.01 chips translates into 2.932 m. of error for E1 signal and into 0.293 m. for E5a (for the sake of fair comparison, we modelled the measurement errors at chip level because E1 and E5a signals have different chip rates).

In Fig. 2 we see the RMSE values for the case of uncorrelated pseudorange measurements and zero mean error. We observe that when the standard deviation of error is smaller than 0.01 chip the dual-frequency methods (LS and CLS) perform the best. However, when the standard deviation increases further, a single-frequency Galileo receiver operating in E5a carrier frequency would perform the best. When the pseudoranges are characterised by a full positive correlation (see Fig. 3), LS and CLS dual-frequency methods perform always better than the single-frequency ones (we notice that while the assumption of fully positively correlated pseudoranges might be extreme, it has been reported in [23]).

When the pseudoranges have a full negative correlation, the relative performance of the dual-frequency methods remains the same than in the case of no correlation (see Fig. 4). On the other hand, we notice that in case of dual-frequency receivers the choice of the best frequency changes. Finally, the performance of the methods for different correlation factors can be see in Fig. 5. In particular, we observe that the RMSE of LS and CLS methods decreases with increasing degree of correlation while in the case of BFC and single-frequency methods, the RMSE is not affected by the varying correlation factor.

4   Conclusions and Future Plans

In this paper, we investigated the performance of single- and dual- frequency receivers in terms of satellite-receiver range estimation and under the assumption that the received signals have been contaminated due to ionospheric and multipath propagation effects. More precisely, we examined the performance of three dual-frequency receiver methods: The first one is the Least Squares (LS) method which estimates the unknown total electron content and range by trying to minimise the sum of squared distances between the observed responses in the observation set, and the responses predicted by the linear approximation. The second method is a variant of LS, called Constraint LS (CLS) which imposes certain constraints on the solution and therefore, it is more complex. The third method is called Brute Force Constraint (BFC) and it was proposed by the authors due to its low computational burden.

The performance of the above-mentioned methods was compared with the performance of single-frequency GPS and Galileo receivers in term of Root Mean Square Error (RMSE). Furthermore, we assumed that the Klobuchar and Ne-Quick model were used to estimate the electron content, in the single- and dual- frequency receivers, respectively. The results showed that when the pseudoranges are uncorrelated, the LS and CLS methods are superior to single frequency methods only in the case when the standard deviation of the error is smaller than 0.01 chip. If the standard deviation is higher than 0.01 chip, a single-frequency Galileo receiver operating at E5a carrier frequency would perform the best. Moreover, dual-frequency methods perform the best when the pseudoranges are characterised by a full positive correlation. Finally, the simulation results showed that the RMSE of LS and CLS methods decreases with increasing correlation factor while the in the case of the other methods, RMSE is only weakly affected.

We notice that our simulations were done under limited assumptions, such as theoretical modelling of multipath errors and equal error variances on E1 and E5a signals; more remains to be investigated about the possible advantage of a dual-frequency receiver under more realistic assumptions.

Acknowledgements. The research leading to these results has received funding from the European Union's Seventh Framework Program (FP7/2007-2013) under grant agreement n°227890 (GRAMMAR project). The work has also been supported by the Tampere Doctoral Program in Information Science and Engineering (TISE) and by the Academy of Finland.

References

1. Galileo: European alternative to GPS needs more funding (2010),
   http://www.europarl.europa.eu/news/public/default_en.htm?redirection
2. Kaplan, E.: Understanding GPS: Principles and Applications. Artech House, Boston (1996)
3. Odijk, D.: Fast Precise GPS Positioning in the Presence of Ionospheric Delays. PhD thesis, Publications on Geodesy 52, Netherlands Geodetic Commission, Delft, Netherlands (2002)
4. Klobuchar, J.A.: Ionospheric effects on GPS. In: GPS World, pp. 48–51 (1991)
5. Radicella, S.M.: The NeQuick model genesis, uses and evolution. Annals of geophysics 52, 417–422 (2009)
6. Belabbas, B., Schlueter, S., Sadeque, M.Z.: Impact of NeQuick correction model to positioning and timing accuracy using the instantaneous pseudo range error of single frequency absolute positioning receivers. In: ION GNSS 2005, Los Angeles, CA USA, pp. 712–722 (2005)
7. Skournetou, D., Lohan, E.S.: Ionosphere-corrected range estimation in dual-frequency GNSS receivers. accepted in IET Radar, Sonar and Navigation (2010) (in press)
8. Engel, U.: A theoretical performance analysis of the modernized GPS signals. In: Position, Location and Navigation Symposium, 2008 IEEE/ION, pp. 1067–1078 (2008)
9. Kay, S.M.: Fundamentals of Statistical Signal Processing. Detection Theory, vol. 2. Prentice Hall, Englewood Cliffs (1993)
10. Klobuchar, J.: Ionospheric time-delay algorithm for single-frequency gps users. IEEE Transactions on Aerospace and Electronic Systems AES-23, 325–331 (1987)
11. Bidaine, B., Prieto-Cerdeira, R., Orus, R.: Nequick: In-depth analysis and new developments. In: Proc. of the 3rd ESA Workshop on Satellite Navigation User Equipment Technologies NAVITEC 2006 [CD-Rom], Noordwijk, The Netherlands (2006)

12. Parkinson, B.W., Spilker, J.J.J.: Global positioning system: Theory and applications. In: American Institute of Aeronautics, 370 L'Enfant Promenade, SW, Washington (1996)
13. Prieto-Cerdeira, R., Orus, R., Arbesser-Rastburg, B.: Assessment of the ionospheric correction algorithm for Galileo single frequency receivers. In: Proceedings of the 3rd ESA Workshop on Satellite Navigation User Equipment Technologies NAVITEC 2006 [CD-Rom], Noordwijk, The Netherlands (2006)
14. Hurskainen, H., Lohan, E.S., Nurmi, J., Sand, S., Mensing, C., Detratti, M.: Optimal dual frequency combination for Galileo mass market receiver baseband. In: IEEE Workshop onSignal Processing Systems, SiPS 2009, pp. 261–266 (2009)
15. Gao, G.X., Datta-Barua, S., Walter, T., Enge, P.: Ionosphere effects for wideband GNSS signals. In: ION Annual Meeting, Cambridge, Massachusetts (2007)
16. Hein, G.W., Avila-Rodriguez, J.A.: Combining Galileo PRS and GPS M-Code. Inside GNSS 1, 48–56 (2006)
17. Mannucci, A.J.: Ionosphere and ionospheric delay,
    http://www.cosmic.ucar.edu/summercamp_2005/
    presentations/Mannucci_Anthony_20050602.pdf (2005) (accessed January 5, 2010)
18. Misra, P., Enge, P.: Global Positioning System: Signals, Measurements, and Performance. Ganga-Jumuna Press, Lincoln (2001)
19. Kuusniemi, H., Lachapelle, G., Takala, J.: Position and velocity reliability testing in degraded GPS signal environments. GPS Solutions 8, 226–237 (2004)
20. Liu, G., Liu, Y.: An methodology of automatic guidance system. In: Proceedings of the 2006 IEEE International Conference on Networking, Sensing and Control, ICNSC 2006, pp. 568–571 (2006)
21. O' Keefe, K., Julien, O., Cannon, M., Lachapelle, G.: Availability, accuracy, reliability, and carrier-phase ambiguity resolution with Galileo and GPS. Acta Astronautica 58, 422–434 (2006)
22. Zhang, J., Lohan, E.S.: Galileo E1 and E5a link-level performances in single and multipath channels. In: International ICST Conference on Personal Satellite Services (PSATS), Malaga, Spain (2011)
23. WorkingGroupC: Combined performances for Open GPS/Galileo receivers (2010), http://ec.europa.eu/enterprise/policies/satnav/documents/index_en.htm, to be presented in the Fifth Meeting of the International Committee on Global Navigation Satellite Systems (ICG) jointly hosted by Italy and the European Union

Author Index